AF754325

Portrait of a Scientific Racist

Portrait of a Scientific Racist

Alfred Holt Stone of Mississippi

James G. Hollandsworth Jr.

LOUISIANA STATE UNIVERSITY PRESS
BATON ROUGE

Published by Louisiana State University Press
Copyright © 2008 by Louisiana State University Press
All rights reserved
Manufactured in the United States of America
First printing

Designer: LAURA ROUBIQUE GLEASON
Typefaces: WHITMAN, TEXT; ROCKWELL EXTRA BOLD, DISPLAY
Typesetter: J. JARRETT ENGINEERING, INC.
Printer and binder: THOMSON-SHORE, INC.

Frontispiece: Alfred Holt Stone, ca. 1935. Courtesy Mississippi Department of Archives and History.

LIBRARY OF CONGRESS CATALOGING-IN-PUBLICATION DATA

Hollandsworth, James G.
Portrait of a scientific racist : Alfred Holt Stone of Mississippi / James G. Hollandsworth, Jr.
p. cm.
Includes bibliographical references and index.
ISBN 978-0-8071-3336-1 (cloth : alk. paper) 1. Stone, Alfred Holt, 1870—1955. 2. Stone, Alfred Holt, 1870-1955—Political and social views. 3. Plantation owners—Mississippi—Biography. 4. Cotton farmers—Mississippi—Biography. 5. Mississippi. State Tax Commission—Officials and employees—Biography. 6. Racism—Mississippi—History—20th century. 7. Eugenics—Mississippi—History—20th century. 8. Mississippi—Race relations—History—20th century. 9. Racism—United States—Philosophy. 10. Eugenics—United States—History—20th century. I. Title.
F341.S78H65 2008
305.80092—dc22
[B]

2007051856

The paper in this book meets the guidelines for permanence and durability of the Committee on Production Guidelines for Book Longevity of the Council on Library Resources. ♾

For Anne Lipscomb Webster, who opened the door

Contents

Preface

I did not intend to write this book. It started out as a biographical sketch to accompany a finder's aid for a large collection of material at the Mississippi Department of Archives and History (MDAH) that I was working on. The subject, Alfred Holt Stone, was well known in the state, but not much was known about his personal life. There were two articles about Stone in the *Journal of Mississippi History*, one written by John David Smith that dealt with Stone's work on the history of slavery, and another, a eulogy, written in 1955 on the occasion of Stone's death. In addition, MDAH had the typescript of Stone's recollections of his childhood, which he wrote when he was in his sixties, and there was also material at MDAH regarding Stone's activities as an executive of the Staple Cotton Cooperative Association and later as Mississippi's tax commissioner. But two articles, some newspaper clippings, the typescript of his recollections, and miscellaneous documents related to his public service were the extent of what I had to work with.

Although this material provided the outlines of Stone's life and some information about his childhood, it did not offer much insight in regard to more important questions concerning his racial views, particularly his motivation for assembling such a large collection of material on black people. Nor were there any obvious leads as to where more information about Stone could be obtained. Stone and his wife, Mary, had only one child, a boy, who died in infancy. Consequently, there were no direct descendants to carry Alfred Holt Stone's memory forward. In addition, I was unable to locate any of Stone's papers, which was surprising given that Stone, of all people, would have appreciated the importance of his personal papers as a testament to his many years of public service. Nevertheless, either they had disappeared, been destroyed, thrown away, or were languishing out of mind in the attic of an unidentified relative.

Nevertheless, something about Stone's life needed to be written. His collection of race-related material at MDAH is a significant historical and sociological resource, and the effort Stone devoted to assembling the collection warranted at least a biographical sketch to accompany the collection's finding aid. In addition, there was a good bit of curiosity among those of us who were familiar with the collection about Stone's motivation for compiling such an eclectic archive of black material. Yet, the lack of primary source material was discouraging. Then, an unexpected breakthrough occurred. I found a collection of Stone's correspondence that immediately propelled the biographical sketch into a full-length biography.

The clue to discovering this new material was a name, Walter F. Willcox. Willcox was a statistician and nationally known economist at the turn of the twentieth century. Born in Massachusetts, Willcox held a position on the faculty at Cornell University and joined the U.S. Census Bureau as a chief statistician for the 1900 census. Willcox's name had surfaced almost immediately when I began my research. In 1908, Stone had published a collection of his writings under the title *Studies in the American Race Problem.* Willcox contributed three papers of his own to complement Stone's nine chapters and wrote a preface in which he praised Stone's work and noted the similarities between the two men's positions on the race problem.

I searched the Internet for references to Willcox and came up with many hits, the most useful being an article by Mark Aldrich in a 1960 issue of *Phylon* entitled "Progressive Economists and Scientific Racism: Walter Willcox and Black Americans, 1895–1910." In this excellent paper, Aldrich mentioned an incident in which Willcox accused the race-baiting governor of Mississippi, James K. Vardaman, of misinterpreting some of his data. In discussing the controversy, Aldrich referred to Alfred Holt Stone, who sided with Willcox. Most importantly, Aldrich cited six letters between Stone and Willcox that he had found in the Library of Congress.

I followed up Aldrich's article with a trip to Washington. The finding aid in the Manuscripts Reading Room of the Library of Congress indicated that there was a folder with correspondence between Willcox and Stone, and I requested it. You can imagine my excitement when the staff member in the reading room handed me the folder. It contained more than two hundred letters between Stone and Willcox beginning in 1900 and continuing without interruption for more than ten years, the period during which Stone studied and wrote about issues involving race. In addition, both sides of the correspondence had been preserved because Willcox had dictated his letters, and his stenographer had made a carbon copy of everything he dictated. It also became obvious as I read these letters that Willcox was Stone's mentor, the one person most responsible for Stone's rapid ascent to the heights of respectability as a racial theorist.

The Willcox–Stone correspondence yielded dozens of clues as to where additional information about Stone could be found. Furthermore, Stone's letters to Willcox offered a behind-the-scenes perspective on Stone's correspondence with other people, such as Booker T. Washington and W. E. B. Du Bois. Although Stone was polite and deferential in his letters to Washington and Du Bois, he expressed less charitable views, presumably his real feelings, in a parallel correspondence with Willcox.

For these reasons, the biographical sketch I had intended to write about Alfred Holt Stone became a biography, but not a biography in the traditional sense since I have focused on that decade during which Stone was spending most of his time developing his ideas as a racial theorist. Of course, I wanted to explore his early life to understand why he had developed a passion for the subject, and I also wanted to give the reader an idea of what happened to Stone after economic reality curtailed his career as a racial theorist. But I have kept the focus of the book on his thinking about issues involving race, and there was plenty of material for me to work with in that regard. In addition to Stone's correspondence with Willcox and other notables of the period, I could rely on the corpus of Stone's writings, as well as the collection of articles, pamphlets, and monographs he assembled, many of which he had underlined and annotated.

Painting the portrait of a racial theorist at work required situating his thinking in a political and social context. To put Stone's thinking into perspective, I have attempted to provide readers a necessarily brief picture of the complicated context in which racial theories developed at the turn of the twentieth century. Readers who are particularly interested in that subject can find more information in four excellent studies.

George M. Fredrickson's *The Black Image in the White Mind: The Debate on Afro-American Character and Destiny, 1817–1914* was first published in 1971 and reissued in 1987. A sweeping exploration of "a neglected aspect of American intellectual history" when Fredrickson wrote it in the mid-1960s—and one that has been followed by much good work—it remains as valuable a resource now as when it was first published.

Those interested in the role of slavery in the development of racial theories that emerged after emancipation will be amply rewarded by reading John

David Smith's *An Old Creed for the New South: Proslavery Ideology and Historiography, 1865–1918*. Smith's work is particularly helpful in understanding the assumptions that colored Stone's ideas in regard to African Americans.

Both *The Black Image in the White Mind* and *An Old Creed for the New South* deal primarily with the white perspective. The third book, Neil McMillen's *Dark Journey: Black Mississippians in the Age of Jim Crow*, explores the racial equation from the black perspective. Along the same lines, racial issues in Mississippi around the turn of the century are examined in a recent work by Stephen Cresswell, *Rednecks, Redeemers, and Race*, in which Cresswell is able to balance the white with the black perspective. All four works offer an overview of issues involving race during the period in which Stone was an active racial theorist, and they go beyond what I have attempted to do here.

An area in which I like to think I am passably competent is tracing the development of a person's ideas and appreciating the factors that may have shaped those ideas over time. My background as a psychologist has provided me with a bent in that direction, but I have not tried to psychoanalyze Stone or delve into the dark mysteries of racial prejudice. Rather, I have tried to let the personae of this work do the talking with words that came either from articles, interviews, comments at professional meetings, or letters that documented their thinking. As far as the letters were concerned, I would like to observe how lucky I was that the telephone was not widely used during the first decade of the twentieth century. How much less we would know about what these men thought had they been able to pick up the phone and call.

As a racial theorist, Alfred Holt Stone is all but forgotten today, and his writings in regard to the race problem are of interest from a historical perspective only. But each generation is faced with recurrent themes that emerge when people from different cultures live in close proximity. I hope that the reader who finds the pursuit of ideas about racial relationships interesting in itself will judge this book to be worth his or her time.

Acknowledgments

Many persons have helped me with the research for this book. I would like to acknowledge in particular the assistance of Margaret Burri in the Milton S. Eisenhower Library at the Johns Hopkins University; Petrina Jackson and Susette Newberry in the Kroch Library at Cornell University; Bruce Kirby at the Library of Congress; Joseph D. Schwartz, who works at the National Archives in College Park, Maryland; John David Smith, the Charles H. Stone Distinguished Professor of American History at University of North Carolina–Charlotte; and John Strom at the Carnegie Institution of Washington. I would also like to thank Tom Henderson for allowing me access to the library at Millsaps. In addition, I would like to acknowledge the steady help, friendly smiles, and sage advice of people at the Mississippi Department of Archives and History in Jackson: Clinton Bagley, Nancy Bounds, Joyce Dixon-Lawson, Elbert Hilliard, Hank Holmes, Grady Howell, Jean Hudspeth, De'Niecechsi Layton, Walterine Robinson, William Thompson, Clay Williams, and Julia Young. I owe a special debt of gratitude to two members of the Department of Archives and History who read the manuscript as I wrote it, giving me valuable feedback and encouragement. Their names are Anne Lipscomb Webster and Susan Johnson.

Abbreviations

AHSC	Alfred H. Stone Collection, University of Mississippi, Oxford
BTWM	Booker T. Washington Papers on microfilm, Library of Congress, Washington, DC
BTWP	Booker T. Washington Papers, University of Illinois Press, www.press.uillinois.edu
CIA	Carnegie Institution archives, Washington, DC
DBP	W. E. B. Du Bois Papers, University of Massachusetts, Amherst (microfilm, 89 reels)
JFJP	J. Franklin Jameson Papers, Library of Congress, Washington, DC
LC	Library of Congress, Washington, DC
MDAH	Mississippi Department of Archives and History, Jackson
NA	National Archives, Washington, DC
NA-CP	National Archives, College Park, MD
PFP	Percy Family Papers, Mississippi Department of Archives and History, Jackson
PMHS	*Publications of the Mississippi Historical Society*
TSC	The Stone Collection, Mississippi Department of Archives and History, Jackson
SARP	*Studies in the American Race Problem,* by Alfred Holt Stone (New York: Doubleday, Page, 1908)
WCCH	Washington County [Mississippi] Court House
WFWP	Walter F. Willcox Papers, Library of Congress, Washington, DC

Portrait of a Scientific Racist

INTRODUCTION: The American Race Problem

The United States has been grappling with questions stemming from the economic, social, and political relationship of white people with black folk since a Dutch trader docked at Jamestown, Virginia, in 1619 and exchanged his cargo of African slaves for food.[1] Although issues concerning relations between the two races in America seem destined to persist indefinitely, the importance of these issues has ebbed and flowed. Race was not the major issue, for example, during the framing of the Constitution. Slavery was taken for granted, and determining the relative strength of big states versus little states in the new government was considered to be more important. In contrast, issues involving race assumed prominence several decades later when abolitionists challenged, and slave owners defended, the moral and economic legitimacy of slavery. That debate continued to generate heat and hatred throughout the crisis over secession, the Civil War, and Reconstruction. The issue of racial equality also transformed the nation's emotional landscape one hundred years later during the modern civil rights movement.[2]

Given the importance of these racial milestones, it is easy to forget that there was another time in the history of the United States when questions concerning the natural and appropriate relationship between the black and white races dominated this country's public and private agenda. Today, people

1. "Chronology on the History of Slavery and Racism," www.innercity.org/holt/slavechron.html. For a history of "race" as a legal concept, see Daniel J. Sharfstein, "The Secret History of Race in the United States," *Yale Law Journal* 112 (March 2003): 1473–510.

2. For a thoughtful narrative of race relations in the United States, see Thomas F. Gossett, *Race: The History of an Idea in America*, rev. ed. (New York: Oxford University Press, 1997). For an extensive treatment of race relations since emancipation, see Joel Williamson, *The Crucible of Race: Black-White Relations in the American South since Emancipation* (New York: Oxford University Press, 1984).

may not recall that period of racial upheaval because it fell between two significant, and thus more memorable, events—Reconstruction and the First World War. Nevertheless, tension over race relations in America peaked toward the end of the nineteenth century, and the intensity of feelings regarding the status of black people vis-à-vis white folk during that period can be illustrated by what happened when the president of the United States invited a black man to dinner at the White House.

William McKinley, a Republican, had been reelected president in November 1900 with Theodore Roosevelt as his running mate. On September 6, 1901, just six months after his inauguration, McKinley was shot and killed by an anarchist, and Roosevelt became president.[3] Five weeks later, Roosevelt invited Booker T. Washington to the White House for dinner to discuss issues that affected black citizens in the United States.

One of Roosevelt's biographers has described Booker T. Washington as "the archetype of the modern, sensible black man." An advocate of black self-sufficiency as well as black acceptance of the South's insistence that African Americans abandon ambitions of political involvement and aspirations for social equality, Washington had invited Roosevelt to visit the Tuskegee Institute in Alabama during the vice president's anticipated trip through the South in the fall of 1901 to see the vocational training programs Washington had established there.[4] McKinley's assassination made canceling the trip necessary, but Roosevelt did not forget Washington's invitation and asked him to stop by the White House the next time he was in the nation's capital. Roosevelt thought that Washington was a spokesperson for the black race that he could consult without fear of offending white voters.

On October 16, 1901, Roosevelt learned that Washington was in town and invited him to dinner at the White House. The group attending dinner was small: Roosevelt, his wife, Edith, and a guest from Colorado in addition to Washington. Washington knew that the invitation might create controversy and slipped into and out of the White House without attracting the press's attention. The private dinner probably would have gone unnoticed had not the Executive Mansion's doorkeeper routinely released a list of visitors to the press. One sentence concerning the event appeared the next morning at the bottom of an inside page in the *Washington Post*. "Booker T. Washington, of

3. Samuel Eliot Morison and Henry Steele Commager, *The Growth of the American Republic* (New York: Oxford University Press, 1962), 2: 476. McKinley died on September 14, 1901.

4. H. W. Brands, *T. R.: The Last Romantic* (New York: Basic Books, 1997), 422. For a description of Washington as white America's "acceptable" black man, see John David Smith, *An Old Creed for the New South: Proslavery Ideology and Historiography, 1865–1918* (Westport, CT: Greenwood Press, 1985), 201–3.

Tuskegee, Ala., dined with the President last evening." The short item was picked up and reprinted throughout the nation.[5]

Many white southerners were aghast when they learned that the president of the United States had invited Washington to dinner. By doing so, they thought, Roosevelt had officially bestowed on Booker T. Washington the honor of social equality, and the promotion of a black man to that status was unacceptable. "The most Damnable outrage which has ever been perpetuated by any citizen of the United States was committed yesterday by the President, when he invited a nigger to dine with him at the White House," an editorial in the *Memphis Scimitar* cried. "It would not be worth more than a passing notice if Theodore Roosevelt had sat down to dinner in his own home with a Pullman car porter, but Roosevelt the individual and Roosevelt the President are not to be viewed in the same light."[6]

Outrage over the incident would have enjoyed a short life span had the invitation been viewed as politically motivated. After all, Roosevelt was a Republican, and, as such, he had a legitimate interest in determining what inroads he could make in the solidly Democratic South. But the incident was not about politics; it was about race. A cartoon in the *Atlanta Constitution* emphasized that point by depicting a black man seated by Roosevelt in a cozy embrace. As Alfred Holt Stone from Mississippi noted several years later, "Within forty-eight hours the President was being denounced for having crossed the 'social equality' dead-line, through breaking bread with a Negro."[7]

Many white southerners at the turn of the twentieth century could work themselves into a frenzy when it came to the question of social equality between the two races. "When Mr. Roosevelt sits down to dinner with a Negro," the *New Orleans Times-Democrat* observed, "he declares that the Negro is the social equal of the white man." That observation may have had some merit, but the sensational conjecture in another paper that Roosevelt's wife and her black guest may have touched thighs under the table did not, nor did the charge that Roosevelt was promoting a "mingling and mongrelization" of the Anglo-Saxon race, as yet another paper characterized the meeting.[8]

By today's standards, these reactions to Roosevelt's inviting Washington

5. Mark Sullivan, *Pre-War America*, vol. 3 of *Our Times: United States, 1900–1925* (New York: Scribner's, 1930), 131–33; Edmund Morris, *Theodore Rex* (New York: Random House, 2001), 52.

6. October 17, 1901, quoted in Morris, *Theodore Rex*, 54.

7. Alfred Holt Stone, *Studies in the American Race Problem* (New York: Doubleday, Page, 1908), 245; hereafter cited as *SARP*.

8. *New Orleans Times-Democrat* quoted in Sullivan, *Pre-War America*, 136; miscellaneous newspaper clippings quoted in Morris, *Theodore Rex*, 55.

to dinner would seem humorous had they not been so serious. Both men received death threats and hate mail over the incident, and the honor of southern womanhood was invoked to exaggerate the controversy's significance. "It means that the President is willing that negroes should mingle freely with whites in the social circle—that white women may receive attentions from negro men," an editorial in the *Richmond Times* warned. "It means that there is no racial reason in his opinion why whites and blacks may not marry and intermarry, why the Anglo-Saxon may not mix negro blood with his blood."[9] This dramatic leap in logic was picked up and repeated in papers throughout the South. "Washington should send his daughter to the White House for Christmas," another paper suggested. "Maybe Roosevelt's son will fall in love with her and marry her."[10]

The antagonism of white people toward black people had been building steadily since the end of Reconstruction. Most students of American history know that Reconstruction ended around 1877, when white supremacists regained control of state governments throughout the South. Typically, they also know that African Americans had been disfranchised, the Republican Party had been dismantled in these states, and that carpetbaggers had been sent packing. What these students might not recall is that, despite the reestablishment of white supremacy, relations between the two races did not settle into a stable, albeit lopsided, status quo. They continued to deteriorate.[11]

The African American historian Rayford W. Logan called attention to this period in his *The Negro in American Life and Thought: The Nadir, 1877–1901.* As the book's title suggests, Logan argued that "the last decade of the nineteenth century and the opening of the twentieth century marked the nadir of the Negro's status in American society."[12] Interestingly, the deterioration in race relations following Reconstruction did not occur because of issues left unresolved by the reestablishment of white supremacy. Those matters, such as the disfranchisement of black voters, had been accomplished. According

9. Quoted in Sullivan, *Pre-War America*, 134.

10. Undated newspaper clipping quoted in Morris, *Theodore Rex*, 55.

11. Stephen Cresswell, "The Persistent Institution: Conflict and Racial Separation," in *Rednecks, Redeemers, and Race: Mississippi after Reconstruction, 1877–1917*, by Cresswell (Jackson: University Press of Mississippi, 2006), 52–68.

12. Williamson has described the period from 1889 to 1915 as being characterized by "the rage of radicalism," defining radicalism as "the concept that Negroes, freed from the restraining influences of slavery, were rapidly 'retrogressing' toward their natural state of bestiality" (*Crucible of Race*, 111). Edward J. Blum explores the evolution of racism from 1865 to 1898 by tracing the contribution of religion, imperialism, and sectional reconciliation to the triumph of white supremacy on a national level (*Reforging the White Republic: Race, Religion, and American Nationalism, 1865–1898* [Baton Rouge: Louisiana State University Press, 2005]).

to Logan, it was the well-intentioned efforts by politicians in the North that prompted white southerners to tighten the screws of racial oppression.[13]

Republicans had recaptured the White House in 1888 with the election of Benjamin Harrison and won the majority of seats in both the House and the Senate. The Grand Old Party was in control of the executive and legislative branches of government for the first time in fourteen years, and some of its members had not forgotten their heritage of belonging to the party that had freed the slaves.[14] In 1890, two Republicans named Henry—Henry Cabot Lodge and Henry W. Blair—individually sponsored legislation that was anathema in the South. Lodge's bill would establish federal supervision for federal elections. Furthermore, the bill would have enforced the Fourteenth Amendment by reducing representation in Congress for any state that systematically prevented an identifiable class of citizens, such as African Americans, from voting in federal elections. Blair's proposal would have provided federal funds to support public schools, regardless of race, with the most money going to those states with the highest rates of illiteracy. Blair's aid would necessarily favor black pupils in the South who were attending schools receiving very little in the way of state funding.[15] Both measures died in the Senate, but the debate in Congress over the two bills served to remind white southerners that the black segment of their population presented a problem with which they had to contend.[16]

The idea that black people in the South were a serious liability instead of a valuable labor force surfaced shortly after the Emancipation Proclamation.[17] Black people had not been a problem during slavery because the legal and social prohibitions of the South's "peculiar institution" kept African Americans in what was assumed to be their rightful place and clearly specified the

13. Rayford W. Logan, *The Negro in American Life and Thought: The Nadir, 1877–1901* (New York: Dial Press, 1954), 53; see also Smith, *Old Creed for the New South*, 287.

14. Thomas Adams Upchurch, *Legislating Racism: The Billion Dollar Congress and the Birth of Jim Crow* (Lexington: University of Kentucky Press, 2004), 2–4; Neil R. McMillen, *Dark Journey: Black Mississippians in the Age of Jim Crow* (Urbana: University of Illinois Press, 1989), 41.

15. For an extensive discussion of each bill, see Upchurch, *Legislating Racism*, 46–65 (Blair's bill) and 85–107 (Lodge's bill).

16. Reasons for the bills' defeat were complex and due only in part to opposition from southern Democrats (Upchurch, *Legislating Racism*, 52–63, 90–107 passim). For an example of southern opposition to Lodge's bill, see J[ames] Z. George, *The Federal Election Bill: Speech of Hon. J. Z. George, of Mississippi, in the Senate of the United States, December 10, 1890* (Washington, DC: n.p., 1890), 50 pages. The Alfred Holt Stone Collection (TSC) in the Mississippi Department of Archives and History (MDAH) has a copy of this speech.

17. George M. Fredrickson, *The Black Image in the White Mind: The Debate on Afro-American Character and Destiny, 1817–1914* (1971; Hanover, NH: Wesleyan University Press, 1987), 324–25.

"rights and duties" of their masters.[18] Such was not the case following emancipation, and stuffing the black genie into her old bottle was problematic, but not impossible.

The first reference to the term "negro problem" in the Library of Congress online catalogue appears in the title of a book published in 1864 by Hollis Read, *The Negro Problem Solved; or Africa as She Was, as She Is and as She Shall Be. Her Curse and Her Cure.* Reed was addressing the question of what to do with former slaves after the Civil War ended, now that they had been freed. His solution was to send them all to Africa.[19] As it turned out, his solution was wildly impracticable, and the "War Amendments" to the Constitution (Thirteenth, Fourteenth, and Fifteenth), which enfranchised former slaves and ensured their rights as citizens, made the question of what to do with them moot during Reconstruction.

Ten years after the publication of Read's book, H. H. Goodloe wrote an article entitled "The Negro Problem" for the *Southern Magazine.* His piece was a pessimistic assessment of the future of race relations in the South and foreshadowed the intense debate over what to do with all of those black people after Reconstruction ended. According to Goodloe, the problem went beyond the usual struggle between labor and management; it encompassed the most basic feelings of revulsion and rejection. "That antagonism exists between the two races in their relations to each other, and in form and degree different and greater than that usually recognized as between capital and labor is so plainly observable," Goodloe wrote, "that we have only to open our eyes to existing and constantly recurring facts to be convinced of its truth."[20]

In 1877, J. R. Ralls published a pamphlet entitled *The Negro Problem. An*

18. For example, see J. H. Thornwell, *The Rights and the Duties of Masters. A Sermon Preached at the Dedication of a Church, Erected in Charleston, S.C., for the Benefit and Instruction of the Colored Population* (Charleston, SC: Steam-Power Press of Walker and James, 1850), 51 pages. There is a copy of this pamphlet in TSC.

19. For other examples of colonization as a solution to the race problem, see Henry A. Scomp, "Can the Race Problem Be Solved?" *Forum* (December 1889): 365–76; Wade Hampton, "The Race Problem," *Arena* (July 1890): 132–38; or "The Negro Problem," *National Criterion* (September 1902): 11–13. TSC has copies of all three articles. See also H[orace] S[mith] Fulkerson, *The Negro; As He Was; As He Is; As He Will Be* (Vicksburg, MS: Commercial Herald, 1887), 118. Fulkerson, a Mississippian, argued that African Americans should be colonized in Cuba and other islands in the Caribbean. Greene C. Chandler, another Mississippian, believed that African Americans should be resettled on public lands in the western United States (Walter Chandler, ed., *Journal and Speeches of Greene Callier Chandler* [N.p., 1953], 162).

20. H. H. Goodloe, "The Negro Problem," *Southern Magazine* 7 (April 1874): 373. The

Essay on the Industrial, Political and Moral Aspects of the Negro Race in the Southern States, in which the author blamed northern do-gooders for stirring up conflict between the two races in the South by encouraging unrealistic expectations among African Americans. "While the great mass of this race [the black race] will very probably remain in a state of ignorance for successive generations," Ralls observed, "numbers will be educated, which will stimulate their pride and awaken new aspirations, and, finding a superior, and the ruling race above them, whose sphere cannot be attained, on account of caste, will cause them to agitate foolish and impracticable questions, which will beget a restive, jealous, and dissatisfied spirit among the masses, that must make society, in the future, more disorderly and insecure than at present."[21]

Use of the term "the negro problem" or its variant, "the race problem," exploded around 1890 as the political gains achieved by white supremacists were being solidified by the adoption of post-Reconstruction constitutions.[22] An online search of the Library of Congress catalogue for books published in the United States will reveal that fifteen entries with "negro problem" in the title were catalogued during a twenty-five-year period spanning the turn of the century (1889–1914); twelve additional books with the term "race problem" in the title were catalogued during almost the same period (1891–1918). In addition, dozens of articles, speeches, and pamphlets were published with titles bearing one of the terms.[23] If the topic is defined

---

Southern Magazine was the official organ of the Southern Historical Society. There is a copy of this article in TSC.

21. J. R. Ralls, *The Negro Problem. An Essay on the Industrial, Political and Moral Aspects of the Negro Race in the Southern States. As Presented under the Late Amendments to the Federal Constitution* (Atlanta: Jas. P. Harrison and Co., 1877), 116 pages. There is a copy of this monograph in TSC.

22. Upchurch has asserted that the Mississippi Constitution of 1890 was "the most significant reaction to the Lodge [federal elections] bill" (*Legislating Racism*, 110).

23. The use of either term in book titles dropped off after the First World War. There is only one book in the online catalogue with "race problem" in its title after 1914, and six with "negro problem" after 1918. Articles, pamphlets, and journal articles also used both terms in their titles extensively during the same period (1889–1909). Thirty of these publications can be found in TSC; eighteen using the term "race problem," and twelve using "the negro problem." The Daniel A. P. Murray Collection of African American pamphlets, which is online at the Library of Congress (memory.loc.gov/ammem/aap/aaphome.html), has nine titles using these terms, all published between 1889 and 1906. Five of these are also in TSC. Finally, the Birney Anti-Slavery Collection of Johns Hopkins University lists an additional five titles with these terms published between 1884 and 1908, as well as one that contains the phrase "the negro issue" (1904) and another that includes "the negro question" (1908). TSC has three addi-

without reference to specific words in the title, the number of books, pamphlets, and articles dealing with the race problem in the United States during this period runs into the hundreds.[24] It is not an exaggeration to say that from the end of Reconstruction to the end of the First World War, the race problem was as significant as any domestic topic facing the nation.[25] In terms of its impact, the race problem was as divisive and contentious in 1900 as the debate over abortion in the United States one hundred years later.

The great majority of the publications dealing with the race problem assumed the white man's point of view, and it was not surprising that African Americans did not like being seen as a problem.[26] "I object to characterizing the relation subsisting between the white and colored people of this country as the Negro problem, as if the Negro had precipitated that problem, and as if he were in any way responsible for the problem," Frederick Douglass wrote in 1890.[27] Another African American, W. S. Scarborough, also objected: "This question, improperly styled the 'Negro problem,' is in reality the white man's question. From the negro's standpoint the conditions that usually enter into a problem are absent and therefore the wonder why all this discussion in regard to the blacks, why this confusion, these sectional differences, this bitter strife concerning the negro's rights,—his citizenship?"[28]

In 1894, a year before he died, Douglass argued in a speech that the "negro problem" was a misnomer. "The marvel is that this old trick of misnaming things, so often played by Southern politicians, should have worked so

---

tional pamphlets with "the negro question" in the title. They were published in 1892, 1906, and 1908.

24. A book published in 1921 that was "intended to be an interpretation of the leading aspects of the Negro problem of today" included a bibliography of 347 works on the topic, almost all of them appearing after 1899 (Julia E[mily] Johnsen, *Selected Articles on the Negro Problem* [New York: Wilson, 1921], 1). The earliest entry cited in Johnsen's bibliography is an article published in 1879 (James G. Blaine, L.Q.C. Lamar, James A. Garfield, Alexander H. Stephens, Wendell Phillips, Montgomery Blair, and Thomas A. Hendricks, "Ought the Negro to Be Disenfranchised? Ought He Have to Have Been Enfranchised?" *North American Review* 268 [March 1879]: 225–83). There is a copy of this article in TSC.

25. John David Smith, "Religion and 'The Negro Problem,'" pt. 2 of *The Biblical and "Scientific" Defense of Slavery*, vol. 6 of *Anti-Black Thought, 1863–1925* (New York: Garland, 1993), xiv.

26. Smith, *Religion and "The Negro Problem,"* xvi.

27. Frederick Douglass, *The Race Problem. Great Speech of Frederick Douglass, Delivered before the Bethel Library and Historical Association, in the Metropolitan A.M.E. Church, Washington, D.C., October 21, 1890* (N.p., [1890?]), 5. There is a copy of this pamphlet in TSC.

28. W. S. Scarborough, "The Race Problem," *Arena* (October 1890): 560. There is a copy of this article in TSC.

well for the bad cause in which it is now employed,—for the Northern people have fallen in with it," he declared. "It is still more surprising that the colored press of the country, and some of the colored orators of the country, insist upon calling it a 'negro problem,' or a Race problem, for by it they mean the negro Race," Douglass added. "Now—there is nothing the matter with the negro. He is all right. Learned or ignorant, he is all right. He is neither a Lyncher, a Mobocrat, or an Anarchist. He is now, what he has ever been, a loyal, law-abiding, hard-working, and peaceable man; so much so, that men have thought him cowardly and spiritless."[29]

If the race problem was not about the inadequacies of black people, as Douglass claimed, then what was it about? In 1909, the Reverend Quincy Ewing tried to answer this question in an article entitled "The Heart of the Race Problem" in the *Atlantic Monthly*.[30]

Ewing knew what he was talking about because he was from the South and had been around black folk all of his life. Born in 1867 on the family plantation near Thibodaux, Louisiana, Ewing had pursued academic and theological studies at the University of the South in Sewanee, Tennessee. He had been ordained in 1891 and became rector of St. James Episcopal Church in Greenville, Mississippi, in 1895. His tenure at St. James had been cut short, however, when he spoke out against the race-baiting of James K. Vardaman when Vardaman ran for governor in 1903. Ewing subsequently assumed the pulpit of a church in Birmingham, Alabama, before relocating in 1906 to Christ Church in Napoleonville, Louisiana, where he remained for twenty-three years.[31]

"During the past decade, newspaper and magazine articles galore, and not a few books, have been written on what is called the 'Race Problem,' the problem caused by the presence in this country of some ten millions of black and variously-shaded colored people known as Negroes," Ewing wrote at the

29. Frederick Douglass, *Address by Hon. Frederick Douglass, Delivered in the Metropolitan A.M.E. Church, Washington, D.C. Tuesday, January 9th, 1894, on the Lessons of the Hour, in which He Discusses the Various Aspects of the So-Called, But Mis-Called, Negro Problem* (Baltimore: Thomas and Evans, 1894), 31. This speech is in TSC.

30. Quincy Ewing, "The Heart of the Race Problem," *Atlantic Monthly* 103 (March 1909): 389–97.

31. Ewing's father was a surgeon in the Confederate Army (Charles E. Wynes, "The Reverend Quincy Ewing: Southern Racial Heretic in the 'Cajun' Country," *Louisiana History* 7 [Summer 1966]: 222–23). Ewing delivered a homily against lynching while he was at St. James and had it published in *Outlook* ("How Can Lynching Be Checked in the South?" *Outlook*, October 12, 1901, 359–61). A copy of this article is in TSC. Napoleonville is about twenty miles from Thibodaux.

beginning of his article. "But, strange as it may sound, the writer has no hesitation in saying that at this date there appears to be no clear conception anywhere, on the part of most people, as to just what the essential problem is which confronts the white inhabitants of the country because they have for fellow-citizens (nominally) ten million Negroes."[32]

Ewing attempted to fill this conceptual void by evaluating the reasons white southerners used to explain the race problem. According to Ewing, white southerners claimed that "The Negroes, as a rule, are very ignorant, are very lazy, are very brutal, are very criminal." In addition, white people in the South often claimed that they experienced a "personal aversion" when in the presence of people as different in appearance and manner as black people. Based on these statements, Ewing identified three characteristics of African Americans that purportedly lay at the root of the race problem—poor work habits, criminal behavior, and cultural differences.[33]

As far as poor work habits were concerned, Ewing pointed out that African Americans were good workers. In fact, they were valued as laborers throughout the South. "In very many Southern communities," he noted, "the vexing problem for employers is not too many, but too few Negroes." To emphasize the point, Ewing referred to what happened to labor recruiters from the North when they ventured into the South. "If any one doubts that Negroes are wanted as laborers in Southern communities, very much wanted," Ewing remarked, "let him go to any such community and attempt to inveigle a few dozen of the laziest away. He will be likely to take his life in his hands, after the usual warning is disregarded!"[34]

As far as the second reason for the race problem was concerned, Ewing reasoned that the high rate of criminal behavior among black males was a social, not a racial, phenomenon. More specifically, "the Negroes occupy everywhere in this country the lowest social and industrial plane, the plane which everywhere else supplies the jail, the penitentiary, the gallows, with the greatest number of their victims." Based on his forty years of residence in the South, Ewing was convinced "that the Negro is not more given to crimes and misdemeanors than the laboring population of any other section of the country." In addition, the incidence of black criminality was complicated by

32. Ewing, "Heart of the Race Problem," 389. Quotations in subsequent paragraphs are from that article. The emphasis was in the original in every case.

33. For a discussion of "differences from the 'normal'" as a basis for racial discrimination, see Smith, *Religion and "The Negro Problem,"* xii–xiii.

34. For specific examples of the white resistance to outsiders wanting to recruit black labor, see Cresswell, *Rednecks, Redeemers, and Race*, 49–50.

the fact that "in the section of the country where these penal statistics are gathered [the southern states], all the machinery of justice is in the hands of white men." According to Ewing, that reality skewed the data on criminal activity.

Ewing also dismissed the third reason southern whites gave for the race problem. How could white southerners claim a personal aversion based on the physical and cultural differences of a race with whom they had lived in close proximity their entire lives? "The Negro is *not* a mystery to people whom he has nursed and waited on, whose language he has spoken, whose ways, good and bad, he has copied for generations," Ewing wrote, "and his personal presence does not render them uncomfortable, not, at any rate, uncomfortable enough to beget the sense of a burden or a problem."

"So much for what the race problem is not," Ewing concluded. The real foundation of the race problem "is the white man's conviction that the Negro as a race, and as an individual, is his inferior: not human in the sense that he is human, not entitled to the exercise of human rights in the sense that he is entitled to the exercise of them."

> The problem itself, the essence of it, the heart of it, is the white man's determination to make good this conviction, coupled with constant anxiety lest, by some means, he should fail to make it good. The race problem, in other words, is *not* that the Negro is what he is in relation to the white man, the white man's inferior; but this, rather: How to keep him what he is in relation to the white man; how to prevent his ever achieving or becoming that which would justify the belief on his part, or on the part of other people, that he and the white man stand on common human ground.

Ewing illustrated his conclusion by remarking on something that was self-evident at the time. "Everywhere in the South," he noted, "friction between the races is entirely absent so long as the Negro justifies the white man's opinion of him as an inferior; is grateful for privileges and lays no claim to *rights*." In other words, Ewing wrote, the expression "Good Nigger" is used to refer to a black person who knows his or her place, while the expression "Bad Nigger" means a black person who has gotten out of line.[35]

35. Ewing's observations would be confirmed in 1944 by Gunnar Myrdal, Richard Sterner, and Arnold Rose's encyclopedic survey of race relations in the South, which was funded by the Carnegie Corporation and published under the title *An American Dilemma* (New York: Harper, 1944). For an assessment of the importance of Myrdal, Sterner, and Rose's findings, see Gene Roberts and Hank Klibanoff, *The Race Beat: The Press, the Civil Rights Struggle, and the Awaken-*

Ewing concluded his article by tracing the race problem back to slavery. According to Ewing, the race problem was a manifestation of the South's struggle to perpetuate an institution that had been abolished. "How to maintain the institution of chattel slavery, ceased at Appomattox," Ewing explained. "The problem, How to maintain the social, industrial, and civic inferiority of the descendants of chattel slaves, succeeded it, and is the race problem of the South at the present time. There is no other."[36]

Ewing's conclusion contradicted the popular belief that white southerners willingly abandoned slavery after the Civil War. In fact, the opposite was true, as historian John David Smith has argued. Slavery continued to serve as the model for race relations in the South. "In its broadest sense," Smith has written, "slavery provided a metaphor that explained and justified race relations in the postwar South. This was a period of severe racial tension. Many race-related questions remained unsolved after Appomattox. What was to be the social, economic, and political role of the freedmen? How would blacks react to their new social status? And how would they fare physically without the alleged protection offered by slavery? Americans glanced backward to slavery for answers to these and other questions. They employed slavery as a constant touchstone, a reference point for contemporary issues."[37]

Because of slavery's relevance to the race problem, arguments used to defend slavery were resurrected to defend white supremacy after Reconstruction.[38] Traditionally, the defense of slavery was based on the Bible, particularly the Old Testament, in which references to involuntary servitude abound.[39] But the Bible could also be used to attack slavery and, by extension, the doctrine of white supremacy, such as when Paul wrote "There is neither

ing of a Nation (New York: Knopf, 2006), 3–11. For additional confirmation of Ewing's observations, see McMillen, *Dark Journey*, 45; and Smith, *Religion and "The Negro Problem,"* xii, xiv.

36. For a similar assessment, see Ulrich B. Phillips, "The Central Theme of Southern History," *American Historical Review* 34 (October 1928): 30–43.

37. Smith, *Old Creed for the New South*, 5.

38. For the evolution of proslavery arguments, see Bertram Wyatt-Brown, "From Piety to Fantasy: Proslavery's Troubled Evolution," in *Yankee Saints and Southern Sinners*, by Wyatt-Brown (Baton Rouge: Louisiana State University Press, 1985), 155–82.

39. For example, see John H. Hopkins, *Bible View of Slavery* (N.p., [1861?]), 1. For a rebuttal, see "A Vermonter," *Review of a "Letter from the Right Rev. John H. Hopkins, D.D.LL.D. Bishop of Vermont on the Bible View of Slavery"* (Burlington, VT: Free Press, 1861), 28 pages. Both pamphlets are in TSC.

Jew nor Greek, there is neither bond nor free, there is neither male nor female: for ye are all one in Christ Jesus" (Galatians 3:28).[40] Consequently, it was not surprising that white supremacists searched for other ways to strengthen their position, and one approach that looked promising appeared in the guise of a new discipline, anthropometry.

Although anthropometry was a primitive science before the Civil War, some would say a pseudoscience, Dr. Samuel George Morton of Philadelphia had established himself as an authority when his book *Crania Americana* was published in 1839. In it, Morton argued that anatomical differences in the average skulls of different races implied the inheritance of different mental abilities.[41] Using those findings as a starting point, it was a short step for Morton to argue that different races were, in fact, different species, a theory that became known as polygenesis. Josiah C. Nott, a physician and amateur ethnologist from Mobile, Alabama, endorsed Morton's theory of polygenesis and tied it neatly to the biblical argument of black inferiority with a notion that the black race had been created separately by God to inhabit Africa and, as a result, people of European stock were the only authentic descendants of Adam.[42]

The real breakthrough for white supremacists interested in scientific theories to "prove" that people of African descent were inherently inferior came just before the Civil War when Charles Darwin published *The Origin of Species*. The book's subtitle said it all: *The Preservation of Favoured Races in the Struggle for Life*. As George M. Fredrickson has noted, "The theory that evolution toward higher forms of life stemmed primarily from the conflict and competition of varieties and species, with the resulting 'survival of the fittest' and disappearance of the unfit, had obvious attraction for those who believed that some human races had a more exalted destiny than others."[43]

Although Darwin's theory of evolution was clearly at odds with the notion of polygenesis, that inconsistency was handled skillfully in 1866 when Nott morphed his belief in different species into the idea of slowly evolving differ-

40. For passages in the New Testament that deal with slavery, see "Slavery in the Bible: Passages from the Christian Scriptures," www.religioustolerance.org/sla_bib12.htm.

41. For a critical assessment of Morton's work, see Stephen Jay Gould, *The Mismeasure of Man* (New York: Norton, 1981), 50–69.

42. Fredrickson, *Black Image in the White Mind*, 74–75; Reginald Horsman, *Josiah Nott of Mobile: Southerner, Physician, and Racial Theorist* (Baton Rouge: Louisiana State University Press, 1987), 83–85, 104–5, 115–16.

43. Fredrickson, *Black Image in the White Mind*, 230.

entiation of races within a single species. White supremacists had nothing to fear. It would be hundreds of thousands of years before persons of African descent caught up, if ever.[44]

For some people, the salient issue concerning African Americans at the turn of the century was not "catching up" but "falling back." Basing their predictions on Darwin's theory of natural selection, some white supremacists claimed that people of African descent would not only dwindle in numbers, but actually regress toward their more natural state of savagery.[45] Slavery, they argued, had protected the inferior black race from the healthy competition with other races to survive and prosper. As a result, African Americans had ridden piggyback on European Americans in terms of their intellectual and cultural progression. With the paternalistic and uplifting protection of slavery gone, the black race in America would revert to savagery.[46]

The idea that African Americans were reverting to savagery complicated the race problem. On one hand, white southerners sought a system of social control that would keep the black segment of their population actively employed as an efficient labor force, primarily in agriculture, just as people of African descent had functioned for hundreds of years during the time of slavery. On the other hand, many white southerners were convinced that something had to be done to prevent African Americans from sliding backward into an unproductive and criminal state of barbarism. Suggestions as to how to accomplish both goals were often contradictory.

The question of black education provides an example. Professor Paul Barringer, professor of medicine and chair of the faculty at the University of Virginia, thought that a regimen of vocational training under the strict discipline of white instructors would reverse the black trend toward savagery.[47] John Roach Straton, a Baptist clergyman and professor at Mercer University in Georgia, was convinced that education would not have much effect on the black race, whom he called "this simple-minded, impressionable, and imita-

44. Ibid., 231–34; Horsman, *Josiah Nott of Mobile*, 249–50; Smith, *Religion and "The Negro Problem,"* xxvi–xxvii.

45. Fredrickson, *Black Image in the White Mind*, 245–47; Williamson, *Crucible of Race*, 111. See Frank A. Fetter, "Social Progress and Race Degeneration," *Forum* (October 1899): 228–38, for an example of the regression hypothesis. There is a copy of this article in TSC.

46. Fulkerson, *The Negro; As He Was; As He Is; As He Will Be*, 17–19, 60, 101.

47. P[aul] B[randon] Barringer, *The American Negro: His Past and Future*, 3rd ed. (Raleigh, NC: Edwards and Broughton, 1900), 23. There is a copy of this article in TSC. The titles and institutional affiliations for Barringer and Straton come from Fredrickson, *Black Image in the White Mind*, 252 and 271, respectively.

tive people."[48] James K. Vardaman, the governor of Mississippi, was bitterly opposed to spending any money at all for the education of African Americans because he was convinced that it would ruin them as productive workers.[49]

These differences of opinion over something as basic as education highlighted the dilemma facing white southerners. Dare we educate the black man in order to make him a more efficient laborer when doing so might foster unrealistic expectations in regard to his opportunity to participate in the country's economic, political, and social life? In addition, there was the question whether an educated black labor force would turn its back on hard work for subsistence wages, the sort of work uneducated workers were forced to accept, and abandon the fields for better jobs in factories. Finally, there was that sticky issue of the impact education might have on black suffrage. Educational requirements to vote had been inserted in the constitutions of many southern states as a mechanism for disfranchising black voters. Would encouraging literacy among black citizens undercut the constitutional mechanism for black disfranchisement?

Answers to complex questions such as these were not to be found in the Bible, nor did theories spawned by Darwin's observations provide much guidance. Rather, finding answers to these questions depended on scholars in the areas of economics, history, political science, and sociology, disciplines that would become known collectively as the social sciences.

Unfortunately, the social sciences were not ready to tackle the race problem when it was presented to them. They were in the midst of a transition from informal groups of independent generalists into professional organizations comprised of specialists based in the nation's finest universities. The discipline of history is a case in point. American historians following the Civil War were often well-educated, independently wealthy enthusiasts who used literature as the model on which to base their studies.[50] As such, their

48. John Roach Straton, "Will Education Solve the Race Problem?" *North American Review* 170 (June 1900): 801. There is a copy of Straton's article in TSC, as well as a copy of Booker T. Washington's response ("Will Education Solve the Race Problem? A Reply," *North American Review* 171 [August 1900]: 221–32).

49. Vardaman's article, "A Governor Bitterly Opposes Negro Education," appeared in *Leslie's Weekly* on February 4, 1904, two weeks after his inauguration. Vardaman began the article with these words: "The race question is one of the most serious problems which confront the civilization of the present country. The entire republic is interested in it; but the South, where the nigger lives in such large numbers, is of course more widely affected and therefore more materially and vitally interested" (104).

50. John Higham, *History: Professional Scholarship in America* (1965; Baltimore: Johns Hopkins University Press, 1983), 156–57.

worldview tended to be idealized, and a romantic depiction of the antebellum South came naturally.[51]

As intelligent and as well-educated as they may have been, the first generation of postbellum historians were at a distinct disadvantage when it came to sorting through the complexity of the issues with which they were confronted. "The typical historian was often an amateur, untrained in methodology, as easily inclined to accept tradition as he was documentary evidence," a student of historiography noted. "Furthermore, history was often considered a vehicle for teaching ethical values rather than as an end in itself."[52]

The alternative to the amateur scholar in the social sciences was the professionally trained academic, and the growth of professional organizations in these disciplines toward the end of the nineteenth century provides evidence of the gradual shift to the new model of scholarship. The American Economic Association was organized in 1885[53] and held its annual meetings for many years with the American Historical Association, which had been founded in 1884.[54] Anthropologists organized the American Anthropology Association in 1902 "to promote the science of anthropology,"[55] and political scientists followed a year later.[56] In 1905, sociologists gathered at a joint meeting of the American Historical, Economic, and Political Science Associations to form the American Sociological Society.[57]

Whereas literature had been the model for amateur scholars, science became the model for the new professionals,[58] and the desideratum for social scientists was objectivity. "The essence of scientific training is accuracy, first of observation, second of statement, third of verification,—all of this to be

51. Ibid., 92–93. For examples of a romanticized view of the South and slavery by an amateur historian, see Robert Cruden, *James Ford Rhodes: The Man, the Historian, and His Work* (Cleveland, OH: Press of Western Reserve University, 1961), 261–77.

52. Robert Reynolds Simpson, "The Origin of State Departments of Archives and History in the South" (Ph.D. diss., University of Mississippi, 1971), 34; Higham, *History*, 156–57.

53. "AEA General Information," www.vanderbilt.edu/AEA/org.htm.

54. "American Historical Association," www.historians.org/.

55. "A Brief History of the American Anthropological Association," www.aaanet.org/history.htm.

56. "About the American Political Science Association," www.apsanet.org/section_21.cfm.

57. "Establishment of the American Sociological Association," www.asanet.org. The American Sociological Society became an "Association" in 1959. The sixth discipline usually considered to be one the social sciences, psychology, formed its national organization, the American Psychological Association, in 1892.

58. Simpson, "Origin of State Departments of Archives and History in the South," 34.

done with the personal bias absolutely eliminated," a historian observed of his discipline in 1907.[59]

The most active center for the promulgation of the scientific method in the social sciences, especially history, and the training of graduate students in its tenets was at Johns Hopkins University under the leadership of Herbert Baxter Adams. Trained at the University of Heidelberg, Adams imported the German tradition of "scientific" history to Johns Hopkins in 1876.[60] According to Adams, conducting historical research was analogous to what naturalists did in the field. As he put it, books were "like mineralogical specimens, passed from hand to hand, examined and tested."[61] Another historian described the process as "the most unwearied search after every scrap of proof that could throw light upon the topic."[62]

The major problem facing this new emphasis on primary sources was the lack of material. Extensive collections that scholars rely on today at major universities did not exist at the time.[63] Furthermore, organizations at the state level for building and maintaining collections, such as state-supported departments of archives and history, were just beginning to be organized in the southern states.[64] As a result, the South lagged behind the North in terms of its ability to address complex issues. "The South, we agree, is seldom fairly presented in the account of any national interest or undertaking," William H. Kilpatrick, a historian from Georgia, stated to an audience at Columbia University in New York. "Why? The reason is, in a nutshell, that the South has not made available its data for history writing." Kilpatrick believed that the

59. [Colyer Meriwether], "Scientific History," *Publications of the Southern History Association* 11 (September–November 1907): 300.

60. Smith, *Old Creed for the New South*, 137–38, 142.

61. Herbert B. Adams, *The Study of History in American Colleges and Universities*, Circular of Information No. 2 (Washington, DC: Bureau of Education, 1887), 175, quoted in Simpson, "Origin of State Departments of Archives and History," 35. Adams was the "chief architect" of the American Historical Association, and he played a prominent role in it until his death in 1901 (Higham, *History*, 11).

62. [Meriwether], "Scientific History," 301.

63. It was not until the 1920s that the development of southern archives began in earnest. For an overview, see John David Smith, "'Keep 'Em in a Fire-Proof Vault': Pioneer Southern Historians Discover Plantation Records," *South Atlantic Quarterly* 78 (Summer 1979): 376–91; reprinted in John David Smith, *Slavery, Race, and American History: Historical Conflict, Trends, and Method, 1866–1953* (Armonk, NY: M. E. Sharpe, 1999), 133–43.

64. Simpson, "Origin of State Departments of Archives and History in the South," 4; John K. Bettersworth, "Mississippi Historiography: Research Materials and Researchers," *Mississippi Quarterly* 10 (Summer 1957): 138–45.

South itself was to blame for this shortcoming. "We must begin at once to make good this deficit," he declared. "The task is not easy. It requires both trained historians and adequate data."[65]

Albert Bushnell Hart, a history professor at Harvard, summarized the new approach in 1910: "What we need is a genuinely scientific school of history, which shall remorselessly examine the sources and separate the wheat from the chaff; which shall critically balance evidence; which shall dispassionately and moderately set forth results. For such a process we have the fortunate analogy of the physical sciences," Hart continued. "Did not Darwin spend twenty years in accumulating data, and in selecting typical phenomena, before he so much as ventured a generalization? . . . In history, too, scattered and apparently unrelated data fall together in harmonious wholes; the mind is led to the discovery of laws; and the explorer into scientific truth is at last able to formulate some of those unsuspected generalizations which explain the whole framework of the universe."[66]

Hart's reference to Darwin can be taken literally because the theory of evolution had been woven into fabric of the new historiography. "The concept of cumulative, on-going change, operating through the endless chain of tangible causes and effects, became for scientific historians the very essence of historical wisdom," John Higham wrote in his study of American historiography. In that regard, the theory of evolution "strengthened the realistic approach to history immensely. It showed that history was not a thing of shreds and patches, and this made the revolt against romantic history coherent and intellectually exciting." Understandably, the scientific historians scorned the "mere narrative" of history and rejected the emphasis that romantic historians had placed on personalities. Scientific historians, instead, prided themselves on studying the evolution of institutions.[67] The new approach opened new avenues to study of the race problem in ways that earlier paradigms could not.

The transformation of the social sciences did not occur overnight.[68] In the field of history, for example, professional historians were able to as-

65. William H. Kilpatrick, *Preserving Southern History Material: An Address before the Southern Club of Columbia University, July 31, 1923* ([New York]: Columbia University Press, 1923), 1, 6.

66. Albert Bushnell Hart, "Imagination in History," *American Historical Review* 15 (January 1910): 232–33.

67. Highham, *History*, 94–97.

68. Ibid., 8; John David Smith, "High Authority or Failed Prophet? Alfred Holt Stone and Racial Thought in Jim Crow America," *Journal of Mississippi History* 68 (Fall 2006): 198.

sume control over the American Historical Association's governance by the mid-1890s, but amateurs were routinely elected by the Executive Council as presidents of the organization well into the next century.[69] Nevertheless, a new path of inquiry had been charted, and the turn of the century was an exciting time for people interested in finding solutions to important issues, such as the American race problem. The need for objective research based on extensive data had been identified, even if the researchers to do the work and the facilities to compile the data were still in the process of being developed. Biblical explanations had been laid aside, and the search for data to help reveal the natural patterns of reality that lay beneath the outward manifestations of social custom was pursued with enthusiasm. Like any topic in a state of flux, the field was open to anyone, amateur or professional, with ideas or theories that could be examined with the newfound rigor of scientific methodology.

One of the new breed of racial theorists to enter the arena was a cotton planter from the Mississippi Delta. His name was Alfred Holt Stone, and his theories integrated an amateur's defense of slavery with the new, scientific approach. The story of his foray into the heated debate over the race problem reveals much about him, but even more about the complexity of racial thinking during the first decade of the twentieth century.

69. Higham, *History*, 13–16. The same could be said for the Southern Historical Association (E. Merton Coulter, "What the South Has Done about Its History," *Journal of Southern History* 2 [February 1936]: 25–28).

1 AN ORDERLY BRAIN

Alfred Holt Stone was a man with many friends, as evidenced by an editorial that appeared in the *Jackson Clarion-Ledger* on July 15, 1949, under the heading "Many Citizens Will Welcome Mr. Stone's Reappointment."

> Many thousands of Mississippians welcomed Governor Fielding Wright's announcement this week that he intends to reappoint Alf H. Stone as Chairman of the State Tax Commission when the latter's current term expires next spring.
>
> Many of these Mississippians welcomed the announcement as personal friends of Mr. Stone's, he having such friends in all sections.
>
> Many others, not knowing Mr. Stone personally, welcomed the announcement because they have some knowledge, as taxpayers, of the record he has set during his 17 years as head of the State Tax Commission, because of personal experience with the department's policy of courtesy and consideration for the public, a policy consistently honored and enforced throughout Mr. Stone's three terms as chairman.
>
> Chairman Stone has earned national recognition as an authority on administration of tax laws, and the administration of the Mississippi State Tax Commission has served as a model for officials of many other states.
>
> With many others, we welcome Gov. Wright's announcement that he will reappoint Mr. Stone, and congratulate both.

Stone was six months shy of his eightieth birthday when his reappointment became official in April 1950, making him one of the oldest office-holders in the history of the state.[1] Of all of Mississippi's public figures over

1. None of Mississippi's governors were close to eighty when they were elected. The oldest, Hugh Lawson White, was seventy when he was inaugurated in 1936. In fact, only seven of the fifty-six men who have served as governor of Mississippi lived to be eighty. They are

the years, only one, Senator John C. Stennis, trumped Stone when it came to holding office at an advanced age.[2]

Stone's reappointment to his fourth consecutive term as state tax commissioner was a tribute to his unimpeachable service in the administrations of seven governors. It is unlikely that a public official in such an important position could have survived the politics and patronage of seven administrations if there was any question about his competence. Furthermore, Stone's durability in office was a credit to his honesty and sincerity when it came to dealing with the public, something that the citizens of the state recognized and appreciated.

This aspect of Stone's personality can be seen in an open letter to "our taxpaying friends," which Stone distributed just before Christmas in 1941. "We come to you this year with a very brief message; just a few words of sympathetic interest and understanding," he began. "We fully realize the weight of your burdens. We would lighten them if we could, but such alleviation is neither within our province or our power. We make no tax laws. Our sole responsibility is as to their administration. It is in the discharge of this duty that we have endeavored to formulate a policy and to observe a procedure which we think of and refer to as a humanizing process."[3]

Stone's letter continued with an explanation of why collecting taxes should be viewed as a public trust. "We believe that the State no less than the taxpayer should have a conscience. The State owes a duty to the taxpayer no less binding and obligatory than the duty of the taxpayer to the State." Stone closed his letter with these words: "In this solemn hour of our Nation's trial and her need, we wish for each and all of you everything which it is seemly and proper that we wish." The letter was dated December 17, 1941, ten days after the Japanese had attacked Pearl Harbor.

Alfred Holt Stone was sixty-one when he was first appointed state tax commissioner in 1932, an age at which most people have begun thinking about retirement. Not surprisingly, he had pursued another career before

---

Henry Whitfield (84), Adelbert Ames (98), Robert Lowry (81), Andrew Houston Longino (88), Hugh Lawson White (84), Ross Robert Barnett (89), and William Forrest Winter (84 in 2007) ("Governors of Mississippi from 1817 to 2004," mshistory.k12.ms.us/features/feature47/governors-first.htm).

2. Born in 1901, Stennis was elected to his final term in the U.S. Senate when he was eighty-one. He died on April 23, 1995, at the age of ninety-three (politicalgraveyard.com).

3. Alfred H. Stone, *A Personal Word to Our Taxpaying Friends with a Christmas Letter to Our Staff* (Jackson: [Mississippi State Tax Commission], 1941).

becoming a tax commissioner. Stone was a cotton planter in the Mississippi Delta.

Growing cotton is an unpredictable business, and like all of the cotton planters in the Delta during the first part of the twentieth century, Stone had to contend with two natural disasters: the boll weevil and flooding from the Mississippi River.[4] Faced with these challenges and other problems related to growing cotton, Stone worked, like other cotton planters, to keep his head above water. Of the many the useful things Stone accomplished in that capacity, his involvement with the Staple Cotton Cooperative Association was probably the most important.

The Staple Cotton Cooperative Association—or Staplcotn, as it became known—was organized in 1920 by cotton growers who wanted to sell their crops directly to buyers without having to go through a middle man. Stone was one of the founders of Staplcotn, a member of its board of directors, and the editor of its monthly newsletter for more than three decades.[5]

Staplcotn was a great success, and Stone was big part of it. "A truly great man" is how an editorial on the front page of the *Staple Cotton Review* described Stone at his death. "In his magnificent editorship of *The Staple Cotton Review,* and in all that he did for the Staple Cotton Cooperative Association and the Staple Cotton Discount Corporation, for more than thirty-two years, those of us who worked with him, who loved and revere him, can say in all sincerity that his work and plans and visions extended well 'beyond the reach of ordinary capacities.'"[6]

Despite his success in two distinct careers, Stone enjoyed a third career that brought him even greater notice on the national scene. This career was relatively brief when compared to his tenure as the tax commissioner and his work with Staplcotn, and it occurred earlier in his life. Yet, it was this career that brought him recognition that exceeded the attention he was to achieve later. This career was based on Stone's success as a self-taught, multidisciplinary theorist of the race problem.

4. Most people have heard about the boll weevil, but they may not be aware of the frequency with which the Mississippi River overflowed its banks before the Federal Flood Control Act in 1928 provided protection from periodic inundations. In the fifty years prior to the great flood of 1927, a period that essentially coincides with the development of the Delta as an important agricultural basin, there were major floods in 1882, 1884, 1890, 1897, 1903, 1912, and 1913 (A[lfred] H[olt] Stone, "'And the Waters Prevailed,'" *Staple Cotton Review* 5 [July 1927]: 3).

5. Stone continued to serve as the editor of the Staplcotn newsletter after he assumed duties as the state tax commissioner (Stone's obituary in the *Jackson Daily News,* May 12, 1955).

6. "Empty Giant Shoes," *Staple Cotton Review* 33 (June 1955): 1.

Stone lived in a part of the state where African Americans greatly outnumbered white people. In fact, Stone often made a point that there were more Negroes, as he referred to African Americans, in Washington County, Mississippi, than in any one of twenty-eight states or territories of the Union.[7] Like every planter in the Mississippi Delta, Stone relied almost entirely on black labor for his success as a cotton grower. It is not surprising, therefore, that he developed a special interest in that segment of the population. What set Stone apart from his white colleagues in the cotton-growing business who also wanted to understand their black workers was Stone's mental discipline and capacity for tedious research. It was these characteristics, coupled with a good education and a quick mind, that enabled Stone to become a recognized authority on questions involving race.

Beginning with a conference on the race problem that he attended in 1900 until the boll weevil arrived in Washington County in 1909, Alfred Holt Stone threw himself into the study of people of African descent. His interests in black folk extended well beyond Mississippi and included not only the rest of the South, but also the North, the Caribbean, and Africa. He wrote extensively, gave lectures at professional meetings and universities, and eventually became known as an expert on the subject.[8] In 1908, Representative Benjamin G. Humphreys from Mississippi referred to Stone during a speech in Congress as "perhaps the most profound student of the race question in this country to-day."[9]

Stone could not have become an expert without accumulating a great deal of information about his topic. An avid reader, Stone collected all sorts of material for his personal library on people of African descent, including census reports, papers presented at professional meetings, speeches delivered by politicians and educators, economic assessments of countries with indigenous black populations, sermons by preachers either defending slavery or attacking it, published narratives written by slaves and freedmen, gov-

7. For example, see Alfred Holt Stone, "Foundations of Our Differences," in *SARP*, 44. The twenty-eight states and territories Stone listed were: West Virginia, Indian Territory (now Oklahoma), Massachusetts, Delaware, Oklahoma, Michigan, Connecticut, Iowa, California, Rhode Island, Colorado, Nebraska, Minnesota, Wisconsin, Washington, Arizona, New Mexico, Montana, Maine, Oregon, Wyoming, Vermont, Utah, New Hampshire, South Dakota, Idaho, North Dakota, and Nevada. His data came from Walter F. Willcox, "Census Statistics of the Negro," *Yale Review* (November 1904): 274–86. There is a copy of this pamphlet in TSC.

8. Smith, "High Authority or Failed Prophet?" 196.

9. Benjamin G. Humphreys, *In Defense of the State of Mississippi against the Charge of Peonage: Speech of Hon. Benjamin G. Humphreys on Mississippi in the House of Representatives, Monday, March 2, 1908* (Washington, DC: n.p., 1908), 2.

ernmental reports, and a host of other sources spanning a period from the late eighteenth century through the first decade of the twentieth. By the time he was finished, Stone had assembled an extraordinary collection of material on "the Negro and cognate subjects." Some of the titles he collected were hardbound books, but most were articles, pamphlets, and paperbound monographs, more than three thousand of them. Fortunately for scholars today, his collection is still intact.

Usually, one can tell a lot about what a person believes by looking at his or her library. Although it is not unusual for the libraries of well-educated people to contain titles that reflect a variety of opinions, the preponderance of books in an individual's personal library will tend to reflect his or her beliefs. As a Harvard law professor remarked about nominees for the U.S. Supreme Court, "Probably the best place to look, if you want to guess his future attitude toward important cases, is not in his file of clients or in his safe-deposit box, but at the books in his private library at home."[10]

You would never guess what Stone believed by looking at his library. Although Stone's writings reflected the strong racial prejudices that were common in both the North and South at the time, he did not limit his collection to works that reinforced his opinions. He wanted it all, even though much of what he read was at variance with his stated beliefs. Stone collected pamphlets written by abolitionists as readily as he collected pamphlets defending slavery. Speeches in Congress dealing with the extension of slavery reflected arguments from both sides in Stone's collection. And pamphlets that criticized the southern states for disfranchising black voters during the early days of Jim Crow were balanced by pamphlets defending the practice.

Stone collected material for his library because he loved books. "While in no sense a recluse, books were his most constant companions, and his private library was filled with thousands of precious and well-thumbed masterpieces of all ages," an editor wrote in the *Jackson Daily News* the day after Stone died in 1955. "He did not depend on memory alone for his learning. All that he read was well assimilated and stored away in an orderly brain. Learning was the greatest comfort of his old age—far more precious to him than any worldly wealth or personal ambition."[11]

Stone gave his collection to the Mississippi Department of Archives and History (MDAH) in the 1930s. His hardbound books were catalogued and

10. Zechariah Chafee, *Free Speech in the United States* (Cambridge: Harvard University Press, 1946), 359.

11. "Alfred Holt Stone," editorial in the *Jackson Daily News*, May 12, 1955.

put on the shelves with the department's other holdings. The articles, pamphlets, and paperbound monographs were kept separate from other collections and stored until a suitable finding aid could be developed. Lil Kirkpatrick wrote an article about Alfred Holt Stone in 1981 that described his collection of articles and pamphlets as follows: "Its value lies in its uniqueness. The encyclopedic compilation of research material covers broad areas of interest and points of view." Silas McCharen, a staff person at the MDAH whom Kirkpatrick interviewed for the article, said: "The Stone Collection is an attempt to bring together in one place the best of the available knowledge of the time. It attempts to give a holistic vision of the place of black people in the world—their history, their problems, their struggles. Its value to the student of black studies cannot be overstated."[12]

Stone's collection of articles and pamphlets at the MDAH is his greatest legacy. Although he was an innovative and honest tax administrator, many tax commissioners have come and gone in Mississippi since Stone's four terms. And although Staplcotn is still the largest producer-owned cotton cooperative in the United States, Stone was only one of many men who have made that organization the success it has become.[13] But Stone's collection of material on black folk stands alone. It is a unique reflection of Stone's personality, an accomplishment that was solely his own, and a treasure for scholars of race relations today.

12. Lil Kirkpatrick, "Archives Receives Stone's Typewriter," *Jackson Clarion-Ledger*, June 14, 1981.

13. Staplcotn Web site, www.staplcotn.com/Index1.htm.

2 ESSENTIALS OF GREATNESS

Sometimes people end up in unexpected places by chance. Eli Evans, the biographer of Judah P. Benjamin and chronicler of Jewish people in the South, told an interesting story to this effect when he spoke at the Old Capitol in Jackson, Mississippi, on a book-signing tour.[1] Evans's first book was entitled *The Provincials: A Personal History of Jews in the South,* and he was fascinated by accounts of how Jewish peddlers ended up as merchants in small towns scattered across the southern landscape. According to Evans, when he asked a question such as, "How did the Goldsteins end up in Griffin?" the reply was often, "That is where the horse died." The story of how Alfred Holt Stone's family ended up in Greenville, Mississippi, was along these lines. The Stones were not Jewish, and the horse did not die, but Stone's father was a merchant, and his path to Greenville involved a horse.

The first member of Stone's family to reach America came from England and settled on the eastern shore of Chesapeake Bay in 1628. His name was William Stone, and although his family grew and prospered in the New World, William Stone's religious beliefs conflicted with the official doctrine of the Anglican Church in Virginia. Joining other Protestants who had immigrated to America with him, William Stone reached an agreement with Lord Baltimore to settle in Maryland and moved his family to the future site of Annapolis in 1648. In August 8 of that year, Lord Baltimore named William Stone governor of Maryland, making him the first Protestant governor of a predominantly Catholic colony. A year later, William Stone signed the Re-

1. Judah P. Benjamin from Louisiana served in Jefferson Davis's Confederate cabinet throughout the Civil War and was easily the highest-ranking officer of the Jewish faith to serve on either side during the conflict (Eli N. Evans, *Judah P. Benjamin: The Jewish Confederate* [New York: Free Press, 1988]). The date of the book signing was October 30, 1988.

ligious Toleration Act, which protected the religious freedoms of Christian denominations in the colony. As such, the act was a precursor to the First Amendment in the Bill of Rights.[2]

William Stone's descendants were counted among Maryland's most prominent leaders by the time of the American Revolution. Thomas Stone, William Stone's great-great-grandson, signed the Declaration of Independence. Another great-great-grandson, Michael Jenifer Stone, represented Maryland in the first United States Congress, and Michael's younger brother, John Hoskins Stone, later became governor.[3]

Alfred Holt Stone's grandfather, Caleb S. Stone, was the first member of the American Stones to be born outside of Maryland or Virginia. Caleb's father, James, had moved west to establish a pioneer farmstead in Madison County, Kentucky, near Boonesboro around the turn of the nineteenth century. Caleb grew up in Kentucky but decided in 1830 to seek his fortune in Missouri, much of which was still a frontier wilderness. Caleb Stone did well in Missouri, first as a merchant and later as a farmer. He became a wealthy man and married Ann Wilson, who also had lived in Kentucky before her family decided to try their luck in Missouri. Twelve children were born from this union, the fifth of whom was Walter Wilson Stone, Stone's father, born in 1840.[4]

Stone claimed in his memoirs that his grandfather, Caleb Stone, did not believe in slavery. "He abhorred it," Stone wrote. "My father often told me that his father left Kentucky and removed to Boone County, Missouri, more because of slavery than for any other reason."[5] But if Caleb Stone hated slavery, he kept strange company, for he became active in the fiercely proslavery wing of the Democratic Party, which was a power to be reckoned with in that part of the slave state.[6] In fact, if Caleb Stone had left Kentucky to avoid

2. Alfred Holt Stone identified William Stone as an ancestor in "Some Recollections of a Southern Planter," chap. 2, p. 4 (typescript in Stone's subject file, MDAH).

3. Information about the Stones in Maryland comes from "The Political Graveyard," politicalgraveyard.com/bio/stone.html; and "William Stone," en.wikipedia.org/wiki/William_Stone.

4. A. H. Stone, "Some Recollections," chap. 2, p. 4; *Biographical and Historical Memoirs of Mississippi: Embracing an Authentic and Comprehensive Account of the Chief Events in the History of the State and a Record of the Lives of Many of the Most Worthy and Illustrious Families and Individuals* (Chicago: Goodspeed Publishing Co., 1891), 2: 855.

5. A. H. Stone, "Some Recollections," chap. 2, p. 4.

6. Jonas Viles, *The University of Missouri: A Centennial History* (Columbia: University of Missouri, 1939), 94.

slavery, he moved to the wrong place. Boone County was one of only 7 counties in Missouri (out of 113), in which more than a quarter of the population was enslaved.[7]

In 1843, Caleb Stone was elected by the Missouri legislature to serve on the Board of Curators for the newly established University of Missouri in Columbia.[8] His was a powerful position, particularly after he was elected president of the board in 1849, and Caleb Stone used his position to go after the university's first president, John H. Lathrop. Caleb Stone and other members of the board objected to Lathrop because he was from New York State, had gone to Yale, and was suspected of harboring antislavery sentiments. Sensing the growing opposition on the board, Lathrop resigned to assume the presidency of another recently organized university in the Midwest, the University of Wisconsin in Madison. His replacement, someone whom Caleb Stone supported enthusiastically, was Dr. James Shannon, an Irish-born Baptist minister and slave owner from Kentucky who was well known for his aggressive use of biblical texts to defend slavery.[9]

It appears that Caleb Stone was a bit of a gadfly who alienated people with his uncompromising nature. This characteristic was evident when some members of the board who had supported of Dr. Lathrop sought to pass a resolution thanking the university's first president after he resigned. As introduced, it read, "Resolved, That we have the highest confidence in the learning, talents, integrity, and upright moral character of President John H. Lathrop, and while we deeply regret the loss of his valuable service to our State University[,] we cordially recommend him to the confidence of the community wherever his lot may be cast." Board members who had been prepared to fire Lathrop if necessary refused to pass the resolution unless the phrase "while we deeply regret the loss of his valuable services to the State University" was struck out. That amendment to the resolution passed by a vote of six to three, but Caleb Stone wanted to do something more to humiliate the erstwhile academic administrator and insisted that the minutes reflect that he, Stone, "did not entertain the highest confidence in the

7. The total number of counties in Missouri in 1860 was extracted from a table posted on the Internet by the National Association of Counties ("Missouri," www.naco.org/Template.cfm?Section=Find_a_County&Template=/cffiles/counties/state.cfm&statecode=mo). The slave statistics came from the 1860 census and are plotted on a map entitled "Civil War Missouri," in T. J. Stiles, *Jesse James: Last Rebel of the Civil War* (New York: Knopf, 2002), 11.

8. "University of Missouri Board of Curators List," muarchives.missouri.edu/uw-rg1a.html.

9. Viles, *University of Missouri*, 24, 50–52, 80–81, 90, 92–94; "Dr. James Shannon, 1799–1859," www.therestorationmovement.com/shannon.htm.

learning, talents, integrity, and upright moral character of President John H. Lathrop."[10]

Although Caleb Stone was reelected to the Board of Curators for more than a decade, support in the legislature for his dogmatic opinions ended in 1859. His failure to be reappointed may have resulted from the polarization of politics in the years just prior to the Civil War and a weakening of the extreme prosouthern faction in Boone County.[11] Whatever the reason, Caleb Stone sided with the South when war broke out two years later and was appointed adjutant-general of the Missouri State Guards, an organization that eventually was incorporated into the Confederate army. Caleb Stone survived the war, primarily because he was assigned to the quartermaster department, and returned to Columbia in 1865, where he resided until his death in 1873 at the age of sixty-three.[12]

Walter Wilson Stone, Stone's father, grew up in Columbia and gravitated naturally to the academic subjects taught at the university over which his father wielded so much influence. Graduating in 1859, Walter began reading law while teaching Latin and Greek at his alma mater to make a living. Like his father, Walter Stone sided with the South and joined a local company. Arrested by federal authorities for trying to recruit men for the Confederate army, he was briefly held in the same building where he had taught. Walter Stone slipped away, however, and made his way south.[13] Eventually, he enlisted as a private in the Ninth Missouri Infantry and served throughout the war in the Confederacy's Trans-Mississippi Department.[14] Promoted to the rank of captain and assigned to a staff position, Walter Wilson Stone participated in the Red River campaign, although it is not clear whether he saw combat.[15]

10. Frank F. Stephens, *A History of the University of Missouri* (Columbia: University of Missouri Press, 1962), 72–73.

11. Viles, *University of Missouri*, 94.

12. *Biographical and Historical Memoirs of Mississippi*, 2: 855.

13. "Captain Stone Death Victim," *Jackson Daily Clarion-Ledger*, June 5, 1930. For Alfred Holt Stone's account of his father's escape, see "Some Recollections," chap, 2, p. 10.

14. There were two regiments designated as the Ninth Missouri Infantry in the Confederate army. Stone's father belonged to the regiment commanded by Colonel, later Brigadier General, John B. Clark in Parson's Division (Jerry Ponder, *The 9th Missouri Infantry Regiment, C.S.A., and the 12th Missouri Regiment, C.S.A.* [Doniphan, MO: Ponder Books, 1996], 32–34).

15. Accounts of Captain Stone's service in the Confederate army vary. The most reliable information about his service comes from the "Biographical Memoranda" he filled out 1903, which is in his subject file at the MDAH. Unfortunately, this account is brief (approximately fifty words). His obituary and the biographical sketch in Goodspeed provide more detailed but

The end of the war found Captain Stone in Shreveport, Louisiana. Although many Confederate soldiers started for home as soon as they learned that Lee and Johnston had surrendered their armies in the East, Captain Stone remained at his post and was paroled on June 10, 1865, the same day that the local newspaper published a federal order abolishing slavery in Louisiana.[16]

Initially, Captain Stone was reluctant to go home and did odd jobs around Shreveport. But his father, Caleb, had gone back to Missouri, and he missed his mother, so Captain Stone started out cross-country on an old gray nag, heading toward the Mississippi River, where he planned to catch a steamboat north. After several days of riding in the midsummer heat, Captain Stone came to the river across from Catfish Point on the Mississippi side. Catfish Point is one of those swampy extensions of land formed by the river's serpentine course and lies in Bolivar County about fifteen miles north of Greenville.[17]

Captain Stone hailed a ferryboat that was tied up on the far side, and the ferryman took him and the old horse across, setting Captain Stone down about three miles above Mound Landing near Scott, Mississippi, his first time to set foot in the Delta. Captain Stone spent the night in the ferryman's cabin and paid the fare with his last money, a five-dollar gold piece. His money gone, Captain Stone decided to sell the horse in order to buy a ticket on a steamboat headed north. Fortunately, he was able to get sixty dollars for the nag from a widow and spent the night with a neighbor. "I shall never forget the kindness of Gus Newman and his good wife during my stay," Captain Stone recalled almost fifty years later. "My acquaintance with the Newmans proved the anticipation of kindly friendships in pioneering the region around Greenville."[18]

Captain Stone caught a boat to St. Louis and took the train to Columbia to

---

conflicting information (*Jackson Daily Clarion-Ledger*, June 5, 1930, and *Biographical and Historical Memoirs of Mississippi*, 2: 856, respectively).

16. "Biographical Memoranda," Walter Wilson Stone subject file, MDAH; John D. Winters, *The Civil War in Louisiana* (Baton Rouge: Louisiana State University Press, 1963), 423–27.

17. W[alter] W[ilson] Stone, "Some Post-War Recollections," in Henry Tillinghast Ireys, *Memoirs of Henry Tillinghast Ireys: Papers of the Washington County Historical Society, 1910–1915*, ed. William D. McCain and Charlotte Capers (Jackson: Mississippi Department of Archives and History, 1954), 237. The description of the horse is verbatim from this source.

18. Ibid., 237–38. Captain Stone read his memoirs to the Washington County Historical Association during the fall and winter of 1912–13.

see his family. It did not take long for him to discover that Confederate gray was out of style in postbellum Missouri, and he headed back south to find more hospitable surroundings. Memphis, Tennessee, was a bustling center of economic activity following the war, and Captain Stone decided to try his luck there. In late August 1865, Captain Stone arrived in Memphis "with nothing to do, and but little to do it with; not altogether aimless, but without a specific object to aim at, a 'creature of circumstance floating upon the current of events,' waiting for something to turn up, but not able to turn it himself."[19]

Despite the modest assessment of his prospects, Captain Stone did have a college degree, he was young, and he was healthy. Consequently, it did not take long before he cast his lot with a group of investors in Memphis who planned to lease an abandoned plantation and recruit black soldiers recently discharged from the Union army to work in the fields. The idea was not novel,[20] but the investors behind this scheme were particularly well suited to run a plantation like a military operation. One was Francis P. Blair Jr., from Missouri, the brother of the postmaster general in Lincoln's cabinet and a major general in the Union army.[21] Another was the former Confederate general Nathan Bedford Forrest.[22]

Given that the investors would be relying on free labor in a country that had been worked recently by slaves, they deemed it advisable to locate the operation "in some remote and inaccessible locality," ostensibly to minimize the distraction of bright lights in a big city. After studying several locations in the Cotton Belt, the investors settled on a spot just west of Lake Village in Chicot County, Arkansas, across the river from Greenville. However, the investors wanted someone to inspect the plantation in person before committing themselves to a course of action, someone who was familiar with the

19. W[alter] W[ilson] Stone, "Greenville in 1865," in Ireys, *Memoirs of Henry Tillinghast Ireys*, 78. Captain Stone read this paper before the Washington County Historical Association in 1910.

20. For example, see James E. Yeatman, *Suggestions of a Plan of Organization for Freed Labor, and the Leasing of Plantations along the Mississippi River, under a Bureau or Commission to be Appointed by the Government* (St. Louis, MO: Western Sanitary Commission, 1864), 8 pages. A copy of this pamphlet is in TSC.

21. Ezra J. Warner, *Generals in Blue: Lives of the Union Commanders* (Baton Rouge: Louisiana State University Press, 1964), 35–36.

22. W. W. Stone, "Greenville in 1865," 81. For an account of Forrest's postbellum commercial ventures, see Jack Hurst, *Nathan Bedford Forrest: A Biography* (New York: Knopf, 1993), 265–76.

area, and someone without anything else to do. Captain Stone fit the bill admirably, and he was dispatched downriver by steamboat.[23]

The morning of the second day out of Memphis, Captain Stone was called from his berth and advised that he was at his destination. "Where am I?" he asked the "mud clerk" who was supervising the discharge of freight as Captain Stone disembarked. "Greenville, Mississippi," the clerk told him as the steamboat pulled away. Captain Stone found himself standing on the riverbank surrounded by nothing "but sky, weeds, and water." The place Captain Stone wanted to go was on the other side of the river, but someone at the makeshift landing told him that if he walked about a mile downriver to Old Greenville, which had been destroyed during the war, he would find a camp of black soldiers. He could get one of them to row him across in a skiff. Despite his bitterness at having to rely on his former enemies, black ones at that, Captain Stone reached the camp and asked the first African American soldier he encountered to take him across the river. Somewhat to his surprise, the soldier was glad to help, and Captain Smith was able to complete his mission.[24]

The cotton-planting scheme in Chicot County did not work out, but Captain Stone's foray downriver was fortuitous. It was his second trip to the Greenville area, and he liked what he had found there. The people were friendly, especially to veterans in gray, and the town looked like a place that was destined to grow. The old town of Greenville had been burned during the war, and the river had encroached on what was left. Consequently, the town's residents had determined to move the site of their municipality to a location that was protected from the mighty Mississippi's indomitable yet fickle course. They found a spot about a mile upriver from the old site where an island acted as a buffer between the main channel and the new town. The land belonged to a widow, Mrs. Harriet B. Theobald, and she obligingly sold lots as a means to recover from the financial ruin that her plantation, Blantonia, had suffered as a result of the war.[25]

23. W. W. Stone, "Greenville in 1865," 78–79. Ironically, the parcel of land the investors were interested in buying was directly across Chicot Lake from another cotton plantation that thirty years later became known as Sunnyside.

24. Ibid., 79–80. Captain Stone described his attitude toward black soldiers as follows, referring to himself in the third person: "Thus the prospect was gruesome indeed, for he [Stone] had been fighting negro soldiers for two years, was envenomed against the whole brood, and now had to appeal to them for help out of distress."

25. W. W. Stone, "Some Post-War Recollections," 272–73; James F. Brieger, "Hometown Mississippi," 2nd ed. (photocopy of a typescript, 1980, MDAH), 514; Henry T. Ireys, "County Seats and Early Railroads of Washington County," *PMHS* 14 (1914): 284–85.

Captain Stone returned to Memphis and formed a partnership with J. M. Maury and E. V. Ferguson to establish a merchandising business in the new town. Captain Stone returned by steamboat later that fall, bringing lumber for their store with him. He was present at the dedication of the new site for Greenville in January 1866, and one of the first buildings to be erected at the foot of Washington Avenue sported a sign reading "Maury, Stone and Company."[26] Doubtless, Captain Stone's experience as a staff officer in the Confederate army suited him for his new career, and he quickly established himself as a successful and respected member of the community.

Young, energetic, and ambitious, Captain Stone sought to enlarge on his success in merchandising by trying his hand at growing cotton. After all, cotton was the staple that drove the Delta economy, and Stone did not see why he should not share in the largesse resulting from the conscientious cultivation of that crop. Fortunately, he had some help for his new venture. Captain Stone had learned from one the passengers on the steamboat that an old friend, Captain H. W. Anderson, who was also an ex-Confederate and expatriate from Missouri, had settled in Washington County to try his hand at growing cotton. Captain Stone renewed his friendship with Captain Anderson, and eventually the two men went into the cotton-growing business together.[27]

The former Confederate captains-turned-cotton-planters leased Ararat Plantation on Deer Creek. There were many cotton plantations lining Deer Creek, and one of them, Whitehall, was owned by Dr. Alfred C. Holt, a physician who resided with his family in New Orleans. Dr. Holt provided the capital, and his brother-in-law ran the operation, which included a large frame dwelling in which Dr. Holt stayed when visiting from the Crescent City. It so happened that another one of Captain Stone's friends was visiting Ararat when Dr. Holt brought his wife and oldest daughter with him to Whitehall. Captain Stone's friend knew the Holt family and suggested that they ride over one day to pay their respects. Captain Stone agreed, and that was how, in May 1868, he met Eleanor Holt. The specifics of the courtship have been lost to history, but Captain Stone's amorous assault prevailed, and the two were married on October 2, 1869, at the Prytania Street Presbyterian Church in New Orleans.[28] Eleanor was nineteen, ten years younger than her

26. W. W. Stone, "Some Post-War Recollections," 243–45.

27. Ibid., 244–45.

28. Ibid., 263–64; A. H. Stone, "Some Recollections," 21. Captain Stone's brother, Owen, also moved to Greenville and became a medical doctor. He married his brother's wife's sister, Mary Holt, in 1881 (Alice Wade and Katherine Branton, "Marriage Records from the Greenville

husband. A year later on October 16, she gave birth to her first child, a son, whom they named Alfred Holt Stone after her father.[29]

Eleanor Holt's family, too, had come to Mississippi from England by way of Kentucky. Her father had moved to Mississippi to practice medicine and to grow cotton before the Civil War. A slaveholder and ardent secessionist, Dr. Holt was elected to the Mississippi Secession Convention from Wilkinson County and served on the committee that framed Mississippi's ordinance of secession.[30] During the war he served as a physician, possibly with the Army of Tennessee, and relocated to New Orleans after the collapse of the Confederacy, although he maintained extensive business interests in Mississippi, particularly as they related to growing cotton.[31] Eventually, he retired from the practice of medicine and died at his home in Summit, Mississippi, in 1881.[32]

Not everyone in the Holt family shared Dr. Holt's secessionist views. His brother, Joseph, provides one of those interesting examples of how the Civil

'Times,'" in *Early Records of Mississippi: Issaquena County, Washington County,* 3: 82 [typescript, 1983, MDAH].

29. Alf Stone was Walter and Eleanor's only son. He had four sisters who lived to adulthood: Anna, Lilian, Daisee, and Ellena (1880 Census, Stoneville Precinct, Washington County, Mississippi). Stone was born in Dr. Holt's house on Polymnia Street in the Garden District in New Orleans (A. H. Stone, "Some Recollections," chap. 2, p. 1).

30. *Biographical and Historical Memoirs of Mississippi,* 2: 134. Holt was not in favor of immediate secession. Rather, he assumed the position that Mississippi should wait until several southern states were ready to secede together and form a new government. "I am in favor of a concert of action among the cotton-growing States," he stated in a broadside that Holt circulated as part of his campaign to gain election to the secession convention. "I am *not* in favor of Mississippi going out alone and *will resist it*" ("The Address of Dr. A. C. Holt," December 10, 1860, oversized broadside file, MDAH; emphasis in original).

31. The various databases of persons who served in the Confederate army do not provide evidence of Dr. Holt's service, including searches of the "Civil War Soldiers and Sailors" site maintained by the National Park Service (www.itd.nps.gov/cwss/index.html); the American Civil War Research Database (www.civilwardata.com/), or H. Grady Howell, *For Dixie Land I'll Take My Stand! A Muster Listing of All Known Mississippi Confederate Soldiers, Sailors and Marines* ([Madison, MS]: Chickasaw Bayou Press, 1998). Reference to his service is based on information from the biographical sketch of Alfred Holt Stone in Dunbar Rowland, *The Official and Statistical Register of the State of Mississippi, Centennial Edition, 1917* (Madison, WI: Democrat Publishing, 1917), 941, and from William D. McCain, ed., *The Story of Jackson: A History of the Capital of Mississippi, 1821–951* (Jackson, MS: Hyer, 1953), 2: 222.

32. The date of Dr. Holt's death was October 5, 1881 ("Died," *New Orleans Daily Picayune,* October 7, 1881). His wife, Mary Williams Stone, died on August 17, 1899, at the home of a son-in-law in Bay St. Louis, Mississippi ("Died," *New Orleans Times-Democrat,* August 19, 1899; also in *New Orleans Daily Picayune,* August 19, 1899).

War turned brother against brother. Joseph Holt read law and gained distinction in Kentucky before also moving to Mississippi. Settling in Port Gibson near Vicksburg, Joseph Holt became active in the Democratic Party and made a speech at the 1836 National Democratic Convention in support of vice-presidential aspirant Richard M. Johnson, a hero of the War of 1812 who was also from Kentucky. Johnson won, but not without controversy, for he was married to a woman of mixed-race ancestry. Johnson defended his marriage by pointing out that it was customary for slave owners to have sexual relations with their black servants. The difference between him and other slave owners was that he assumed responsibility for his behavior and sanctioned the relationship in the eyes of God through their marriage vows.[33]

For reasons that are unclear, possibly because of his support for Johnson, Joseph Holt moved back to Louisville in 1842 and continued his involvement in the Democratic Party, albeit the northern branch. In 1856, he campaigned for James Buchanan and was rewarded with an appointment as the commissioner of patents and later as the postmaster-general. When the southerner John B. Floyd resigned as the secretary of war in the final months of the Buchanan administration, Joseph Holt was appointed to that post, serving as the interim secretary of war from January 1, 1861, until March 4, when Abraham Lincoln was inaugurated.[34]

Joseph Holt pursued a course that was diametrically opposed to his brother's when war loomed on the horizon. Speaking out vigorously against secession in Kentucky, Joseph Holt's staunch Unionism was rewarded when President Lincoln appointed him in September 1862 as the first judge advocate general of the army. His service in that capacity revealed a keen mind, common sense, and an iron will. In 1864, for example, Major General William T. Sherman sought his opinion in the following matter. "The question arises daily, and I expect to execute a good many spies and guerrillas under that law without bothering the President," Sherman wrote from Nashville, Tennessee. "Spies and guerrillas, murderers under the assumed title of Confederate soldiers, deserters on leave, should be hung quick, of course after a trial," he explained, "for the number of escapes made easy by the changes on guard during the long time consumed by trial and reference have made that class of

33. "Joseph Holt," www.famousamericans.net/josephholt/; "Richard M. Johnson, 1780–1850," http://www.juntosociety.com/vp/rmjohnson.html. Johnson served as the vice president to Martin Van Buren.

34. Holt's appointment was dated December 31, 1860, and signed by James Buchanan (*War of the Rebellion: A Compilation of the Official Records of the Union and Confederate Armies*, 128 vols. [Washington, DC: Government Printing Office, 1880–91], ser. 3, 1: 21; hereafter cited as *OR*).

men bold and dangerous, and our own scouts and detachments have so little faith in the punishment of known desperadoes that a habit is growing of 'losing prisoners in the swamp,' the meaning of which you know." Judge Holt responded to Sherman the next day with one sentence. "You have no authority to carry into execution a death sentence pronounced by a military commission."[35] Judge Holt was not a man to quibble. Among other things, he implemented the administration's policy of keeping prisoners in jail without recourse to habeas corpus and trying individuals by military commissions instead of civil courts. Judge Holt also prosecuted Peace Democrats in 1864 and later tried to link Jefferson Davis and the Confederate government to the conspiracy to assassinate President Lincoln.[36]

As could be imagined, Judge Holt was not popular among the Holts who sided with the South. Although Stone wrote many years later that his wife's uncle "was one of the greatest lawyers of his time," he also observed that Judge Holt "was the only renegade member of the family, as we regarded it when I was a youngster. In my grandfather's [Dr. Alfred Holt's] home," he added, "his name was not allowed to be mentioned in the evening talks about the family fireside."[37]

Politics aside, Alfred Holt Stone inherited a number of favorable characteristics from his family: integrity, industry, and intelligence. Both the Stones and the Holts were impressive ancestors, and having strong opinions appears to have been a trait that was shared by both sides of the family. Stone was thus the proud son of a proud family. "To say that Alf Stone had within him the essentials of greatness is a poor and inadequate tribute to the man," the editor of the *Jackson Daily News* wrote the day after Stone died. "Human greatness is something hard to define within a few phrases," he continued, "but one had to be in the presence of Alf Stone only a few minutes to realize that he was a man set apart, a figure above the multitude."[38]

35. W. T. Sherman to Joseph Holt, April 6, 1864; and J. Holt to W. T. Sherman, April 7, 1864, *OR*, ser. 2, 7: 18–20.

36. Holt was the government's chief prosecutor at the trial of conspirators in the Lincoln assassination ("Joseph Holt," www.spartacus.schoolnet.co.uk/USACWholt.htm; Warner, *Generals in Blue*, 232–33).

37. A. H. Stone, "Some Recollections," chap. 2, p. 2.

38. "Alfred Holt Stone," *Jackson Daily News*, May 12, 1955.

3

NO TWILIGHT ZONE

The first glimpse we have of Alfred Holt Stone's early life comes from the pages of an autobiographical sketch he wrote on a train in 1938 while traveling home after a meeting with the National Association of Tax Administrators in New York.[1] He was sixty-seven years old at the time, and recalling early years in the Mississippi Delta was customary for the region's prominent citizens. Stone's father had recorded his recollections twenty-five years before, and Stone's father-in-law two years before that.[2]

Stone began his memories with the account of an incident that still shone brightly in his mind's eye. His recollection of this incident is important, not only because it sheds light on a memorable event in a significant life, but also because it prefigures a paradox that complicates our understanding of Stone's thinking when he assembled his huge collection of publications dealing with "the Negro and cognate subjects."

There is no question that Stone thought of the incident as a metaphor for his later work on the race problem. "I suppose that each of us has some earliest recollection, some incident or event which is so emphasized by time or circumstance as to tie his subsequent life to its beginnings," Stone recalled.[3] "Doubtless these first impressions are sometimes more imagination than

1. A. H. Stone, "Some Recollections," chap. 2, p. 20.

2. W. W. Stone, "Some Post-War Recollections," 231 (originally read before the Washington County Historical Association in 1912 and 1913), and H. T. Ireys, "Autobiography," 1 (originally read before the Washington County Historical Association in 1910), both in respective subject files at MDAH and later published in William D. McCain and Charlotte Capers, eds., *Memoirs of Henry Tillinghast Ireys: Papers of the Washington County Historical Society, 1910–1915* (Jackson: Mississippi Department of Archives and History, 1954), 231–89 and 6–36, respectively.

3. A. H. Stone, "Some Recollections," chap. 1, p. 1. Stone referred to the incident publicly in a speech to the Jackson Lions Club on May 30, 1947 (typescript, 1947, MDAH), 2.

memory. Long association or frequent repetition has clothed them with the attributes of life for those to whom they become a cherished possession. But there is no border line of doubt, no twilight zone, in my own case," Stone added. "If every incident of my life were as vivid and as real as this, I would have no misgivings as to the safety or significance of the joint adventure upon which you and I, as passenger and pilot, are presently about to embark."[4]

The incident Stone was referring to occurred in the fall of 1876 when he was six years old. As it turned out, 1876 was an important year in Mississippi's history. Seven years earlier, African Americans had gained the right to vote when the state's constitution, framed during the early days of Reconstruction, was ratified. With blacks outnumbering white Mississippians 54 percent to 46 percent, black voters and their white allies in the Republican Party had instantly become a political power to be reckoned with.[5] As could be expected, Democrats were not willing to accept their status as a minority party lying down and began to fight back. Initially, their opposition proceeded along legal paths, such as fielding strong candidates and courting black voters. But this strategy failed, and the statewide election of 1873 saw a rejection of the Democratic Party's conservative platform when black Mississippians and white Republicans captured both houses of the legislature. In addition, a former Union general from Maine, Adelbert Ames, defeated a native Mississippian, James L. Alcorn, in the race for governor. Ames's victory was particularly important because, under the provisions of the state's Reconstruction constitution, the governor had the authority to appoint judges, and all three branches of the state government were now firmly in Republican hands.[6]

At this point, racial attitudes in the conservative white community shifted to the nasty side. Previously, criticism of black politicians had been mild or mocking but generally sympathetic in its portrayal of African Americans because Democratic leaders wanted to garner black political support. But, according to Vernon Lane Wharton, the chronicler of the black Mississippians from 1865 to 1890, the tone of the white press changed dramatically following the election of 1873. By the summer of 1874, sympathy had evolved into

4. Ibid.

5. "Census Data for the Year 1870," fisher.lib.virginia.edu/cgi-local/censusbin/census/cen.pl?year=870.

6. Nicholas Lemann, *Redemption: The Last Battle of the Civil War* (New York: Farrar, Straus, and Giroux, 2006), 55–61; Vernon Lane Wharton, *The Negro in Mississippi, 1865–1890* (New York: Harper and Row, 1965), 176. Ames defeated Alcorn by a vote of 69,870 (58 percent) to 50,490 (42 percent).

open dislike and eventually hatred. By 1875, a distinct and rigid color line in Mississippi politics had been drawn.[7]

With the polarization of politics along racial lines, white opposition to the black dominance at the polls took a rougher form. One of the first indications of the Democrat's get-tough campaign strategy was the application of economic pressure. Across the state, planters favoring Democratic candidates announced that any black man who voted for the Republican ticket would be out of a job when contracts were issued for the next growing season.[8]

It did not take long for this sort of pressure on the black electorate to assume a more direct and often violent form. White Mississippians, many of whom had served in the war as Confederate soldiers, joined so-called "Democratic Clubs" that were organized expressly for the purpose of intimidating prospective black voters. According to Wharton, Democratic leaders may have denied the existence of violence or tried to blame it on African Americans whenever it could not be ignored, but they never repudiated its use, and in some cases, encouraged it. The slogan "Carry the election peaceably if we can, forcibly if we must" became a byword in the conservative press.[9]

Open conflict between Republican and Democratic forces increased through the fall of 1874 and into the spring and summer of 1875 as the statewide elections of that year approached. Racial disturbances erupted in Austin, Vicksburg, Louisville, Macon, Yazoo City, Clinton, Satartia, and Coahoma County. Dozens of African Americans as well as a handful of white club members were killed in these clashes, and although conservative leaders tended to downplay the role of violence in the election, the evidence of white coercion was overwhelming.[10]

An investigative committee in Congress collected enough testimony about election irregularities in Mississippi to fill two volumes. Among other things, the committee found that men in the Democratic Clubs had been "furnished with the best arms that could be procured in the country," brand-new, lever-action Winchester rifles, according to Hiram B. Putman, an election

7. Wharton, *Negro in Mississippi,* 183; William C. Harris, *The Day of the Carpetbagger: Republican Reconstruction in Mississippi* (Baton Rouge: Louisiana State University Press, 1979), 656–57.

8. Wharton, *Negro in Mississippi,* 185–86.

9. Ibid., 187. See also the resolutions passed at a "Mass Democratic and Conservative Meeting" in Hinds County, *Hinds County Gazette,* July 28, 1875.

10. Cresswell, *Rednecks, Redeemers, and Race,* 5–6; Lemann, *Redemption,* 80–99; Wharton, *Negro in Mississippi,* 190–92.

registrar in Washington County.[11] The committee also found that these clubs used violence or threats of violence to force the withdrawal of Republican candidates from the election and to prevent black voters from either showing up at the polls or casting a vote for a Republican candidate if they did. Violence and intimidation directed against black voters was widespread, the committee noted, especially in twenty-two counties, one of which was Washington County, where Stone lived.[12]

Apparently, intimidation worked. The statewide elections of 1875 reversed the results of the Republican-dominated election two years before. Democrats captured both houses of the legislature and then used the power of impeachment to remove the Republican lieutenant governor, an African American.[13] With the lieutenant governor out of the way, the legislature turned its attention to Governor Ames. Claiming that he was unfit to hold office, the Democrats began impeachment proceedings against him too. Initially Ames fought back, but he faced unanimous opposition from the Democrats as well as some Republicans who objected to the high taxes levied during Ames's administration.[14] Faced with overwhelming odds, Ames resigned on March 28, 1876, and left for Maine, never to return.[15] Now that the state had no governor or lieutenant governor, the president pro tempore of the Senate, John W. Stone (no relation to Alfred Holt Stone) became governor of Mississippi.[16] Using his constitutional authority, Governor Stone dismissed Republican judges and replaced them with Democrats.[17] The conservatives were now in control of all three branches of government, and the aggressive

11. "Force, Fraud, and Intimidation: 'The Mississippi Plan,' 1875," in Robert W. Johannsen, ed., *Reconstruction, 1865–1877* (New York: Free Press, 1970), 177; U.S. Congress, Select Committee, *Mississippi in 1875. Report of the Select Committee to Inquire into the Mississippi Election of 1875, with the Testimony and Documentary Evidence* (Washington, DC: Government Printing Office, 1876), 1430–31. Putman's name was misspelled "Putnam" in the committee's report.

12. "Force, Fraud, and Intimidation," 178–79; Select Committee, *Mississippi in 1875*, 1431. Nineteen of the twenty-two counties had majority black populations. These counties were: Amite, Chickasaw, Claiborne, Clay, Copiah, De Soto, Grenada, Hinds, Holmes, Kemper, Lowndes, Madison, Marshall, Monroe, Noxubee, Rankin, Warren, Washington, and Yazoo ("Census Data for the Year 1870"). The three counties with white majorities were Alcorn, Lee, and Scott. The figures for Clay County are from the 1880 census because that county was not formed until 1871.

13. Lemann, *Redemption*, 163; Harris, *Day of the Carpet Bagger*, 694.

14. Lemann, *Redemption*, 164; Harris, *Day of the Carpet Bagger*, 624–25.

15. A congressional committee investigating the election determined that the charges against Ames were without merit (Johannsen, *Reconstruction, 1865–1877*, 173).

16. Harris, *Day of the Carpet Bagger*, 697.

17. Ibid., 698–99.

tactics they had used to regain power, the "Mississippi Plan" as it was known, became the prototype for similar efforts in other southern states.[18]

Alfred Stone grew up with the Mississippi Plan, and his earliest memory involved an incident leading up to the national election in 1876. That election was important because it would determine the state's six representatives to the U.S. Congress and thus consolidate the Democratic Party's victory achieved in the election for state offices the year before. In addition, voters would be voting for a new president of the United States, an office that Republicans had held since 1861. For white Mississippians, Democratic success in this election would mark an end to Reconstruction, an outcome they had been waiting for since the end of the war. "Our people had sworn to themselves that the year 1876 would witness a resumption of their independence by the Southern States,—the restoration of home rule and the end of carpetbag, scalawag and negro domination, regardless of the results of the national election," Stone noted in his recollections. "They wanted peace; wanted it desperately; wanted it enough to die for it. But, peace or no peace, they had made up their minds that the travesty of government, under which they had lived for ten years, had come to its closing hour. If they had to sell their own lives in order to end it,—well and good. They were ready to do just that. But they were also prepared to take just as many as possible with them, when they went down in defeat, if defeat should be their portion."[19]

In 1876, Stone's family lived on a plantation on the banks of Deer Creek about three miles north of Arcola, a small community southeast of Greenville. The name of the plantation was Sligo, and the house in which the Stones lived was built of hewn logs floored with whipsawed planks cut from trees on the property.[20]

One night about a month before the election, Stone's father rode to Arcola to meet with some forty or fifty white men from the surrounding countryside to talk about "the ways and means of carrying the coming elections." Stone's mother stayed at home. "My mother knew the purpose of the meeting that night," he recalled. "She knew also that every man was armed, with either a Winchester rifle or a Colt's Army or Navy revolver." Both weapons fired .44 caliber cartridges, and Stone's father was in charge of their ammu-

18. Lemann, *Redemption*, 170–209; "Force, Fraud, and Intimidation," 171.

19. A. H. Stone, "Some Recollections," chap. 1, p. 2.

20. The location of Sligo Plantation and the description of the house come from Walter W. Stone's "Some Post-War Recollections," 266. Alfred Holt Stone referred to his home as a "rude house" in his memoirs.

nition. Consequently, he had stored about two gross of .44 caliber ammunition (14,400 cartridges) in the house.[21]

A fire broke out sometime after Stone's father left for his meeting. More than sixty years later, Stone remembered vividly how "the blazing glare from our burning home lit up the Heavens and shut out the autumn stars from my awestricken eyes." Stone's mother gathered Alfred and the two other children around her on a half-burnt mattress beside the road. She held a rifle across her lap as she watched the house burn to the ground. A hundred or more African Americans, workers on the Stone plantation and surrounding farms, stood behind her and around the building, watching the fire. According to Stone, she was not afraid of the "dark faces visible in the reflected light," but she was concerned that her husband would misinterpret the situation when he came galloping up the road from Arcola with dozens of other white men who had seen the flames from a distance.[22]

Stone recalled in his memoirs that "all our [white] people, women and men, were doing all in their power to keep the peace, to do nothing to furnish excuse for continued alien control." But there had been trouble between the races in the neighborhood before. According to Stone, a white man had been taken from his home one night "and horribly tortured to death." In reprisal, a group of white men from the community had killed sixteen African Americans. For this reason alone, Stone's mother was concerned that the white men converging on the burning house would see the black crowd and assume the worst. Her concerns were heightened considerably when fire reached the ammunition and set it off with a cascade of explosions.[23]

What happened next was branded in Stone's mind. An old and dignified African American, Uncle Jim Weaver, stood nearby. His complexion was jet-black, and his hair and full beard were as white as snow, a commanding presence, as Stone remembered him. "Like hundreds of his kind throughout the South," Stone wrote, "he was more, much more, than a faithful servant. He was our companion and friend." Stone's mother caught up a sheet that had been saved from the fire as her husband and the other men approached. Holding one corner, she gave another to Uncle Jim. Together, "they held it between them," Stone recalled, "high as they could above their heads,—held it across the open road, where the dying flames could bring it into view;—not a white flag of surrender, but an emblem of safety and peace, telling the com-

21. A. H. "Some Recollections," chap. 1, pp. 4–5.

22. Ibid.

23. Ibid., chap. 1, p. 5.

ing men that no one had been harmed, that there was no need of vengeance or reprisal. Not a shot was fired."[24]

The peaceful resolution of the potentially violent confrontation at the Stone place anticipated the generally quiet conditions that existed in Washington County during the election a month later. Perhaps the peace was due to a tradition in the county of office sharing between races—"fusion" politics, as it was called—that worked best in counties with extremely large black majorities.[25] Eighty-five percent of Washington County was black in the 1870s, and white politicians and black voters may have agreed to an uneasy truce. On the other hand, it is possible that the black electorate in Washington County had been subdued by a continuation of the violence surrounding elections the year before, in which case Uncle Jim Weaver's gesture may indeed have signified surrender.[26]

For whatever reason, a large number of African Americans in Washington County voted for the Democratic ticket in November 1876. The Democratic candidate for Congress, James R. Chalmers, a Confederate general who had served under Nathan Bedford Forrest, received 2,905 votes as compared to 1,598 for the black incumbent, John R. Lynch. The large turnout for the Democratic candidate was remarkable given that there were only 2,164 white men, women, and children in the entire county as compared to 12,3895 black citizens.[27]

Registrar Putman claimed that the Democratic majority had been achieved by fraud, although he did acknowledge that a number of Republican voters had become disenchanted with the arrogance of their party leaders following the election of 1873.[28] But the *Greenville Times* offered another explanation. "This gratifying termination to the recent canvass is attributable to the earnest and consistent efforts of the Democratic leaders to convince the colored population that their true political interests are identical with ours," the editor observed. "Every planter, merchant, and lawyer, was an active canvasser during the week preceding the election. The work of dividing

24. Ibid., chap. 1, pp. 5–6.

25. Harris, *Day of the Carpet Bagger,* 666–67; McMillen, *Dark Journey,* 60; Wharton, *Negro in Mississippi,* 203.

26. Lemann, *Redemption,* 171. Lemann also noted that in 1876 Democrats issued "certificates of loyalty" to black registrants who promised to vote for the Democratic Party. These certificates protected the bearer and his family from violence or the loss of employment.

27. "The Election in Our County and District," *Greenville Times,* November 11, 1876.

28. Select Committee, *Mississippi in 1875,* 1434–35, 1439. See also Harris, *Day of the Carpet Bagger,* 660, 681.

the colored vote was rendered all the easier by the absolute absence of opposition."[29]

"Absolute absence of opposition." What did that mean? One indication that the threat of violence was always just beneath the surface appeared in the *Greenville Times* two days later in a political cartoon under the heading "Mississippi Democratic by 40,000 Majority." The cartoon depicted two figures in silhouette, both running. The first one was carrying a carpetbag, and the one behind bore the stereotypic features attributed to persons of African ancestry. The caption—apparently referring to the large number of African Americans who voted for Democratic candidates—read, "The Carpet Bagger's parting speech. 'That's what makes me 'spies a Nigger!'." To help explain why they were running, the cartoonist punctured the two figures with thirteen bullet holes.[30]

Elsewhere in Mississippi, the Democratic victory in 1876 had been achieved by keeping black voters away from the polls. The extent of the success of Democratic intimidation can be seen clearly in the voting patterns in most of the state's predominantly black counties. In 1873, for example, 4,709 voters cast votes for Republican candidates in Warren County, which was 70 percent black. In 1876, the number of Republican voters had dropped to 623, a decline of 87 percent. Amite County, which was 62 percent black, experienced a 93 percent reduction in Republican voters between 1873 and 1876. Two counties, Lowndes and Madison, both more than 70 percent black, saw the number of Republican voters decline by 99 percent. In Yazoo County, which was 72 percent black, 2,433 Republicans cast ballots in 1873. In 1876, only two showed up at the polls to vote.[31] Not surprisingly, the national election of 1876 in Mississippi resulted in a Democratic landslide. Democratic candidates captured all six of the state's congressional seats, and Samuel J. Tilden, the Democratic presidential candidate, received the state's electoral votes by a margin of 56,853.[32]

Stone did not dwell on the outcome of the election in his recollections, nor did he mention the violence and intimidation that were an essential part of it. Instead, he offered a rationale for white paternalism and restated

29. "The Election in Our County and District," *Greenville Times*, November 11, 1875.

30. "Mississippi Democratic by 40,000 Votes," *Greenville Times*, November 13, 1876.

31. *Jackson Daily Times*, December 8, 1876, quoted in Wharton, *Negro in Mississippi*, 201. The election figures comparing 1873 to 1876 for Amite, Lowndes, and Madison Counties were 1,093 to 73; 2,723 to 13; and 2,323 to 13, respectively. For a statement concerning the general impact of white coercion on black voting patterns, see McMillen, *Dark Journey*, 39.

32. Wharton, *Negro in Mississippi*, 200.

his contention that the white and black southerners would have solved the problem peacefully if agitators from the North had not interfered.

> During many of the years which have lengthened and passed into the shadows since that autumn night, I have been a student of negro life and character. I have written and spoken much of the relations between the two races, destined, it seems, to work out their fate upon a common ground. But not one word has ever come in harshness from my lips or pen. I have set down naught in uncharity or malice. I dealt with negroes for more than forty years in the intimate contacts of plantation life. I have a conscience void of offence toward them, individually and as a race. I have never abused one of them nor taken advantage of one in any transaction. It was not in the blood which I inherited to act otherwise. But no matter how sorely I might have been tempted, the indelible vision of my mother and our faithful black friend, of these two that night, at opposite racial extremes but unified through a common purpose, would have made impossible any act unkind or mean. And I have often thought of the lesson it held for both races, if only it could have been learned and taken to heart in time; if only we could have been left alone.[33]

Stone's metaphor reflected a belief that he articulated consistently in his writings on the race problem. Namely, the passive and unconditional acceptance by African Americans of white supremacy was the only viable path to racial harmony.[34] Stone's point of view was not original, for it was widely accepted in both the North and the South during his lifetime. Nor should we expect this son of a Confederate veteran to have adopted a racial doctrine that varied significantly from that of the culture in which he grew up. Nevertheless, it is ironic that Alfred Holt Stone, defender of the Old South, would systematically assemble one of the country's most comprehensive and valuable collections of material dealing with the potential for progress and promise among African Americans.

33. A. H. Stone, "Some Recollections," chap. 1, p. 6.

34. See J. William Harris, *Deep Souths: Delta, Piedmont, and Sea Island Society in the Age of Segregation* (Baltimore: Johns Hopkins University Press, 2001), 126 and 202, for an assessment of Stone's views on white paternalism.

4

FOOD FOR THE HUMAN MIND

Alfred Stone received most of his early education from his father, much of it on horseback. Stone recalled that he would ride behind his father on a heavy osnaburg saddle blanket when Captain Stone toured the plantation. "[James A.] Garfield's often quoted description of a university, as a log with Mark Hopkins on one end and a boy on the other," Stone wrote more than fifty years later, "has always reminded me of some of the most helpful and instructive hours of my life,—those thus spent [behind the saddle] with my father."[1]

Not many planters in the Mississippi Delta were better prepared to teach their sons than Captain Walter Wilson Stone. "My father was a man of scholarly attainments, with a fine mind and a wonderfully accurate memory," Stone wrote in his memoirs. "My first serious reading of history was through a six volume set of Gibbons' 'Decline and Fall of the Roman Empire,' with Milman's notes, which was awarded him [my father] as a prize in his junior year at the University of Missouri, for an original essay in Greek."[2]

Captain Stone appreciated the importance of the printed word and filled his household with books, magazines, and newspapers, even if the furniture was meager. "We took the old Scribner's Magazine and Harpers Weekly,—though Thomas Nast's political cartoons more often stung and irritated than pleased," Stone recalled. The family also read the *North American Review*, the *Edinburgh Review*, and the *Eclectic Magazine of Foreign Literature, Science, and Art*, as well as subscribing to newspapers from Saint Louis and Columbia,

1. A. H. Stone, "Some Recollections," chap. 2, pp. 19–20.

2. Ibid., chap. 2, p. 18. Editions of Gibbons's classic work on the Roman Empire with annotations by Henry Hart Milman, Dean of St. Paul's Cathedral in London, were issued almost yearly between 1843 and 1855.

Missouri, and the *New York Tribune*. Stone remembered that they were well informed about events occurring in distant places, even though there were no telegraphs, telephones, or railroads in that part of the Delta at the time.[3]

Captain Stone also took his son on trips to see some of the world they read about in magazines and newspapers. In fact, excursions beyond the confines of the Delta became a family routine because relatives on both sides lived in distant places; Captain Stone's kin in Columbia, Missouri, and Ella Stone's family in New Orleans. The family visited their relatives every year during Stone's childhood. In the hot summer months, they would take a steamboat upriver from Greenville to Saint Louis and then travel to Columbia by train. During the winter months, the Stone family would travel downriver to visit grandparents in New Orleans.[4]

It was during a visit to see the folks in Missouri that young Alf Stone went to see the circus, and not just any circus, but the circus of the great showman Phineas T. Barnum. There were two "really stupendous attractions" that Stone remembered. One was "the largest aggregation of elephants ever seen in captivity." The other was the electric illumination for the circus's tents and grounds. Interestingly, Stone reported that "I was much more fascinated by the electric lighting machine than by the elephants." "I was of a mechanical turn of mind," he explained, "and was never really much interested in animals."[5]

In 1882, when Alf was twelve years old, Captain Stone was elected to the Mississippi legislature, and they traveled together to Jackson for the session.[6] The same year, Captain Stone took his young son to New Orleans to attend a convention promoting the construction of a transisthmus canal through Nicaragua. They stayed at the historic Saint Charles Hotel, and Stone remembered attending the convention's keynote address by Senator John T. Morgan from Alabama. Senator Morgan believed that a canal through Nicaragua would help southerners compete in Pacific markets, and he spoke for more than four hours.[7] His speech must have been a good one, for "I do not believe that a man left the audience," Stone wrote many years later. But what

3. Ibid., chap. 2, p. 21.
4. Ibid., chap. 2, pp. 6–7.
5. Ibid., chap. 2, p. 7.
6. Ibid., chap. 1, p. 20.
7. Ezra J. Warner, *Generals in Gray: Lives of the Confederate Commanders* (Baton Rouge: Louisiana State University Press, 1959), 222. See also "The Nicaragua Canal Convention at New Orleans," *Manufacturer and Builder* (April 1893): 78–79.

Alf remembered most about the convention was one of the delegates from Texas, Frank James, Jesse's brother.[8]

Captain Stone was an active member of the Democratic Party and had been for many years. As a result, he was offered the position of state auditor on the Democratic ticket in 1885. The Mississippi Plan was still in force, and its effectiveness was such that a demoralized Republican Party failed to field a slate of candidates. Consequently, the Democrats swept to an almost unanimous majority in the legislature and captured all of the statewide offices. Robert Lowry was reelected governor with 99 percent of the votes cast. Captain Stone garnered 92,499 votes, a total that was exceeded only by the Democratic candidate for superintendent of education. Remarkably, only two people in the entire state voted for someone other than Walter Wilson Stone in the state auditor's race.[9]

Captain Stone traveled to Jackson in January 1886 to assume his new office,[10] and his fifteen-year-old son came with him. The move allowed Alf to pursue his education at another level when he enrolled at the Jackson Collegiate Academy under the tutelage of Dr. George S. Roudebush.[11] Dr. Roudebush was from Ohio but had moved to Mississippi in 1854 at the age of twenty-six. He attended Jefferson College outside of Natchez and was ordained a minister in the Presbyterian Church in 1860.[12] Shortly after the war, he established an academy on north State Street to prepare young Jacksonians for study at the university. Over the years, many of the state's most prominent citizens received their secondary education under the guidance of Dr. Roudebush.[13]

Dr. Roudebush also served as the moderator of the State's Presbyterian

8. A. H. Stone, "Some Recollections," chap. 2, pp. 12–13. Jesse James died on April 3, 1882, but Frank James lived until 1915. The Panama Canal opened for traffic in 1914. (Dates from www.answers.com/.)

9. *Jackson Clarion*, December 9, 1865; McCain, *Story of Jackson*, 247; see also *The 1868 Constitution of the State of Mississippi*, art, V, sec. 20; and Cresswell, *Rednecks, Redeemers, and Race*, 93.

10. *Jackson Clarion*, January 5, 1886.

11. Handwritten notes in George S. Roudebush's subject file, MDAH; McCain, *Story of Jackson*, 247.

12. George S. Roudebush, *An Address Delivered on the One Hundredth Anniversary of the Organization of the Presbytery of Mississippi Held in Pine Ridge Church, October 11, 1916* (Jackson, MS: Tucker Printing, 1916), 1.

13. For example, Judge William Hemingway, for whom the football stadium at the University of Mississippi is named, attended Dr. Roudebush's academy (W. Eugene Morse, "'Judge' William Hemingway, 1869–1937," *Journal of Mississippi History* 36 [November 1974]: 339).

Synod and became a familiar yet impressive figure in Jackson until his death in 1921 at the age of ninety-three. "Tall, white-haired, slightly stooped, with keenly intelligent face," is how a columnist for the *Jackson Daily News* remembered him. "One might well imagine he had just stepped out of a life-sized portrait of an old-fashioned Southern gentleman of the scholarly type. Indeed, he could have played a role of that sort on the stage or screen without the slightest makeup."[14]

An anecdote about Dr. Roudebush may tell us something about his contribution to Stone's love of books. After his retirement, Dr. Roudebush spent most of his time visiting friends and associates, sharing his voluminous knowledge, and reminiscing about his early days in Mississippi. Sometimes these visits took him out of state. Such was the case around 1920 when Dr. J. B. Hutton, pastor of the First Presbyterian church in Jackson, decided to collect money from among Dr. Roudebush's friends to buy him some new clothes for one of his trips north. It was an easy calling, for Dr. Roudebush was highly esteemed, and the good pastor soon had more than a one hundred dollars, which he took to Madison County, where Dr. Roudebush lived with one of his daughters. It was several months later before Dr. Hutton encountered Dr. Roudebush after his trip, and he was surprised to find that the old scholar was still wearing his threadbare clothes. What had Dr. Roudebush done with the money? "Well," Dr. Hutton reported, "he bought books with it—books he said he had been wanting for a long time to read."[15]

The writer for the *Jackson Daily News* used this anecdote as the centerpiece for his column and compared Dr. Roudebush to Desiderius Erasmus, the great German humanist. "Dear old Dr. Roudebush—genial, gentle, lovable, scholarly, filled with the milk of human kindness—had within him the spirit of Erasmus," the columnist wrote. "He loved books. They meant more to him than sustenance for the body or the costliest raiment. Many years of living had taught him the true value of food for the human mind. He knew that the silent power of books is the great moving power of the world; that there is a joy in reading books which those alone can know who read them with keen interest, enthusiasm and understanding."[16]

Dr. Roudebush's contribution to Stone's education was anchored on the foundation laid down by his father, a foundation that apparently was very solid, for it was during this period that Stone's first written work was pub-

14. Frederick Sullens, "Like Erasmus, Dr. George Roundebush [*sic*] Preferred Books to Food and Clothes," *Jackson Daily News*, July 26, 1953.

15. Ibid.

16. Ibid.

lished. The work was a report on the Interstate Dairymen's and Fruit Growers' Convention held in Jackson over three days in February 1887. The convention was the brainchild of Captain John Festus Merry of Manchester, Iowa, a Union veteran and public relations agent for the Illinois Central Railroad who had a penchant for organizing exhibitions to promote agricultural expertise and products from the Midwest. The idea behind the Jackson convention was to invigorate dairy farming in the Deep South by introducing cattle and dairy methods developed in the Midwest and to improve the horticultural habits of southerners with the infusion of knowledge from northern agronomists.[17]

Two hundred delegates from Iowa and Wisconsin arrived in Jackson by train on Wednesday, February 16, and were feted at a formal reception that evening in the chamber of the House of Representatives. Over the next two days, delegates, farmers, and agriculturalists filled the chamber and listened to more than a dozen papers dealing with dairy farming and horticultural topics.[18] By the time the delegates pulled out of the train depot on Saturday morning to carry their crusade for creameries to New Orleans, everyone agreed that the convention had been a great success. Mississippi's climate, soil, and native grasses all boded well for the establishment of farms stocked with Jerseys and other breeds that had made the dairy industry in the Midwest world-famous.[19]

The Illinois Central Railroad had been instrumental in promoting the Dairymen's and Fruit Growers' Convention, for its directors were eager to do whatever they could to attract commerce and new passengers to Mississippi. To that end, the railroad had provided a special train to carry the delegates to the meeting in Jackson. Hoping to capitalize on the publicity resulting from the convention, the Illinois Central sponsored a competition at Dr. Roudebush's school for reports covering the convention's activities. The assignment must have been formidable given that the convention consisted almost en-

17. "Dairying in the South," *Jackson Clarion*, December 22, 1886; "The Fruit Growers and Dairymens' [*sic*] Convention," *Jackson Clarion*, February 2, 1887. In 1885, Captain Merry had worked with other veterans to establish a national military park at Vicksburg, Mississippi, which, not incidentally, would promote tourism on the Illinois Central. Merry's ambition was realized on February 21, 1899, when President McKinley signed legislation making the battlefield at Vicksburg a national military park (Christopher Waldrep, *Vicksburg's Long Shadow: The Civil War Legacy of Race and Remembrance* [New York: Rowman and Littlefield, 2005], 141–49, 166).

18. "The Interstate Dairymen's and Fruit Grower's Convention," *Jackson Clarion*, February 23, 1887.

19. "Dairy Farming in Mississippi," *Jackson Clarion*, March 9, 1887.

tirely of papers dealing with topics such as "How I Fed the Cows and Made the Butter that Took the Sweepstakes at the Interstate Fair."[20] Nevertheless, Alf gave it a try, and after all the reports had been read, his "was judged the best of the lot." According to an article in the *Illinois Central Magazine* many years later, the railroad printed Alf's report in pamphlet form and distributed thousands of copies. Unfortunately, not a single copy is known to have survived. Alf's paper on the convention would have been of interest today because it dealt with the interface of farming and economic development, two topics in which he would remain involved for most of his life.[21]

Shortly after writing his report on the Dairymen's and Fruit Growers' Convention, Alf Stone left Jackson for Oxford, Mississippi, to attend the university. There Stone was drawn to a professor in the English Department, later its chair, William Rice Sims. Stone remembered Sims as "a Mississippi poet, philosopher and scholar" and probably liked him best because Sims wanted to move the department away from its traditional emphasis on linguistic science and toward the more contemporary study of literature.[22] Whatever the reason, Stone referred to Sims in his recollections as "one of my earlier teachers, and probably the greatest of them all."[23] Stone spent almost three years at the University of Mississippi studying history, political science, economics, and literature. However, his course of study was not a regular curriculum, and consequently he did not receive a degree when he departed to let his career assume a different path.[24]

Sometime toward the end of 1889 Stone decided to study law. At that time it was customary for aspiring attorneys to read law under the supervision of a practicing lawyer, and the more prominent the lawyer, the better. Stone's father was very well connected, having been elected state auditor for a second term in November. Doubtless, it was through some of those contacts that young Stone was invited to read law in the office of Judge Wiley P. Harris.[25]

Judge Harris was born in Pike County, Mississippi, and attended the Uni-

20. The paper on fodder and butter was presented by C. L. Beach of Whitewater, Wisconsin.

21. "Alfred H. Stone Recalls Railroad Pioneers," *Illinois Central Magazine* 27 (September 1938): 5.

22. "History of the English Department," www.olemiss.edu/depts/english/history.html. Professor Sims became chair of the department in 1890.

23. A. H. Stone, "Some Recollections," chap. 1, p. 18.

24. Alfred Holt Stone to "Dear Mr. Willcox," January 10, 1902, WFWP.

25. "Railroad Pioneers," 3.

versity of Virginia before obtaining a law degree from Transylvania College in Lexington, Kentucky. Returning to Mississippi, he was admitted to the bar in 1840 and became a circuit court judge four years later. In 1852, Judge Harris was elected to the U.S. Congress, where he served one term as an archly conservative but undistinguished representative.[26] In 1861, he was elected to the state's secession convention and subsequently served in the Provisional Congress of the Confederate States, where he sat on the committee charged with writing a constitution for the new nation. He returned to Jackson at the end of his term and practiced law for the next thirty years.[27]

Judge Harris was a good lawyer, but more of a plodder than a slasher. "He did not jump to conclusions, but reasoned to them," a colleague remembered. "He was a thinker, and required time for thought—reflection. He did not, like some, see everything he could ever see in a subject, at a glance, but with opportunity for reflection, he saw more than they."[28] Another colleague recalled that "Judge Harris's career has not been characterized by the dash, or the startling brilliant achievements, which bring some men sometimes into notoriety. . . . If the infinite capacity for taking pains is genius then Judge Harris possesses genius in the highest degree."[29]

One of the most important events to occur during Alfred Holt Stone's long life happened while he was studying law in Judge Harris's office. The election of 1889 that returned Alf's father to the post of state auditor also resulted in the election of a state legislature determined to rewrite the state's Reconstruction Constitution of 1868. There were several things in the old constitution that the Democrats wanted to change, such as the governor's power to appoint judges and the apportionment of seats in the legislature

26. Judge Harris went against the state's Democratic Party and voted against the Kansas Nebraska Bill in 1854 because he opposed any compromise on the issue of slavery, believing that accepting a compromise acknowledged the right of the federal government to impinge on the right of slaveholders to take their slaves anywhere they wanted ("Kansas and Nebraska: Speech of Hon. W. P. Harris, of Mississippi, in the House of Representatives, April 24 and 26, 1854," *Congressional Globe*, 33rd Cong., 1st sess., Appendix, 547–50).

27. "Harris Obituary," *Jackson State Ledger*, December 7, 1891, reprinted in *Memorials of the Life and Character of Wiley P. Harris, of Mississippi* (Jackson, MS: Clarion Printing, 1892), 35–40; "Wiley Pope Harris (1818–1891)," in the "Biographical Directory of the United States Congress," bioguide.congress.gov/scripts/biodisplay.pl?index=H000256; Dunbar Rowland, "Wiley P. Harris," in *Courts, Judges, and Lawyers of Mississippi, 1798–1935* (Jackson: Press of Hederman Bros. for State Department of Archives and History and the Mississippi Historical Society, 1935), 267–70.

28. "Remarks of Chief Justice Campbell," *Memorials of Wiley P. Harris*, 19–20.

29. Harris obituary, in *Memorials of Wiley P. Harris*, 39.

that favored counties with large black populations, but it was the disfranchisement of African Americans that topped their list of priorities. In February 1890, the legislature issued a call for a constitutional convention to convene in Jackson on August 12 of that year.[30]

The reemergence of concern over black suffrage despite almost fifteen years of white dominance at the polls resulted from the resurgence of the Republican Party at the national level. In 1888, Benjamin Harrison, a Republican, had defeated the Democratic incumbent, Grover Cleveland. In addition, the Republican Party captured both houses in Congress. Republicans had not enjoyed such success since 1874, and they looked to black voters in the South to help them solidify their position, even if getting them to the polls meant federal supervision of congressional elections.[31]

Influential Democrats in Mississippi, such as Senator James Z. George, sounded the alarm. "The history of the Republican party shows that it dares to do whatever it deems necessary for party success. The predominance of white influence in the States of the South is an obstacle to Republican success. . . . Our duty, therefore, when we meet in convention, is to devise such measures, consistent with the constitution of the United States, as will enable us to maintain a home government under the control of the white people of the State."[32]

Other leaders joined Senator George in his call for white supremacy. "This [black suffrage] was the great and constantly irritating evil," the constitutional convention's president recalled a decade later. "The absolute necessity of excluding ignorance and crime from the ballot box, and of purifying it from fraud and violence, was apparent to all good and patriotic citizens."[33] They did not attempt to hide their intent to disfranchise African Americans. Rather, they proclaimed it openly throughout the state.[34] "There was no ef-

30. William Alexander Mabry, "Disfranchisement of the Negro in Mississippi," *Journal of Southern History* 4 (August 1938): 319–21; Wharton, *Negro in Mississippi*, 208.

31. Henry Cabot Lodge of Massachusetts introduced legislation on June 19, 1890, to ensure black participation in national elections. Although the "Force Bill," as it was referred to in the South, made its way through the House, eventually it was defeated in the Senate (Mabry, "Disfranchisement of the Negro," 321).

32. "Senator J. Z. George. He Addresses a Large Audience at His Old Home," *Jackson Clarion-Ledger*, October 24, 1889. See also E[dward] C. Walthall, *The Federal Election Bill. Speech of Hon. E. C. Walthall, of Mississippi, in the Senate of the United States, Friday, December 12, 1890* (Washington, DC: n.p., 1890). There is a copy of this speech in TSC.

33. S. S. Calhoon, "The Causes and Events that Led to the Calling of the Constitutional Convention of 1890," *PMHS* 6 (1902): 110.

34. McMillen, *Dark Journey*, 40–41.

fort to hide or conceal this great purpose," a delegate noted almost forty years later. "It was given the widest publicity from the rostrum and in the public press long before the members of the convention were even elected."[35]

With such an obvious plan of action, one might have expected the Republicans to organize a resistance, which initially they did. However, efforts to mobilize black voters to elect delegates to the constitutional convention failed in the face of hostile white resistance.[36] Thinly veiled threats warning Republican leaders to mind their own business appeared in the Democratically controlled newspapers, and when a white Republican candidate in Jasper County, Marsh Cook, embarked on an active campaign, he was shot from ambush and riddled with twenty-seven shots. "Cook was slain because of his inflammatory speeches and efforts to stir up strife and bad blood between the races," an editorial in the *Jackson Daily Clarion-Ledger* explained. "The *Clarion-Ledger* regrets the assassination of Cook, but knows that those who did it felt they were doing their country a service in removing a man who had become so offensive as Cook had proven himself to be."[37]

The Mississippi Plan still worked, and when the delegates to the Constitutional Convention convened in Jackson on August 12, 1890, they were white Democrats almost to a man, among them Judge Wiley P. Harris.[38] Nevertheless, there were two challenges facing the delegates committed to the disfranchisement of African Americans. The first was a restriction placed on the state by the U.S. Congress when Mississippi was readmitted to the Union in 1870. The restriction stipulated that Mississippi could never deny suffrage to citizens enfranchised by the Constitution of 1868. The delegates asked a committee headed by Judge Harris to study the issue, and he reported that the congressional prohibition was unconstitutional because it interfered with the right of the state to regulate the franchise. Pleased with his report, the delegates set that obstacle aside and got down to work.[39]

35. Remarks of William C. McLean from Grenada County, in *Proceedings of a Reunion of the Survivors of the Constitutional Convention of 1890 on the Thirty-Seventh Anniversary of the Adoption of the Constitution Held in the Senate Chamber of the New Capitol at Jackson, Mississippi, November 1, 1927* (Jackson: Mississippi Department of Archives and History, 1928), 28; see also Frank B. Williams, "The Poll Tax as a Suffrage Requirement in the South, 1870–1901," *Journal of Southern History* 18 (November 1952): 487.

36. Cresswell, *Rednecks, Redeemers, and Race,* 113–14.

37. "Killing of Marsh Cook" and "A Natural Ebullition," in *Jackson Daily Clarion-Ledger,* July 26 and July 30, 1890, respectively. Cook made only two speeches before he was killed.

38. "Complexion of the Convention," *Jackson Clarion-Ledger,* August 14, 1890; Cresswell, *Rednecks, Redeemers, and Race,* 113–14; Wharton, *Negro in Mississippi,* 211–12.

39. Albert D. Kirwan, *Revolt of the Rednecks: Mississippi Politics, 1876–1925* (New York:

The second challenge was more difficult to overcome. The Fourteenth and Fifteenth Amendments to the U.S. Constitution prohibited states from disfranchising voters on the grounds of race, color, or previous condition of servitude, and they could hardly be dismissed as being "unconstitutional." For the delegates to disfranchise African Americans in Mississippi, they had to come up with language to limit suffrage that did not mention race.[40]

A number of race-neutral disfranchisement schemes were floated. The Australian (that is, secret) ballot, for example, was proposed because delegates assumed that illiterate voters would be unable to complete the ballot without assistance. Other ideas included a poll tax, longer residency requirements, and property or educational prerequisites.[41] One especially creative citizen came up with a plan that would award multiple ballots based on property ownership with the wealthiest citizens getting a handful of votes to cast while those who did not own property would get none.[42]

The problem with all of these schemes was that they fell heavily on poor white voters as well. As can be imagined, any plan that would disfranchise poor white citizens encountered strenuous opposition, even if its intent was to keep black people from voting. Distrust of the elite ruling class among the poor dirt farmers from the hill country was beginning to emerge and would later break into an open revolt after the black suffrage problem was resolved.[43]

Judge Harris was appointed as one of thirty-five members of the convention's "Elective Franchise, Apportionment and Elections" committee.[44] The size of the committee—more than one-fourth of the delegates—attested to the importance of this issue as far as the convention was concerned.[45] After two weeks of deliberation, the committee finally released its report. The re-

Harper and Row, 1951), 66–67; Mabry, "Disfranchisement of the Negro," 324–25; Cresswell, *Rednecks, Redeemers, and Race*, 118.

40. Mabry, "Disfranchisement of the Negro," 325.

41. Ibid., 323.

42. "Who Can Answer This? Desperate Diseases Require Desperate Remedies," *Jackson Clarion-Ledger*, June 26, 1890.

43. Mabry, "Disfranchisement of the Negro," 324–25; Kirwan, *Revolt of the Rednecks*, 79–84. For a discussion of the poll tax as it affected poor whites, see Williams, "Poll Tax as a Suffrage Requirement," 469–96, esp. 487, 495–96.

44. Mississippi Constitutional Convention, *Journal of the Proceedings of the Constitutional Convention, of the State of Mississippi. Begun at the City of Jackson on August 12, 1890, and Concluded November 1, 1890* (Jackson, MS: E. L. Martin, 1890), 22.

45. This committee was also provided with its own staff and given top priority on print jobs (Cresswell, *Rednecks, Redeemers, and Race*, 116).

port contained three provisions in regard to suffrage: a residency requirement (two years), a poll tax (two dollars), and an educational restriction. Both the residency requirement and the poll tax were considered to be a good way of discouraging black voters, but they were far from foolproof. A well-heeled candidate, for example, could come up with the money to pay the poll tax for black voters he wanted to court. Consequently, the educational restriction became the centerpiece of the effort to disfranchise African Americans.[46]

The educational restriction was the handiwork of several delegates, most notably Judge Harris.[47] It stated that qualified voters must be able, on or after January 1, 1896 (later amended to January 1, 1892), "to be able to read any section of the constitution of this State; or he shall be able to understand the same when read to him, or give a reasonable interpretation thereof."[48] The restriction would come into play only after the other electoral qualifications (that is, the residency requirement and poll tax) had been satisfied. Almost immediately, the second part of educational restriction became known as the understanding clause.[49]

Opposition to the educational restriction was intense among poor whites who were convinced that the planter class would use this device to disfranchise them too. The argument swirled back and forth until a delegate representing Lincoln County, Judge J. B. Chrisman, arose on September 8 and cut through the fog of debate. After reviewing the various reasons for drafting a new constitution and dismissing them as of secondary importance because they could have been easily accomplished by amending the present constitution, Judge Chrisman asked what really lay behind the insistence among white people in Mississippi to have a new constitution. "What was the shadow that hung over us," he asked, "that darkened our future and alarmed our people?" His answer was surprisingly straightforward for the venue in which he spoke.

46. Mabry, "Disfranchisement of the Negro," 327. According to Cresswell and Kirwan, the poll tax had more of an impact on rural white voters than expected (*Rednecks, Redeemers, and Race,* 128, and *Revolt of the Rednecks,* 74–75, respectively).

47. There has been considerable controversy over the authorship of the understanding clause. Both Senator James Z. George and Judge Harris have been given credit, although Alfred Holt Stone told Kirwan in 1945 that Harris actually opposed educational restrictions and consented to it only to appease Senator George (Kirwan, *Revolt of the Rednecks,* 69–70). Mabry credits Harris in "Disfranchisement of the Negro in Mississippi," 327.

48. *Journal of the Constitutional Convention,* 229–30.

49. Mabry, "Disfranchisement of the Negro," 327.

> Sir, it is no secret that there has not been a full vote and a fair count in Mississippi since 1875—that we have been preserving the ascendancy of the white people by revolutionary methods. In plain words, we have been stuffing ballot-boxes, committing perjury[,] and here and there in the State carrying the elections by fraud and violence until the whole machinery for elections was about to rot down. That which had a beginning in despair at the situation, and which seemed to justify any means for public preservation, has become a chronic ulcer upon the body politic—and threatened to disintegrate the morals of the people. Thoughtful men every where foresaw that there was disaster some where along the line of such policy as certainly as there is righteous judgment for nations as well as man. And I say, Mr. President, no man can be in favor of perpetuating the election methods which have prevailed in Mississippi since 1875, who is not a moral idiot, and no statesman believes that a government can be perpetuated by violence and fraud. The dullest intellect must see that it leads to political convulsions of some sort dangerous to life, liberty, and property.[50]

Having stripped the franchise issue of its hypocrisy, Judge Chrisman argued that both a property qualification and the educational restriction must be adopted regardless of how much they frightened poor white folks.

Judge Chrisman's speech was printed in the newspapers, and it upset some people terribly. You have got this fraud thing all wrong, they said. "We get indignant when any northern journal or northern senator or representative talks of fraud in southern elections," a citizen noted in the *Port Gibson Reveille*. "We accuse them of slandering our people. In *them* we earnestly deprecate what we call wholesale denunciations. Could there be a more wholesale denunciation than the declaration that we have been engaged since 1875 in 'ballot-box stuffing and perjury and carrying elections by fraud and violence?' So then what we denounce in the northern senator[,] in the southern judge we applaud. Is there any consistency in this?"[51]

But Judge Chrisman had exposed the lie, and many people, including most of the delegates, recognized that something had to be done. They knew that tactics used to suppress the black vote could be turned against white voters when opposing factions within the Democratic Party competed for

50. "Two Good Speeches," *Jackson Clarion-Ledger*, September 11, 1890.

51. See, for example, "Letter from Mr. Stiles," *Port Gibson Reveille*, September 19, 1890; emphasis in original.

power.[52] For the restoration of white supremacy to be complete, Mississippi had to stabilize the electoral system. "Judge Chrisman's are the first bold and honest utterances on the suffrage question that have yet been heard on the floor of the convention," the editor in the *Port Gibson Reveille* wrote. "Every member who heard him knew that he was speaking the truth; how many of them will have the manliness to join him in his efforts to put Mississippi politics on a basis of decency?"[53]

Everyone in Mississippi, black and white, understood that the educational restriction would cut both ways if it was fairly enforced, and such an outcome was unacceptable to politicians depending on the rural white vote. However, the understanding clause offered a loophole. It provided a way by which the educational restriction could be enforced preferentially by favoring the illiterate white applicant over the literate black one.[54] In fact, Judge Chrisman had noted this possibility in his speech when he pointed out that the county registrar would be the only person making a determination of literacy. "It looks as if it was intended that if the registrar wanted the man to vote he would read him some such clause as: Slavery except as a punishment for crime shall be forever prohibited. 'Do you understand that?' 'Oh, yes,'" Judge Chrisman explained. "But if he did not want him to vote he would read him the interstate clause or the section forbidding the Legislature to pass *ex post facto* laws and demand a construction." Yet, despite the obvious potential for its abuse, Judge Chrisman argued that the understanding clause was needed, unless "we are to go on carrying the election with Winchesters or by stuffing the ballot boxes."[55]

Despite continued opposition from skeptical delegates, the convention adopted the suffrage committee's report on September 18 with the educational restriction intact.[56] Opposition flared up again on October 22, when the convention adopted a report submitted by Judge Harris's Judiciary Committee stating that ratification of the new constitution by a popular vote was "unnecessary and inexpedient,"[57] and a week later opponents tried to reconsider the educational restriction once more. Judge Harris was picked to defend the provision and spoke for the entire twenty minutes allotted, calling it "a chief prop of the whole structure." The delegates agreed, and the motion

52. Cresswell, *Rednecks, Redeemers, and Race*, 111.
53. "A Manly Speech," *Port Gibson Reveille*, September 12, 1890.
54. McMillen, *Dark Journey*, 42.
55. "Two Good Speeches," *Jackson Clarion-Ledger*, September 11, 1890.
56. Mabry, "Disfranchisement of the Negro," 331.
57. Cresswell, *Rednecks, Redeemers, and Race*, 122; Wharton, *Negro in Mississippi*, 214.

to reconsider was soundly defeated by a vote of 67 to 34.[58] Three days later, on November 1, 1890, the constitution was adopted with only 8 dissenting votes.[59]

Delegates to the constitutional convention accomplished what they had set out to do. Mechanisms for disfranchising African Americans in Mississippi had been put in place, and this objective had been achieved without violating the Fourteenth or Fifteenth Amendments.[60] In 1892, the suffrage provisions of the constitution were upheld by the Mississippi Supreme Court, and the U.S. Supreme Court added its blessing six years later.[61] Within a decade, the black political power base in the state ceased to exist. In his book on African Americans in Mississippi, Wharton noted: "Rebuffed by unfriendly registrars, frowned on by the mass of the white population, and absolutely forbidden to support any candidates save those of a party based on white supremacy, the Negro voters found it, in the words of one of their leaders, 'a mighty discouraging proposition.' More and more of them, as time went on, simply abandoned the effort."[62]

Alfred Holt Stone had taken a keen interest in the proceedings of the constitutional convention and had attended many its sessions, either sitting in the gallery or chatting with delegates during recesses.[63] However, it was not until 1944 that he published an account of the convention in an article entitled "The Basis of White Political Control in Mississippi." The focus of his article was the convention's reapportionment of legislative districts and establishment of an electoral college for statewide offices. Despite the reality that over the years county registrars had relied heavily on the understanding clause to restrict black suffrage, Stone argued that the combination of having both the legislature and the electoral college dominated by coun-

58. "Understanding Clause: Convention Refuses to Reconsider It," *Jackson Clarion-Ledger*, October 30, 1890.

59. Mabry, "Disfranchisement of the Negro," 331. Judge Harris's successful defense of the educational restriction was the last episode in his long career of public service. Harris died on December 3, 1891 (Rowland, "Wiley P. Harris," 269).

60. Stephen Cresswell, *Multiparty Politics in Mississippi, 1877–1902* (Jackson: University Press of Mississippi, 1995), 104–9.

61. Cresswell, *Rednecks, Redeemers, and Race*, 124; Wharton, *Negro in Mississippi*, 215.

62. Wharton, *Negro in Mississippi*, 215.

63. Alfred Holt Stone, "The Constitution of 1890: A Speech by the Hon. Alfred H. Stone before the Jackson Lions Club, May 30, 1947," passim, and "'As a Man Thinketh in His Heart, So Is He:' Address to the Graduating Class of 1937 of Millsaps College, June 8, 1937," 16, both typescripts in Stone's "Articles and Speeches" subject file, MDAH. See also "Alfred Holt Stone" in McCain, *Story of Jackson*, 2: 222.

ties with white majorities was the primary mechanism for maintaining white supremacy. "The Constitution of 1890 provided certain suffrage qualifications," he explained, "but the framers of that document realized that these were really no more than palliative gestures, in so far as permanent white control was concerned. This was, and is, true because the negro can ultimately meet any qualifications which the white man is willing to erect and accept for himself."[64]

In 1947, Stone began speaking to civic groups about the Constitution of 1890, probably because he was one of the few people still alive who had actually witnessed the deliberations and knew the participants firsthand.[65] Once again, he downplayed the significance of the understanding clause as a mechanism for disfranchising African Americans. "It [the constitutional convention] was not in any sense of the word called for the sole purpose or the express purpose of accomplishing or seeking to accomplish the wholesale disfranchisement of our negroes," he told the Jackson Lions Club. "That was incidental to the chief objective of the convention," which was reapportionment, although Stone did credit Judge Harris with the authorship of the understanding clause. "It was written by Judge Harris. I saw the original—written on the stationery of Judge Harris and in lead pencil." Stone insisted that the two-year residency requirement and the poll tax, which had to be paid well in advance, were the most effective measures for suppressing the black vote because "they legislated against his [the black voter's] racial characteristics," namely "his migratory habits" and lack of foresight.[66]

As in the case of the violence attendant to the elections of 1875 and 1876, Stone refused to acknowledge that after 1890 the preferential enforcement of the understanding clause by county registrars became the chief mechanism for suppressing the black vote. Ironically, the clause did not have to be enforced very often to achieve its political goal because it functioned continuously as a formidable psychological barrier. The fact that the understanding clause was a final, insuperable obstacle between African Americans and their right to vote was known throughout the black community and served to discourage prospective voters from initiating the process. More

64. Alfred H. Stone, "The Basis of White Political Control in Mississippi," *Journal of Mississippi History* 6 (October 1944): 225–36.

65. See, for example, "Alf Stone Speaks at Rotary," "Alf Stone Guest Speaker Tuesday of Hinds Alumni," and "Alf Stone Addresses Pioneers," in the *Jackson Daily News*, April 17, 1947, May 1, 1949, and October 10, 1949, respectively.

66. Stone, "Constitution of 1890," 1, 7, 11, 12. In regard to Harris's authorship of the understanding clause, see Williams, "Poll Tax as a Suffrage Requirement," 487.

specifically, black voters realized that paying the poll tax was a waste of money if they were sure to be disqualified when they went to register. Consequently, it was disingenuous of Stone to say that African Americans were not allowed to vote because they did not pay their poll tax. Nevertheless, he asserted to the end of his life that "things the Negro did, not what he was, cost him his vote."[67]

67. "Things Negro Did, Not What He Was, Cost Him His Vote," *Memphis Commercial Appeal*, June 17, 1948. The article quoted remarks Stone had made in his presentation to students at the University of Mississippi the evening before.

5

THE BUSINESS OF RAISING COTTON

Alfred Holt Stone read law in Judge Harris's office for eighteen months before returning to the University of Mississippi, where he completed requirements for a law degree at the end of the 1890–91 session.[1] Nevertheless, Stone did not practice law because he had something else on his mind—cotton.[2]

Stone's father may have been the state auditor, but he was a speculator in Delta land too. Captain Stone's career in real estate had begun in 1866 when he bought a lot in Greenville shortly after arriving in Mississippi.[3] Within three years, he had leased Ararat Plantation on Deer Creek near Leland and started growing cotton. By 1874, Captain Stone moved his family to a new plantation, Sligo, which is where they lived when their house burned down two years later. It is unclear from surviving records whether Captain Stone was making much money, but it is clear that he was very active in leasing, buying, and selling real estate. By 1882, the family was situated on Camellia Plantation, a larger version of Ararat located at the same place. Although

1. Stone to Willcox, January 10, 1902, WFWP; *Historical Catalogue of the University of Mississippi, 1849–1909* (Nashville, TN: Marshall and Bruce, 1910), 228. The formal title of Stone's degree was bachelor of laws. In March 1955, two months before his death, Stone was recognized as the oldest living alumnus of the law school in conjunction with the school's centennial (Michael De L. Landon, *The University of Mississippi School of Law: A Sesquicentennial History* [Jackson: University Press of Mississippi for the University of Mississippi School of Law, 2006], 90).

2. Stone, "The Constitution of 1890," 4. The Mississippi Bar has no record of Stone's being admitted to membership (personal communication [e-mail] from Theresa S. Robinson, membership records administrator, June 24, 2005).

3. Purchase of lot on Blantonia Plantation from H[arriet] B. Theobald by Captain Stone and Ferguson Maury, January 27, 1866 (Deed book S, 478, WCCH).

Captain Stone's election as state auditor in 1885 ended his direct involvement in growing cotton, he continued to speculate in Delta real estate.[4]

The end of Captain Stone's career as a cotton grower marked a beginning for his son. On April 21, 1892, Walter Wilson Stone deeded Alfred Holt Stone a half interest in a section of land ten miles east of Greenville.[5] The section was a valuable piece of property because the Greenville, Columbus, and Birmingham Railroad ran right through it. It was to become the nucleus for Dunleith Plantation, which Stone was to call home for more than forty years.

Dunleith was not an established plantation when Captain Stone and his wife, Ella, started acquiring land for it in 1884. It grew piece by piece, extending from the railroad in sections and parts of sections until it stretched along the east bank of Bogue Phalia to Hollyknowe, another plantation to the south.[6] By 1896, it was more or less complete, and on January 10 of that year, Captain Stone helped Alf buy all 2,506 acres of Dunleith with money borrowed from the bank.[7]

The Delta was not like the older, more established parts of Mississippi. Except for relatively narrow strips of land along the major tributaries, the Delta had not been settled prior to the Civil War. The foliage was too dense and the water was too high to invite any but the backwoods trapper.[8] But that all changed after the Civil War, when the price of cotton was high and the demand for new cotton-producing acreage was intense. Levees were built, and new projects allowed the swamps to be drained. All that was needed was la-

4. Various transactions recorded in deed books U (pp. 97, 490), E-2 (p. 272), F-2 (p. 517), G-2 (p. 484), M-2 (p. 135), and N-2 (p. 654), WCCH. W. W. Stone returned to the real estate business after his tenure as the state auditor (see letterhead of W. W. Stone's letter to a state representative from Glen Allan, Mississippi, dated May 9, 1906, PFP [box 1, folder 9].)

5. Section 4 of township 18 north, range 6 west (Deed book G-3, 381, WCCH).

6. The intricate trail of transactions that allowed the Stones to develop Dunleith as a plantation can be found in the deed books for Washington County, notably books P-2 (p. 790), R-2 (p. 754), G-3 (pp. 381, 531), F-3 (pp. 410, 569), I-3 (pp. 30, 136), L-3 (p. 660), and M-3 (p. 541). Anne Lipscomb Webster reviewed the plat record for section 4 and found fifty-eight entries for that section alone between 1888 and 1913.

7. By the time Stone became the state tax commissioner in 1932, Dunleith had grown to 3,900 acres (Deed book 245, 71 [WCCH]). Hollyknowe was owned by William B. Swain, one of Captain Stone's real estate partners (*History of Mississippi: The Heart of the South* [Spartanburg, SC: Reprint Co., 1978], 3: 76). See also deed book D-3, 430 for a transaction involving Ella Stone, Mr. Swain, and his mother, Harriett.

8. James C. Cobb, *The Most Southern Place on Earth: The Mississippi Delta and the Roots of Regional Identity* (New York: Oxford University Press, 1992), 43.

bor to clear the trees and brush so that cotton farmers could plant their crop in the rich, alluvial soil.[9]

A situation such as this was ideal for an aggressive entrepreneur like Captain Stone. His was a business of buying underdeveloped land and selling or trading it as its value rose. Captain Stone was not a planter per se, not one of those aristocratic barons of the soil who accumulated wealth. Captain Stone was a speculator, apparently short on cash but long on property. He also must have had wits of steel to deal in reclaimed swampland to the extent that he did.

By the time Alfred Holt Stone got into the business, the tide of speculation had passed. Most of the swamps in the Delta had been drained and the land cleared. If money was to be made, it was to be made by the growing of cotton, a labor-intensive endeavor that depended almost entirely on African Americans.[10] It was this link between growing cotton and black labor that intrigued Stone and eventually led him to devote much of his early life to the study of race and its impact on literally everyone with whom he came into contact.[11]

The cost of growing cotton was high, even if the price it fetched after it was harvested was high too, and nothing added more to the cost of producing a bale of cotton than the wages of the men and women who planted the seed, hoed the rows to keep the weeds down, and picked the bolls when the time to pick them came. And there were other expenses associated with labor that went into the cost of growing cotton. For one, there was food for the workers and places for them to live. They needed clothes to wear and tools to till the ground. Mules to pull wagons and ploughs were an expense too, and the mules had to be fed with corn grown by workers in the same fields that

9. For a good account of the Delta during this period, see John C. Willis's *Forgotten Time: The Yazoo-Mississippi Delta after the Civil War* (Charlottesville: University Press of Virginia, 2000). Robert L. Brandfon's *Cotton Kingdom of the New South: A History of the Yazoo Mississippi Delta from Reconstruction to the Twentieth Century* (Cambridge: Harvard University Press, 1967) is also good, although it focuses primarily on the role of the Illinois Central railroad in the economic development of the Delta.

10. Alfred Holt Stone, "The Italian Cotton Grower: The Negro's Problem," *South Atlantic Quarterly* 4 (January 1905): 42–47; Alfred Holt Stone, "The Negro in the Yazoo-Mississippi Delta," in *SARP*, 87–88, 123.

11. For a statement of Stone's self-professed interest, see "Race Problem Contrasts and Parallels," in *SARP*, 6. According to Stone, his interest in racial theories emerged as early as 1893, when he was only twenty-three years old (Alfred Holt Stone, "Italian Cotton-Growers in Arkansas," *American Monthly Review of Reviews* 35 [February 1907]: 209, and Alfred Holt Stone, "The Negro Farmer in the Mississippi Delta," *Southern Workman* [October 1903]: 457).

produced cotton. Consequently, to make money growing cotton meant controlling the cost of labor.[12] It would be almost fifty years before mechanization reached the Delta, and producing a cotton crop at the turn of the century was done with unskilled, back-breaking labor, just as it had been done for more than a century.[13]

By the time that Stone became a cotton grower, demand for the staple was such that labor held the upper hand. Finding reliable men and women to plant, hoe, and pick was becoming more and more of a problem in the Delta. "An adequate supply of labor is the first essential in the business of raising cotton," Stone noted a few years later when he wrote about his early experiences as a cotton grower. "To secure it constitutes the most serious problem confronting the plantation management," he observed. "Not for forty years has the supply equaled the demand in the alluvial section of Mississippi. Here the Negro still has almost a monopoly of the field of manual labor."[14]

There were several types of arrangements that bound black labor to the rich Delta soil. One was black ownership, and, as John C. Willis has documented in his book about the Delta after the Civil War, that reality had ceased to be an option for African Americans following economic and racial reversals in the 1880s.[15] A second arrangement was that of wage earner. Working for a fixed rate of pay was common, but this arrangement was used primarily for short periods of time, such as during the cotton-picking season in the fall. What most planters needed was a source of labor that they could depend on for the entire year.[16]

Yearly workers were hired according to two basic arrangements, although there were variations depending on the worker's assets and the needs of the

12. For an extended discussion of economic factors affecting cotton production, see Alfred Holt Stone, "The Negro in the Yazoo-Mississippi Delta" and "A Plantation Experiment," both in *SARP*, 81–124, and 125–48, respectively.

13. Cobb, *Most Southern Place on Earth*, 204–6; Donald Holley, *The Second Great Emancipation: The Mechanical Cotton Picker, Black Migration, and How They Shaped the Modern South* (Fayetteville: University of Arkansas Press, 2000), passim. See also "Farmers at the Technological Crossroads: Post–WWII Diffusion of the Mechanical Cotton Harvester," www.eh.net/Clio/Conferences/ASSA/Jan_97/Grove.shtml.

14. Stone, "A Plantation Experiment," 125; see also Willis, *Forgotten Time*, 158–61; and Bertram Wyatt-Brown, "Leroy Percy and Sunnyside: Planter Mentality and Italian Peonage in the Mississippi Delta," *Arkansas Historical Quarterly* (Spring 1991): 60–84; reprinted in Jeannie M. Whayne, ed., *Shadows over Sunnyside: An Arkansas Plantation in Transition, 1830–1945* (Fayetteville: University of Arkansas Press, 1993), 78.

15. Willis, *Forgotten Time*, 144.

16. Brandfon, *Cotton Kingdom of the New South*, 133.

planter.[17] The most popular arrangement after the war was tenant farming, in which the worker paid a predetermined rent in either cash or kind for a field to cultivate and a house to live in. The tenant may have his own implements and livestock, or he may have to buy or rent these items from the planter. Whatever the case, the tenant farmer agreed at the beginning of the year to pay a fixed price to work the land, although collection would necessarily be delayed until the crop was harvested and the renter had money or staple to pay his rent.[18]

The tenant farmer got to keep everything he made in excess of what he had obligated to pay as rent. That arrangement worked well for the tenant when yields or prices were high. But there were advantages for the planter too. Much of the risk associated with bad times was assumed by the tenant, and the planter could count on a predictable return, unless, of course, the tenant did so poorly that he could not pay his rent or skipped town.[19]

A second type of arrangement for employing yearly labor was sharecropping, and, over time, it came to predominate. As in tenant farming, the planter provided a farmer with a field to cultivate and a house to live in. But unlike in tenant farming, the sharecropper divided the result of his labor with the planter. If the planter had to supply livestock, implements, and other supplies, he would normally split the crop with the worker half and half. If the worker supplied his own mule and implements, he might negotiate a better deal, sometimes keeping as much as three-quarters of the crop. In either case, risks associated with poor yields or prices were shared by the tenant and the planter alike.[20]

A key difference between the two arrangements for yearly labor was the degree of supervision the worker could expect from the planter. The share-

17. For an extensive discussion of the contractual arrangements available to Stone when he began business as a cotton planter, see Lee J. Alston and Robert Higgs, "Contractual Mix in Southern Agriculture since the Civil War: Facts, Hypotheses, and Test," *Journal of Economic History* 42 (June 1982): 327–53.

18. James D. Anderson, "The Southern Improvement Company: Northern Reformers' Investment in Negro Cotton Tenancy, 1900–1920," *Agricultural History* 52 (January 1978): 114; Willis, *Forgotten Time*, 36. The going rate for the noncash payment of rent was one-quarter to one-third of the crop (Brandfon, *Cotton Kingdom of the New South*, 132).

19. Cobb, *Most Southern Place on Earth*, 71–72; Willis, *Forgotten Time*, 37; Anderson, "Southern Improvement Company," 114.

20. Cobb, *Most Southern Place on Earth*, 71; Cresswell, *Rednecks, Redeemers, and Race*, 39; Willis, *Forgotten Time*, 34; see also Holland Thompson, *The New South: A Chronicle of Social and Industrial Evolution* (New Haven, CT: Yale University Press, 1919), chap. 4.

cropper was considered to be an employee and was thus liable to closer supervision than the tenant, who, as a renter, was considered to be his own boss. Although the landowner might attempt to supervise the tenant too, the incentive for close supervision lay clearly with sharecropping because if the sharecropper did not do his job properly, the planter lost money too.[21]

Overall, it might be said that tenant farming favored the worker, and sharecropping generally worked out better for the planter. Consequently, there was always a tension between the two arrangements, and which one prevailed depended on the availability of labor. When the demand for labor was high, the worker sought to rent. Over time, however, planters moved exclusively to sharecropping as conditions allowed.[22]

Stone clearly understood the costs and benefits of both arrangements when he embarked on his career as a cotton planter, and it did not take him long to decide that he liked the arrangement that generated rent. Within two years of having acquired Dunleith, he had conceived of a labor program that would establish a system of more or less permanent tenants.[23]

> It was to the desire and hope of building some such "assured tenantry" that our experiment largely owed its inception. Its salient features were: Uniformity of tenant system, all land being rented at a fixed cash rental; the sale of stock, implements, and wagons to tenants upon exceptionally favorable terms; the exercise of proper supervision over the crop; the use of a contract defining in detail the undertakings of each party; the handling and disposition of the gathered crop by the plantation management. Let there be no misunderstanding of the motives behind all this. There was nothing philanthropic about it. It was a business proposition, pure and simple, but certainly one with two sides to it. The plan was to select a number of Negro families, offer them the best terms and most advantageous tenant relation, and so handle them and their affairs as to make them reach a condition approaching as nearly as possible that of independence. The hope was that, having accomplished this our pose, we would thereby also have in large measure solved the labour problem, having at-

21. Willis, *Forgotten Time*, 34, 168; Holley, *The Second Great Emancipation*, 6–7.

22. The proportion of farms in the Delta cultivated by renters decreased from 44 to 19 percent between 1900 and 1920, while the proportion cultivated by sharecroppers rose from 46 to 74 percent over the same period (Willis, *Forgotten Time*, 164). Willis asserts that the boll weevil was primarily responsible for the shift to sharecropping between 1910 and 1920.

23. Stone eventually converted his entire operation to sharecropping and exercised even closer supervision over his labor force (Harris, *Deep Souths*, 132).

tached to the plantation by ties of self-interest a sufficient number of these independent renters to make us in turn measurably independent of the general labour situation.[24]

To be "independent of the general labour situation" was Stone's goal. A steady income generated by rent collected from reliable workers would allow Stone to pay off the notes for Dunleith as they came due, improve the value of the property, and pursue other interests. Certainly, he would be responsible for the "proper supervision" of the renters to make sure that they could pay their rent at the end of the year, but Stone had things on his mind other than growing cotton.

One of the other things that attracted Alf's attention was the active social life to be had in Greenville. Greenville was a sophisticated city—more sophisticated than its population of less than ten thousand might have led one to expect—because it attracted planters, merchants, lawyers, and bankers who congregated there as a result of the wealth that was being created from growing cotton on that rich Delta land.[25]

One of the most prestigious social organizations in Greenville during this period was the Hypatia Club, a group "composed of ladies and devoted to the combined pursuits of literary and social pleasures."[26] Stone was extended an invitation to address this exclusive body on May 15, 1894. The setting was gracious, as it was described in the local newspaper by a reporter, possibly a club member. "The house decorations were sumptuous, such a wilderness of roses that the senses were oppressed by their beauty and fragrance." It is not clear whether the reporter meant to write "oppressed," but she continued: "Full dress was of course de rigeur [*sic*], and the parlors presented a brilliant appearance with the combination of handsome toilets, lights and flowers."[27]

Alf, who had yet to celebrate his twenty-fourth birthday, was there to speak on women's suffrage. "This talented young gentleman's name is a suf-

24. Stone, "A Plantation Experiment," 127–28.

25. Harris, *Deep Souths*, 168.

26. Mary Louise Merideth, "The Mississippi Woman's Rights Movement, 1889–1923: The Leadership Role of Nellie Nugent Somerville and Greenville in Suffrage Reform" (master's thesis, Delta State University, 1974), 32–33.

27. "The Hypatia Reception," *Greenville Times*, May 19, 1894. Hypatia (ca. 370–414) was a neo-Platonist philosopher, mathematician, and astronomer in Alexandria whose dedication to pagan scholasticism resulted in her being attacked by a mob of Christians, stripped naked, and pummeled with pieces of broken pottery until she was dead ("Hypatia," www.scottlan.edu/lriddle/women/Hypatia.htm). It is possible that the members of the Hypatia Club in Greenville were not fully aware of their namesake's religious views or fate.

ficient guarantee of the interest and excellence of his paper," the reporter noted, "which was added to by a running fire of extemporaneous comment." His was a typically male view. "Women should rule by love and kindness and not by force," he argued. If a woman wanted to enter "the cesspool of politics," he warned, she "would divest herself of those tender charms that make her so dear to men."[28]

Apparently, Stone's views were well received because he was greeted with applause when he had finished. The ladies were then served ices and strawberries while a group of musicians provided music for the occasion from the anteroom. It had been a grand evening for "one of the most successful and charming of Greenville's social organizations."[29]

Social life for prominent men in Greenville took a different turn. No ices and strawberries in the parlor for them. They liked to be outside, and one activity that passed social muster was guarding the levee. It was a respectable way for the guys to hang out before golf courses came to the Delta. The levees along the Mississippi River were not very high, about four feet, according to William Alexander Percy in *Lanterns on the Levee.* Built by workers with wheelbarrows, they were prone to break in the spring. To spot breaks as soon as they occurred so that repairs could be made before they widened, prominent citizens from Greenville volunteered to ride along the levee, and Stone joined LeRoy Percy's levee guard when he came back to Greenville after reading law in Jackson.[30]

Will Percy wrote about his father's experience with the levee guard in 1940. Because he was only eight years old in 1893, Will turned to Alfred Holt Stone to resurrect the memories of his father's service on the levee near Greenville.

> I asked Mr. Alf the other day if he remembered old Camp Cousens, and, as I expected, he remembered it all, in detail. While off duty the guards played freeze-out poker for licks administered by the winner with

28. "The Hypatia Reception," *Greenville Times,* May 19, 1894; "The Hypatia," *Greenville Democrat,* May 17, 1894. Stone's talk was entitled "Women and Her Sphere."

29. *Greenville Times,* May 19, 1894.

30. William Alexander Percy, *Lanterns on the Levee: Recollections of a Planter's Son* (New York: Knopf, 1941), 242–43. For a detailed account of the levee system in the Delta before 1893, see Mikko Saikku, *This Delta, This Land: An Environmental History of the Yazoo-Mississippi Floodplain* (Athens: University of Georgia Press, 2005), 144–51. Saikku states that the levees had reached the height of four feet before the Civil War and had been increased to about seven feet by 1882 after the Mississippi Levee District for Bolivar, Washington, and Issaquena Counties was created in 1877.

Mr. Cousens's broad, water-soaked razor strop to the squinched posterior of the loser as he bent miserably over a log. No winner ever stayed his arm because of friendship or compassion. The game was gay and brutal. Hotspur and the heroes beneath Troy would have joined it and have been companionable in that company. It attracted participants from the whole neighborhood. One planter, affable and tipsy, rode up on his fine five-hundred-dollar mare as the players were recovering from the last game in a swim. Without dismounting he whipped his snorting mount into the thick of the swimmers, roaring out the while: "Gotta head like a fish and a tail like a man. I'm a mare-maid."[31]

According to Will Percy, all of the members of his father's levee guard went on to pursue careers of distinction. Individually, they became "a captain in the Spanish-American War, a circuit judge, a leader in the state legislature, two sheriffs, a distinguished Cincinnati lawyer, and a United States Senator [Will's father, LeRoy]." The levee guard who became a captain in the Spanish-American War was Henry T. Ireys Jr., and he had a sister.[32]

Her name was Mary Bailey Ireys, and the specifics of how and when Alfred Holt Stone courted Mary and won her hand were not recorded. Nevertheless, they were wed on June 25, 1896.[33] Theirs was to be a long marriage, almost sixty years, and although Alfred Holt Stone was long-lived, Mary Stone lived even longer, dying in 1969 at the age of ninety-six.[34] She bore one child, Alfred Holt Jr., who died in infancy a year after they were married.[35]

Mary Bailey Ireys Stone came from a prominent Greenville family. Her father, Henry T. Ireys, had come to the Delta shortly after the war. He had several things in common with Alf's father. Both were young when they reached Greenville in 1865 (Henry Ireys was twenty-eight, and Captain Stone, twenty-five), and both played an active part in laying out the streets for the new town when it was moved upriver from where it had been de-

31. Percy, *Lanterns on the Levee*, 243–44.

32. Ibid., 242–43; Princella W. Nowell, *A Closer Look: A History and Guide to the Greenville Cemetery* (Jackson, MS: Hederman, 2003), 63.

33. "Stone-Ireys," *Greenville Democrat*, July 2, 1896, and "Married," *Greenville Times*, June 27, 1896. They were married in the First Presbyterian Church, and Henry T. Ireys Jr. was the best man.

34. The dates on Mary Bailey Stone's headstone in the Greenville Cemetery are June 23, 1873, to December 3, 1969. The dates for Mary Stone life in Nowell's book on the Greenville Cemetery, *A Closer Look*, are given as 1879–1960 (63).

35. Caledonia Jackson Payne, comp., "Old Greenville Cemetery, Greenville, Mississippi, 1880–1982" (photocopy of a typescript, 1983, MDAH).

stroyed during the war. But they differed in the resources they had to rely on when they arrived. Unlike Captain Stone, Henry Ireys's father had left him an inheritance in the Delta, a large tract of productive land along Williams Bayou.

Henry Ireys's father, Henry T. Irish (the family name was changed to Ireys in 1881), had come to Mississippi from Rhode Island in 1821 and settled near Port Gibson. Four years later, he moved to Washington County and began clearing land to grow cotton. He did extremely well and extended his holdings in Mississippi to include fertile land in seven different counties.[36] In 1834, Henry Irish returned to Rhode Island to wed Mary Bailey. Henry Ireys was their eldest child.[37]

Henry's father came back to Mississippi to look after his investments, but Henry stayed in Rhode Island to go to school. He was still there in 1846 when his father died. Although Henry Ireys did not grow up in the South, he did make a trip to New Orleans in the fall of 1852, when he was fifteen years old, to visit his uncle, a wealthy cotton factor (broker) in the city. It was during this visit that Henry went upriver to visit the family plantation on Williams Bayou.[38] Returning north to complete high school in Brooklyn, New York, he worked as a clerk in New York City until his mother called him to join her in Edinburgh, Scotland, where she had taken his sister Kate for her health. Henry spent much of the Civil War in Scotland, but he returned before it was over and headed South to visit his inheritance on Williams Bayou. However, the captain of the gunboat that took him to Greenville warned him not to go ashore because Rebel guerrillas were lurking about, and Henry retreated to New York City for the remainder of the war.[39]

36. At the time of his death, Irish owned land in Bolivar, Carroll, Choctaw, Copiah, Tallahachie, Washington, and Yalobusha Counties ("Henry T. Irish's Will," November 27, 1846, Record of Wills, vol. 1, 1839–1894, WCCH).

37. *Biographical and Historical Memoirs of Mississippi*, 2: 1004–5.

38. In his will, Irish instructed the executor of his estate to sell all of his holdings except Mound Pleasant, the plantation on Williams Bayou. He wanted that plantation to be passed on to his heirs debt-free (Irish's will, WCCH). He also instructed the executor to invest the remainder of his wealth in "stocks or property of any kind or description ay any time & at any point he may deem advisable always providing that he shall never make such investment in any slave holding State or South of the Potomac River." Irish himself owned many slaves but instructed the executor to sell all of them except fifty of the most productive. Perhaps Henry T. Ireys's father foresaw the possibility of sectional conflict and did not want to risk his inheritance to the vicissitudes of war.

39. Ireys, "Life and Times of Henry Tillinghast Ireys," 3–29; Death certificate for Henry Tillinghast Ireys, December 24, 1923, on microfilm at MDAH. See also Stone's eulogistic obituary of his father-in-law in the *Greenville Daily Democrat-Times*, January 8, 1924.

Henry Ireys returned to Greenville in May 1865 as soon as the hostilities ended. Although his visit to the plantation on Williams Bayou in 1865 was short, he returned with his brother John a year later to take possession.[40] In 1868, Henry Ireys married Elizabeth Taylor, called Betty, who came from a prominent Delta family. They were to have six children who survived into adulthood. Alf's buddy in the levee guard, Henry T. Ireys Jr., was the oldest. Mary Bailey Ireys was the second oldest.[41]

In 1878, the railroad magnate Collis P. Huntington appointed Henry Ireys secretary and treasurer of a new railroad, the Greenville, Columbus, and Birmingham, which had been chartered to connect the iron-rich region of central Alabama to the Mississippi River. Starting at Greenville, it never got as far as Birmingham, but it did open up the interior of the Delta and allowed cotton growers to transport their crop to the river.[42] The new position required Ireys's presence in Greenville, so he moved the family into town. It was not long before he used his leverage as a railroad executive to become a banker and later the city's first cotton factor. By the time his daughter married Stone, Henry T. Ireys was a very wealthy man.[43]

Alfred Holt Stone's marriage to Mary Ireys completed the circle, and by 1898 he had found his niche. Approaching thirty, he had a steady income, a wife from a prominent family, and good friends. He also had something else he wanted—time to pursue his academic interests. His father had been a promising scholar before the war interrupted that career, and Stone had grown up in a family in which learning and inquiry were highly regarded. Although at that point he could not quite be called a gentleman farmer, Stone could still distance himself sufficiently from the mundane tasks associated with running a plantation to engage in more intellectual pursuits.

The first avenue Stone took in his search for knowledge was history. The Mississippi Historical Society had been organized on May 1, 1890, in the Chancellor's Office at the University of Mississippi. Stone had already left the university to study law with Judge Harris at that time, but he was familiar

40. Ireys, "Life and Times," 29; Ireys, "County Seats and Early Railroads," 284.

41. *Memoirs of Mississippi*, 2: 1005. Henry T. Ireys Jr. was more than Alf's buddy. Ireys and Stone briefly went into the insurance business together after Stone returned to Greenville from college (Brodie Crump, "Mostly Old Stuff," *Greenville Delta Democrat-Times*, January 8, 1962). Their advertisement appeared on the front page of the June 27, 1896, issue of the *Greenville Times*.

42. Willis, *Forgotten Time*, 101.

43. *Memoirs of Mississippi*, 2: 1004; Ireys, "County Seats and Early Railroads," 267; "Life and Times," 5; "Ireys-Archer Cotton Co.," undated article (ca. 1906) in the *Greenville Times*, souvenir edition.

with the participants. His favorite professor, William R. Sims, was present at the organizational meeting and became secretary and treasurer of the society. Possibly through his contact with Professor Sims, Stone became an active member almost immediately after its formation.[44]

Within two years of joining the Mississippi Historical Society, Stone undertook a project that would foreshadow his interest in collecting primary source material a decade later. On February 23, 1893, Stone wrote a letter to the editor of the *Jackson Clarion-Ledger* and other papers in the state under the heading "The Name of Each Asked For as a Matter of History."[45]

> The undersigned [Alfred H. Stone] wishes to get the names of every book ever written on Mississippi soil, to be used in the preparation of a hand-book of Southern authors. He wants the name of the author or authoress, and the name of his or her works, and date of publication; and when possible such personal data as date of authors birth and death. The only limitation is that the author shall have written at least one book, and have been a Mississippian when it was written. Works of fiction, poetry, politics, law and religion, are all wanted. Parties possessing any such information are requested to communicate with the undersigned [in Greenville, Mississippi], and by so doing will confer a great favor upon him.[46]

Responses to Stone's request arrived in a steady stream from February through June until he had about 120 in all.[47] Unfortunately, most of the authors identified in these letters would be unfamiliar to the reader today because the state's remarkable legacy in literature would not be established until the twentieth century. But Stone was on the right track. Unfortunately, nothing came of the project, and the information contained in the responses to Stone's request never found its way into print.

It is not clear why the project to identify and catalog Mississippi authors was not completed. Possibly it was because the Mississippi Historical Society floundered four years after it was organized for a want of interest,

44. *Memoirs of Mississippi,* 2: 53–54. Judge Wiley P. Harris is also on the 1891 membership roll. Professor Sims's role comes from "Development of Historical Work in Mississippi," *Publications of the Southern History Association* 6 (July 1902): 335.

45. The letter appeared in the February 26, 1893, issue of the *Natchez Daily Democrat* and the February 27, 1893, issue of the *Jackson Clarion-Ledger.*

46. "The Name of Each Asked For as a Matter of History," *Jackson Clarion-Ledger,* February 27, 1893.

47. Stone literary papers (1 box), MDAH.

and sponsorship of Stone's project may have evaporated. Whatever the case, a recent graduate of the scientifically oriented doctoral program in history at Johns Hopkins University, Franklin L. Riley, was appointed as the University of Mississippi's first professor of history in 1897. Riley was keen on state and local history and wanted to revive the recumbent organization. "What do you think of the following plan for resurrecting a defunct society?" Riley asked his major professor at Johns Hopkins University, Herbert Baxter Adams.[48]

The twenty-nine-year-old scholar had a vision and the energy to get things moving, and the response to his invitations to join met with success. In January 1898, Riley chaired the first meeting of the reconstituted society in the chamber of the House of Representatives in the state Capitol. Although Professor Adams could not attend, the father of scientific history submitted a paper to be read by a proxy.[49] Adams had a reputation for inspiring his students, and he did not let this opportunity pass unheeded.[50] "Begin to collect a library for yourselves," he told them. "Students and teachers do not always appreciate the opportunities they enjoy of acquiring good books of History. I would strongly urge students to save their money instead of spending it on poor theaters and variety shows. Buy standard books of literature, art, and history; devote your leisure hours to good reading, always with pen and pencil in hand, and with a dictionary and an atlas beside you. *Seize the moment of excited curiosity* and look up every point on which you need exact information."[51]

Stone was an active ally in Riley's attempt to reinvigorate the Mississippi Historical Society and read a paper at the 1898 meeting entitled "Views

48. Charles S. Sydnor, "Letters from Franklin L. Riley to Herbert B. Adams, 1894–1901," *Journal of Mississippi History* 2 (April 1940): 100–101. The quotation comes from Riley to Adams, October 23, 1897, p. 103.

49. Riley to Adams, December 7, 1897, in Sydnor, "Letters from Riley to Adams," 104 6. A copy of Riley's letter of invitation to prospective members can be found on page 104. The location of the meeting comes from Franklin L. Riley, "The Work of the Mississippi Historical Society," *PMHS* 10 (1909): 36.

50. One of Adams's students, John Spencer Bassett, wrote the entry for his mentor in the *Dictionary of American Biography* and described Adams as follows: "Many of his students became greater scholars than their master, and this despite his excellent record in college and university; but none of them had better than he the power to convert a new provincial into an aspiring and assured research student and writer of history" ([New York: Scribner, 1928)], 1: 70).

51. Herbert Baxter Adams, "The Study and Teaching of History," *PMHS* 1 (1898): 79–80; emphasis in original.

of Dr. Alfred Charles Holt [Stone's grandfather] on the Secession Movement."[52] Unfortunately, Stone's first professional presentation has not survived, but his second effort did. In 1899, the Mississippi Historical Society met in Natchez, and Stone delivered a paper entitled "The Early Slave Laws of Mississippi," which was printed in *Publications of the Mississippi Historical Society*. Compared to the many articles that Stone was to write over the next ten years, this work was brief and superficial. Nevertheless, the paper introduced three themes that were to appear consistently in his writings for the rest of his life: (1) slavery was good, (2) Reconstruction was a disaster, and (3) unless you lived in the Deep South, you did not have a clue as to what the race problem was all about.[53]

Although Stone was later to backtrack on his initial endorsement of slavery as the best and most humane system for regulating the interaction of two races living in close proximity (he came to believe that the strict segregation of races could accomplish the same goal), he never wavered in his denunciation of Reconstruction as "an ineradicable scar on the heart [of the South]." Nor did he change his opinion about the futility of Yankees trying to tell southerners what they should do in regard to matters involving race. "Ignorant" and "blind" is how he described them.[54] His address for the Mississippi Historical Society not only set the tone for all that was to follow, it also became the first step in what was to become his rapid ascent to the most prestigious ranks of scholars thinking about the American race problem at the turn of the century.

52. Riley sent a copy of the program for the meeting to Adams, and it is still in the Herbert B. Adams Papers (box 51) at the Milton S. Eisenhower Library, Johns Hopkins University, Baltimore. Riley also sent Adams a copy of the survey he developed to identify historical material and records held by private citizens in Mississippi.

53. Alfred Holt Stone, "Mississippi's Constitution and Statutes in Reference to Freedmen, and Their Alleged Relation to the Reconstruction Acts and War Amendments," *PMHS* 4 (1901): 143–226. See also Alfred Holt Stone, "A Mississippian's View of Civil Rights, States Rights and the Reconstruction Background," *Journal of Mississippi History* 10 (October 1948): 181–239, which appeared initially in the *Staple Cotton Review* 26 (April 1948). Stone's tripartite theme was anticipated in 1887 by another Mississippian, Fulkerson, when he published a pamphlet entitled *The Negro; As He Was; As He Is; As He Will Be.*

54. Both quotations come from Stone, "A Mississippian's View of Civil Rights," 192. Stone wrote the article in response to a report issued by the President's Committee on Civil Rights in December 1947. The report foreshadowed the civil rights movement, and Stone's response laid out a strategy for the continuation of segregation in the South that became known as "massive resistance."

6 CONVICTIONS OF SOUTHERN MEN

In January 1900, Alfred Holt Stone heard about a conference that caught his attention. A group of twenty-five prominent white leaders in Montgomery, Alabama, had formed the Southern Society for the Promotion of the Study of Race Conditions and Problems in the South. Despite its long name, the society had a simple goal: to get people talking openly about the race problem in the South. To that end, the society organized a three-day conference in Montgomery for May 8 to 10, 1900.[1] It was to be "an open parliament" in which a variety of opinions could be expressed, as Edgar Gardner Murphy, the rector of St. John's Episcopal Church in Montgomery and the society's secretary, described it in a speech on January 10. "The value of the Montgomery Conference," he noted, "will largely lie in the fact that it will be representative of the varied and even antagonistic convictions of Southern men."[2]

Murphy's speech struck a responsive chord with his audience and was reported widely in the press, particularly the part about "Southern men."[3] The intent of the conference was to have white southerners talk about the race problem in the South, not northerners who failed to understand the unique and often contradictory relationships between men and women of African

1. *Race Problems of the South: Report of the Proceedings of the First Annual Conference Held under the Auspices of the Southern Society for the Promotion of the Study of Race Conditions and Problems in the South at Montgomery, Alabama, May 8, 9, 10, A.D. 1900* (Richmond, VA: Johnson, 1900), 7–13. There is a copy of this monograph in TSC.

2. Edgar Gardner Murphy, "An Address at Tuskegee" (N.p., [1900?]), 7. The speech was for the dedication of the Slater-Armstrong Memorial Building at the Tuskegee Institute. There is a copy of this monograph in TSC.

3. Hugh C. Bailey, *Edgar Gardner Murphy: Gentle Progressive* (Coral Gables, FL: University of Miami Press, 1968), 33. See also Booker T. Washington to Francis Jackson Garrison (William Lloyd Garrison's son), February 3, 1900, BTWP.

and European descent. "We feel that much of true progress in connection with our racial difficulties has been embarrassed by the fact that the leadership of Southern opinion has been too largely attempted merely from the North," read a statement in the society's constitution adopted on January 20. "The solution of our problems in the South must come from the Southern [white] people themselves. . . . Suggestions from the North, offered with the best motives, have frequently been based upon inadequate acquaintance with our conditions."[4] Although black people from the South could attend if they sat in the balcony, there were no African Americans on the program.[5]

The society organized the conference program around four topics concerning African Americans in the South. The first of these was the franchise. Should it be limited [should African Americans be excluded] by law? Second, what about black education? Should it be "wholly or chiefly industrial [vocational]?" Next, to what degree should European Americans involve themselves with the religious practices of African Americans? For example, "Should we advise the raising of the standard of ordination for the Negro clergy?" Finally, and perhaps most importantly, what should be the African American's status in relation to the social order? "Is there antipathy to the Negro in the South? If so, is it industrial or racial, or both? Is race antipathy a curse, or a blessing to both races?"[6]

The governor of Alabama, Joseph F. Johnston, opened the conference with introductory remarks on the first evening. The conference organizers had assembled an impressive slate of speakers. With the exception of W. Bourke Cockran, a Democratic congressman from New York, they were all white southerners. Among them, a former governor of West Virginia delivered an address on the black franchise. Also addressing that topic were the mayor of Wilmington, North Carolina, and a prominent newspaper editor from Atlanta, Georgia. Speakers for the session on education included the

4. *Race Problems of the South*, 10.

5. A number of African Americans attended the conference, including Booker T. Washington ("The Montgomery Race Conference," *Century Magazine*, [August 1900], 630–32). Washington wrote, "In my opinion, the greatest value of the conference is in the opportunity which it furnishes in the heart of the South for free speech" (631). There is a copy of Washington's article in TSC. The audience was segregated in the auditorium, with African Americans sitting in the balcony (John David Smith, "'No negro is upon the program': Blacks and the Montgomery Race Conference on 1900," in *A Mythic Land Apart: Reassessing Southerners and Their History*, ed. John David Smith and Thomas H. Appleton [Westport, CT: Greenwood Press, 1997], 131–32).

6. *Race Problems of the South*, 11. For a comprehensive account of the conference, see Smith, "'No negro is upon the program,'" 125–50.

white principal of Hampton Institute in Virginia, the president of Roanoke College in Virginia, and a former congressman from Alabama who currently was the primary agent for two large philanthropic funds dedicated to supporting education in the South.[7]

The session dealing with "The Negro in Relation to Religion" was scheduled for the last day, and five southern churchmen expressed their views before the floor was opened for discussion. Under the rules of the conference, participants wishing to speak during the discussion period had to indicate their intent in writing before the session at which they wished to speak and submit a brief memorandum outlining what they wanted to say. If recognized, their remarks would be limited to ten minutes.[8]

The third speaker to take the podium during the discussion period was not a southerner, but he embraced the southern point of view. His name was Walter F. Willcox. Born in Reading, Massachusetts, the thirty-nine-year-old Willcox was a professor of economics at Cornell University who had taken a leave to work as one of five chief statisticians of the U.S. Census Bureau for the census of 1900.[9] Willcox was also a recent past-secretary of the American Economic Association. Given his work with the Census Bureau and his position in the American Economic Association, Willcox was well known and well connected.[10]

Willcox's ten-minute address had nothing to do with religion. Rules for

7. *Race Problems of the South,* 14–16. The former governor was William A. McCorkle; the mayor of Wilmington was Alfred Moore Waddell; and the editor was John Temple Graves. The educators were Dr. Hollis Burke Frissell, Dr. Julius D. Dreher, and Dr. J.L.M. Curry. The two philanthropic funds were the Peabody and the Slater. Mayor Waddell was a primary instigator of Wilmington race riot two years before ("1898 Wilmington Race Riot Report [draft]," www.ah.dcr.state.nc.us/1898-wrrc/report/report.htm).

8. *Race Problems of the South,* 15–16. Four of the five churchmen were the Rev. D. Clay Lilly, secretary of the Southern Presbyterian Board of Negro Evangelization (Tuscaloosa, Alabama); the Rev. W. A. Guerry, chaplain of the University of the South (Sewanee, Tennessee); the Rev. C. C. Brown, [pastor] (Clinton, South Carolina); and the Very Rev. J. R. Slattery, St. Joseph's Seminary (Baltimore). The fifth speaker was identified only as "a representative Methodist."

9. Stuart A. Rice, "Walter Francis Willcox," *American Statistician* 18 (December 1964): 26.

10. Fredrickson, *Black Image in the White Mind,* 251; Mark Aldrich, "Progressive Economists and Scientific Racism: Walter Willcox and Black Americans, 1895–1910," *Phylon* 40 (1st Quarter 1979): 2; "American Economic Association: Past Officers," www.vanderbilt.edu/AEA/officerspast.htm. Willcox was elected president of the American Economic Association in 1915 and continued to serve on the faculty of Cornell until 1931. He was active in the American Economic Association until his death in 1964 at the age of 104 (Rice, "Walter Francis Willcox," 25–26).

the discussion period simply allowed him to gain the floor, as he explained in his opening remarks:

> Mr. President, Ladies and Gentlemen:—On coming here from Washington, it was my firm intention not to speak. I felt that in a Conference like the present, composed of persons from the Southern States, it might be unwise for a New Englander by birth and ancestry to address you, and it would certainly be ill advised to speak without the careful preparation which such an audience and such a theme require. But at the request of some friends whom I have had the pleasure of meeting, and also in recognition of the generous hospitality I have received in your city and your State, I have consented to occupy a few moments.[11]

The ten-minute rule did not allow Willcox to expound on his theories, but it did allow him to state his basic position in regard to the future of African Americans. As a race, he said, they would become extinct: "The final outcome, though its realization may be postponed for centuries, will be, I believe, that the [black] race will follow the fate of the Indians, that the great majority will disappear before the whites, and that the remnant found capable of elevation to the level of the white man's civilization will ultimately be merged and lost in the lower classes of the whites, leaving almost no trace to mark their former existence."[12]

Predictions that the end of slavery would result in the extinction of the black race in America were hardly new, but to hear it said by a chief statistician for the U.S. Census Bureau impressed Alfred Holt Stone, who sat in the audience.[13] There was something about Willcox's talk that rang true for this cotton grower from Dunleith, for Willcox had touched on the three themes that Stone had articulated or was about to articulate in two addresses before the Mississippi Historical Society.

As far as slavery was concerned, Willcox argued that slavery was good because it had offered African Americans a shelter. "During the period of slavery," he said, "the Negro race in the United States was protected from competition with the whites, somewhat as it would have been by local isolation, or somewhat as domesticated animals are protected from the dangers nature throws about them." Not surprisingly, Willcox was convinced that Reconstruction had been a terrible mistake. "Only since emancipation

11. *Race Problems of the South,* 152.

12. Ibid., 156.

13. Smith, *An Old Creed for the New South,* 44, 49–50; Fredrickson, *Black Image in the White Mind,* 154–59, 161–64, 245–52.

has genuine competition between the races in this country existed," he said, "and during the early years after the Civil War the conditions were such as to favor the Negro race and to handicap the whites. I do not wonder that the reconstruction period has left bitter memories. It was cruel of the victors to aim at uprooting the tree of your Southern civilization, and, reversing the order of nature to expose the roots to the sunshine and bury the flowers and leaves in the soil."[14]

It must have pleased the mostly white audience to hear a New Englander affirm the southern point of view, but Willcox took his endorsement a step further. He admitted that northerners did not know what they were talking about when they criticized the South. "During these two days I have heard much with which I could agree, and in only one case do I wish to register a protest," he stated. "In the North, as I know it, there are few, if any 'implacable enemies' of the South. The danger you have to fear in that quarter is the danger from ignorance, not from enmity, but in its results the one may be as serious as the other."[15]

Stone liked what he heard and made a point to talk with Willcox before the conference ended. The conversation must have been a productive one, for Willcox followed with a letter as soon as he got back to Washington. Stone replied by return mail. "I have your favor of 12th inst.," he wrote, "advising me of my name having been placed upon the mailing list for Cuban Census bulletins. Will you kindly give me the address of the Secretary of the American Social Science Association? I wish to obtain the various Journals of the association and presume they may be had for a reasonable price."[16] Willcox responded with the secretary's address and added, "If I can be of any service to you in pursuing your studies, it would give me much pleasure to do so."[17] Stone was prompt in his reply. "I appreciate very much your kind proffer of assistance," he acknowledged, "and some time shall avail myself of it."[18] Alfred Holt Stone had found a mentor.

Two weeks later, Willcox sent Stone the typescript of a newspaper clipping. "I was chatting with a friend of mine a few days ago," he explained, "a reporter for the New York Evening Post and Boston Transcript, and he made an article based on my conversation, which appeared in the New York Evening Post of May 30. As part of my conversation referred to you, I enclose a

14. *Race Problems of the South,* 154.
15. Ibid., 152.
16. Stone to Willcox from Greenville, May 18, 1900, WFWP.
17. Willcox to Stone, May 22, 1900, WFWP.
18. Stone to Willcox from Greenville, May 29, 1900, WFWP.

copy of that part of the article, thinking it might interest you to see it." Willcox noted that there may be "one or two slight errors, but the gist of the matter, I think, you will find correct."[19]

New York Evening Post, May 30 [1900]
. . . I got quite as much information from a man in Montgomery who was not a speaker at the conference as from any one who was. This man told me his history and experiences with the negroes, and it seemed to me most valuable testimony. He was born in Rhode Island and went, soon after the close of the war, with his parents to the South, where they engaged in planting. He now has an immense estate, upon which 300 negroes live, renting of him what might be called small farms, and turning over the produce. Among his tenantry crime is absolutely unknown, and between the two races—the few whites and the many blacks—there is not the slightest friction. While this is characteristic of the region in which he lives, in the lower part of the state, it is particularly true of his own place, and he attributes it to certain rules which he long ago put in force. Two classes of negroes he will not allow upon his estate, and as soon as he finds any of them he terminates their lease at the earliest possible fate [*sic*]; these are the criminal *per se*—gamblers, idlers, thieves—and the "exhorters." He makes a distinction between colored ministers and a particular type of negroes who take to "exhorting" at the religious gatherings which are the units of social life on the plantation. He finds these men dangerous to the peace and safety of the community.[20]

Another rule which he insists upon in all his leases that this tenantry shall have no business dealings with any other white man except himself. They must sell their cotton to him, and he pays the market price for it. They are also obliged to accept his directions as to the handling of the crop and the management of their lands. While these requirements would not be tolerated by white tenants in the North, they are designed to prevent the negro from being victimized by unscrupulous white men. An inferior class of Jewish traders have gone into the lower part of the South, and are provoking great hostility on the part of such white men as

19. Willcox to Stone, June 12, 1900, WFWP.

20. Originally, "exhorters" were black lay preachers who spoke to the congregation during the segregated worship services for slaves in antebellum churches controlled by white pastors (Walter H. Brooks, "The Evolution of the Negro Baptist Church, 1851," *Journal of Negro History* 7 [January 1922]: 14). It is possible that the function of exhorters had assumed overtones of black activism by the turn of the century.

> the one I talked with. The negro is much taken with them because they call him "Mr." and inquire for the health of "the madame." They buy his cotton for 6 5/8 cents a pound when ordinary men are paying only 6 1/2 cents, but so cast the figures as to leave the negro less money than 6 1/2 at an ordinary computation would bring.[21]

Stone was flattered. "I appreciate the compliment in the reference to myself," he wrote in his reply to Willcox. "However, you will pardon me for correcting an impression under which I see that I left you; I was born in the South,—in the City of New Orleans,—and all my family are of this section. The mistake arose from my referring to my father-in-law—who came here from Newport, R.I., just after the war." After sharing some anecdotes about the black workers on his plantation, Stone spelled out, for the first time, his plans to become a student of the race problem: "It is my purpose to seek membership in the various bodies engaged in treating of questions bearing directly or indirectly on the work in which I am engaged,—and, when possible, I shall attend their meetings and participate in their discussions." In a postscript, Stone asked, "Did you receive a pamphlet from me on the old slave laws of Miss?"[22] Old interests and new opportunities had merged.

Willcox was eager to help. "I shall be glad to render you any assistance in my power regarding membership in associations interested in the subject you are working upon," he wrote. However, Willcox had one particular association in mind. "You would probably find it worth while to join the American Economic Association," he suggested. "The Association published Hoffman's book on Race Traits and Tendencies of the American Negro and the book on the Cotton Industry by M. B. Hammond. Most of the teachers of political economy in our colleges and universities are members, and I think membership in it might be helpful."[23]

The American Economic Association had been founded in 1885 and was "composed mainly of persons interested in the study of political economy or the economic phases of political and social questions." It had over six hundred members by 1900.[24] One of the most notable projects the association

21. Typescript with letter from Willcox to Stone, June 12, 1900, WFWP. For a description of the use of titles and other social rules that defined race relations in Mississippi at the turn of the century, see Cresswell, *Rednecks, Redeemers, and Race,* 53–55.

22. Stone to Willcox from Greenville, June 20, 1900, WFWP.

23. Willcox to Stone, July 26, 1900, WFWP.

24. "AEA: General Information," www.vanderbilt.edu/AEA/org.htm; "Papers and Proceedings of the Twelfth Annual Meeting, Ithaca, N.Y., December 27–29, 1899," *Publications of the American Economic Association,* 3rd ser., 1 (February 1900): 7, 19–36.

had undertaken in its fifteen years of existence was the publication of "Race Traits and Tendencies of the American Negro" by Frederick L. Hoffman, as Willcox had noted in his letter.[25]

Hoffman was a German-born insurance statistician whose expertise lay in the area of predicting black mortality. As Fredrickson has pointed out in *The Black Image in the White Mind,* Hoffman's foreign birth and the imprimatur of the American Economic Association lent his pronouncements an air of objectivity that perhaps they did not deserve.[26] Nevertheless, according to Fredrickson, Hoffman's monograph became "the most influential discussion of the race question to appear in the late nineteenth century."[27]

Hoffman was convinced that the black race in the United States, especially in the South, was slowly dying out and would eventually become extinct. His theory was based on an analysis of the 1890 census data, which showed that the number of African Americans had not increased at the same rate as white people since the last census ten years before. Hoffman considered the possibility that the 1890 census had undercounted African Americans, but he dismissed that possibility out of hand. He was convinced that the black race was dying out and that its excessive mortality was due to inherent deficiencies, both physical and mental. "It is not in the *conditionals of life,* but in *the race traits and tendencies* that we find the causes of the excessive mortality," he argued. "So long as these tendencies are persisted in, so long as immorality and vice are a habit of life of the vast majority of the colored population, the effect will be to increase the mortality by hereditary transmission of weak constitutions, and to lower still further the rate of natural increase, until the births fall below the deaths, and gradual extinction results."[28]

Frederickson has noted that "Hoffman drew inspiration from the hardest school of social Darwinism and condemned philanthropists who would interfere with the struggle for existence by seeking what amounted to the artificial preservation of the unfit."[29] Hoffman framed this point in these terms: "After nearly sixty years of freedom in the [British] West Indies and after

25. Hoffman's monograph appeared as a special issue of *Publications of the American Economic Association* entitled "The Race Traits and Tendencies of the American Negro" (11 [January–March–May 1896]: 1–329).

26. Hoffman made this claim in the preface to his monograph. "Being of foreign birth, a German, I was fortunately free from a personal bias which might have made an impartial treatment of the subject difficult" ("Race Traits and Tendencies," v).

27. Fredrickson, *Black Image in the White Mind,* 249.

28. Hoffman, "Race Traits and Tendencies," 1–3, 95; emphasis in original.

29. Frederickson, *Black Image in the White Mind,* 251

thirty years of freedom in this country, during which the most elaborate efforts have been made to improve the moral and social condition of the race, we find that its physical and moral tendency is downwards." According to Hoffman, "the downward tendencies of the colored race, therefore, can only be arrested by radical and far-reaching changes in their moral nature. Instead of clamoring for aid and assistance from the white race the negro himself should sternly refuse every offer of direct interference in his own evolution. The more difficult his upward struggle, the more enduring will be the qualities developed."[30]

By reading Hoffman you can see where Willcox got the ideas he presented at the conference in Montgomery. But more importantly, the publication of Hoffman's monograph by the American Economic Association was an indication of the degree to which scientific racism had become institutionalized in the United States by the turn of the century.

The association had been able to publish Hoffman's monograph by collapsing three issues of its quarterly, *Publications of the American Economic Association*, into one. Anyone who is familiar with the competitive world of academic publishing knows that journal space is at a premium, and that devoting an entire issue to the work of a single author is unusual. Devoting three of the year's four issues to the work of a single author is almost unheard of. Yet, the American Economic Association did that for Hoffman, and by so doing, endorsed what he had to say. This was the world that Stone was being invited to join.

To this point, the correspondence between Stone and Willcox had been cordial, but polite, the sort of professional chit-chat that can go on forever. But that was to change suddenly because of something that happened unexpectedly in New Orleans. It had nothing to do directly with either Willcox or Stone, but Stone's reaction to it would move his relationship with Willcox to a more personal level.

On Monday night, July 23, 1900, three days before Willcox wrote the letter in which he invited Stone to join the American Economic Association, a black man had resisted three white policemen who tried to roust him as he waited on a doorstep in New Orleans for a girlfriend who was dressing inside. In the scuffle that followed, the black man, Robert Charles, shot one of the policemen in the thigh and took off down the street. The policemen reported the incident, and the precinct captain set out with six men before dawn on Tuesday morning to apprehend the culprit. Arriving at the house where

30. Hoffman, *Race Traits and Tendencies*, 241, 328.

Charles rented a room, the captain posted three of his men in the street and told the other three to come with him. Charles was waiting inside and shot the captain and one of his men, killing them instantly.[31]

Robert Charles escaped into the darkness and disappeared. The city was in an uproar. A black man had killed two white officers and gotten away. Racial tension was already high in the city, and the deaths of two officers opened a floodgate of hatred. All through the day on Tuesday a mob estimated to be as large as five thousand white men and boys roamed the streets of New Orleans, attacking African Americans whom they encountered by chance. Fortunately, no one was killed, but that state of affairs ended on Wednesday. About two thousand people assembled that night at the Robert E. Lee monument on St. Charles. They had been incited by an editorial in the day's edition of the *New Orleans Daily States*. "Under the dark, seething mass of humanity that surrounds us and is in our midst," the editor had written, "all appears peaceful and delightful; we know not, it seems, what hellish dreams are arising underneath; we know not what schemes of hate, of arson, of murder and rape are being hatched in the dark depths. We are, and we should recognize it, under the regime of the free negro, in the midst of a dangerous element of servile uprising, not for any real cause, but from the native race hatred of the negro, inflamed by our Northern philanthropists."[32]

The mob was eager to snuff out the purported threat of a black insurrection and took to the streets. Before the night was over, three African Americans had been killed, six injured seriously enough to be hospitalized, and over fifty beaten, treated, and released. The violence had been so widespread that hundreds of citizens had to be deputized to augment the police force, and contingents of the Louisiana militia had to be posted at known trouble spots throughout the city. Peace was finally restored by nightfall on Thursday.[33]

The police learned of Robert Charles's whereabouts late Friday morning and cornered him in a house on Saratoga Street. Charles fought back for two hours, killing two more policemen, until the house was set on fire. He was killed attempting to flee, and the police dragged the body outside, where the mob shot and stomped it to a bloody pulp. But the mob was not satisfied. All evening they surged through the streets, assaulting African Americans wherever they were to be found. In separate incidents, two black workers who had

31. William Ivy Hair, *Carnival of Fury: Robert Charles and the New Orleans Race Riot of 1900* (Baton Rouge: Louisiana State University Press, 1976), 117–27.

32. *New Orleans Daily States*, July 25, 1900, quoted in Hair, *Carnival of Fury*, 148–49.

33. Hair, *Carnival of Fury*, 137–55.

nothing to do with Robert Charles were chased and beaten to death. Later that night, a black school, which some described as "the best Negro school house in Louisiana," was burned to the ground by white men seeking vengeance.[34]

Articles about Robert Charles and the mob's behavior appeared in newspapers and magazines through the country. Stone read as many of these pieces as he could put his hands on, and he did not like much of what he read.[35] "I have been much worried over the troubles in New Orleans and the attitude assumed by the Northern and eastern press," he wrote to Willcox on August 16. "No one deplores such occurrences more than myself, but I am well enough acquainted with the relations between the two races in the South to know, beyond the shadow of a doubt, that such things contain no deep significance, but are merely sporadic outbursts, precipitated by some such casualty as started this one." Stone blamed the violence directed toward African Americans on "the race hatred of a small but vicious element" of the white population. The same was true in northern cities too, he argued, including Boston. Stone emphasized that the better elements among the white citizens of New Orleans took an active part in suppressing the riot once it had begun. "I had friends who did duty on the special police force, on which the best businessmen of the city served willingly, and who risked their lives in personal encounters with white 'toughs' in rescuing negro men."[36]

Stone's letter to Willcox was longer and more passionate than any that had preceded it. He knew that he was writing a northerner who understood the frustrations of southern men when they were slandered by the northern press. "Side by side before me, I have the editorial expressions of secular and religious editors from Boston, New York, Chicago, Philadelphia and Canada, and the resolutions of negro pastors; associations, editorials of negro editors, sermons of negro preachers and addresses of negro lawyers and physicians and educators, all resident in New Orleans," he explained. "Without one single exception the latter condemn in unmeasured terms the conduct of the white mob and the negro ruffian who started the killing and the negroes who shielded him from the law." So far, so good, but there was more. "Yet I find

34. Ibid., 156–77. The mob had intended to set fire to a black neighborhood but burned the school instead.

35. For a summary of contemporary accounts, see "As the Northern Religious Press Saw It," *Southwestern Presbyterian*, August 16, 1900. The same issue also has an article entitled "As Our Negroes Saw and See It." The *Southwestern Presbyterian* was published in New Orleans.

36. Stone to Willcox from Greenville, August 16, 1900, WFWP.

in the other papers before me, wholesale abuse of the people of N.O. and the South, heaped indiscriminately. They argue with singular unanimity that the attack indicates, beyond question, a 'depraved condition of Southern sentiment,' and a 'low state of public morality in the South.' They are unjust and I am unable to think them ignorant, for I believe them malicious."[37]

Stone ended his letter to Willcox by asking his new friend for help. "Just so long as this well nigh universal vilification and misrepresentation continue, just so long must we feel that righteous indignation which instinctively fills a man who knows himself innocent of charges maliciously preferred,—and just so long must be postponed what I shall always hope for—a complete understanding and reconciliation between the people of the two sections," Stone wrote. "It is to you, my dear sir, and gentlemen like you, who seek the truth, determined to find and proclaim it," he continued, "that I look for the working out of this 'consummation devoutly to be wished.'"[38]

Willcox did not respond immediately to Stone's letter because he was trying to get a report on the Cuban census ready for publication. However, he did reply on September 28 and noted that a disturbance similar to the one in New Orleans had occurred in Akron, Ohio, where his wife's family lived, shortly after Stone had written his letter. Willcox was convinced that events like the riots in New Orleans and Akron "pointed [to] the general conclusion that strained relations between the two races and the tendency to wreck vengeance outside the law under extreme provocation are in nowise confined to the southern states."[39]

When it came to the issue of the northern presses' malicious attacks on southern values, Willcox saw things differently. "It has seemed to me that these papers whose views you deplore probably do little else than express the conviction of a large proportion of southern negroes," he observed, "and that the much more influential factor is that sullen resentment among the southern negroes, rather than the expression of it in quarters remote from the occurrences." Willcox cited Booker T. Washington in particular. "Mr. Booker Washington, for example, when I asked him what was the best representative paper of conservative negro opinion was inclined to select the

37. Ibid.

38. Ibid. The quotation comes from Hamlet's soliloquy, "To Be, or Not to Be."

39. Willcox to Stone, September 28, 1900, WFWP. The riot to which Willcox referred erupted after a black man abducted and raped a six-year-old white girl. In that riot, the mob burned two municipal buildings and broke into the courthouse. The culprit, who plead guilty to the crime, had been moved to Cleveland before the mob could lynch him ("History of Akron: The Riot of 1900," ci.akron.oh.us/history/Riot_1900.htm).

New York AGE and to say that it represented that opinion partly because it was in a place where it could speak frankly."[40]

Willcox's mentioning Booker T. Washington must have struck Stone as coincidental because Stone had just written an editorial about Washington in his hometown newspaper, the *Greenville Times*. In June that year Stone had agreed to assume editorial duties of the weekly paper. On September 22, he had written a favorable piece regarding Booker T. Washington, spokesperson of the black conservative point of view and darling of white southern paternalists. "During his whole career he has preached to his people the abandonment of politics," Stone noted in the editorial, "the dignity of labor and the supreme importance of cultivating, by proper conduct, the friendship and respect of the dominant race."[41] (For verification of Stone's authorship of this editorial, see appendix A.)

Despite Stone's admiration of Washington, it is possible that Willcox's comment in his letter set Stone to second-guessing his enthusiastic endorsement of the black leader. Stone's reassessment of Washington also may have been affected by the appearance of Washington's autobiography, *Up from Slavery*, which was being serialized at the time in *Outlook* magazine.[42] The truth was that Washington presented something of a predicament for white paternalists like Stone. On one hand, Washington preached accommodation with and subservience to the white race, which paternalists liked. On the other hand, the more prominent Washington became, the more he demonstrated what a sincere and hard-working black man could accomplish if he set his mind to it. In this manner, Washington's accomplishments challenged the stereotype of black inferiority that white southerners attributed to African Americans.

Stone was aware of this contradiction and wrote about it in another editorial that appeared in the *Greenville Times* on November 3: "The negro seems at last to have broken into 'literature,' with Booker T. Washington indulging in autobiography, and forecasting, through the medium of a book, the future of his race in America; with Paul Lawrence Dunbar, producing some excellent verse, and Charles W. Chestnutt [*sic*] doing some pretty good work in the way of 'color line' sketches, we are assured by the New England critics that the negro race is proving itself capable of worthy achievements in this field." But such a conclusion was unwarranted, Stone argued, because the people

40. Willcox to Stone, September 28, 1900, WFWP.

41. "Booker T. Washington," *Greenville Times*, September 22, 1900.

42. "Booker T. Washington, 1856–1915, Up from Slavery: An Autobiography," docsouth.unc.edu/Washington/summary.html.

who were producing these works were mulattoes, not true representatives of the negro race. "In the case of ninety-five per cent.," he wrote, "practically all, of the so-called negroes, whose achievements, even where they are most ordinary, are constantly cited by a certain class in support of the theory of the dormant capacity of the race, such infusion of white blood will be found to exist as to utterly destroy the value of the evidence relied on."[43] According to Stone, white blood made black people smarter.

Stone got around to writing Willcox two weeks after the second editorial appeared. Like Willcox, he too had been busy. "Your appreciated favor of Sept. 28th [a newspaper clipping on the New Orleans race riot in Willcox's letter] came duly to hand," he wrote. "It reached me at the beginning of my busiest season; the gathering and marketing of my cotton crop." Stone commented on Willcox's observations about press coverage of the New Orleans race riot but moved on to another matter. "My negroes have all done exceedingly well this year," he reported. "Through close attention on my part, and in some instances, compulsion to constant work, they have made a great crop." Stone reported that all of his workers were out of debt to him and that many had been able to save some money. But there were still some problems with his black labor force that had to be resolved.

> The greatest difficulty I have so far experienced is in directing the expenditure of their money; in endeavoring to make them invest it profitably, instead of following the bent of their inclinations and buying fancy nothings, trinkets and gew-gaws. It is plainly observable that the "better off" they become the more my suggestions are resented. I am now chiefly concerned as to the future. I have no hold or claim on these people. They are as free to move about as I am. They have already begun to manifest (many of them) a desire for a "change," either to leave me entirely or stay only under some other arrangement. I have so far done nothing but discuss the situation with them from the standpoint of their own good. I have pointed out to them that by saving their money now they will be able to supply their own needs next year,—without calling on me at all. I am confident that most, probably two-thirds, of them will remain.[44]

Alfred Holt Stone had reason to be confident. The year 1900 had been a good one for him. His plantation was profitable, and his experiment in "as-

43. "A Lame Conclusion," *Greenville Times*, November 3, 1900. Chesnutt's name was misspelled.

44. Stone to Willcox from Greenville, December 17, 1900, WFWP.

sured tenantry" was bearing fruit. In addition, he had taken his first steps as a racial theorist. The relationship he had established with Walter F. Willcox was promising, and Willcox could open doors. But it would be up to Stone to take advantage of the opportunities Willcox provided. Nevertheless, Stone was ready, for he possessed both the intellect and the writing skills necessary to engage in the national debate on the race problem. He also knew that it would take concentration and hard work to be successful. Stone was capable of both, but the business of growing cotton demanded the same, and it was the competing pulls of two distinct careers that would define Stone's life over the next decade.

7

MY LIFE WORK

Alfred Holt Stone's budding career as a racial theorist received a big boost on February 25, 1901, when Walter F. Willcox invited him to join a group of scholars who had been handpicked to study the economic condition of African Americans. "For some time I have known that plans for the twelfth [1900] census included the collection and tabulation of information regarding the agricultural status of the south and carrying the distinction of race into the various tables," Willcox explained in a letter to Stone. "I have felt that this information should throw somewhat of light upon the present condition of the negroes and that the co-operation of private organizations with the Census Office in interpreting this material would be helpful. Accordingly I proposed to the American Economic Association that a special committee, probably of five members, should be appointed and authorized to prepare for presentation to the Association a report upon the economic condition of the American negro. This recommendation was adopted at the recent Detroit meeting and I was authorized to proceed to constitute the committee."[1]

Willcox would be one of the five members, and he had already recruited two others. One was Harry T. Newcomb, who worked with Willcox in the Census Office as chief of the Agricultural Division. "He is a very able, energetic, and careful man," Willcox observed, "and I have the highest opinion of his ability to interpret the agricultural figures of the twelfth census impartially and successfully." The third member would be William Z. Ripley, an economist and anthropologist from the Massachusetts Institute of Technology who had published an influential book entitled *The Races of Europe: A Sociological Study.* Professor Ripley was among the first social scientists to

1. This and subsequent quotations in the next three paragraphs are from Willcox to Stone, February 25, 1901, WFWP.

delineate subdivisions of geographic races. For example, he divided European Caucasians into three groups: northern (Teutonic), central (Alpine), and southern (Mediterranean).[2] "Ripley is about the only young economist I know who has interested himself in race questions from an anthropological anthropometrical standpoint," Willcox noted. "His ability and reputation would be of material assistance to the committee[,] and he is deeply interested in the proposal."

Stone's qualifications for inclusion in this select group were slim. They consisted of a conversation with Willcox at the Montgomery conference in May 1900, six letters to Willcox, and a copy of "The Early Slave Laws of Mississippi," a marginal piece of scholarship at best. Stone did not even have a college degree. But Willcox had his reasons for selecting Stone. "The other [remaining] two members of the committee are still under consideration," Willcox noted by way of introducing the topic.

> My preference would be to have them be southern men, and if such an arrangement could be made, to have one of them from each race. It of course is desired to have conservative and impartial men on the committee. . . . I have talked with Newcomb and Ripley about the matter, and have read them parts of your letters. Both of them expressed deep interest in your work and thought if you were disposed to take a place on such a committee your experience, your point of view, and your work, would enable you to be a help to us and to the object we all desire to further. I ask, therefore, whether you would be willing to take a position on a committee of five of this sort, the three other members to be those already named [Willcox, Newcomb, and Ripley].

Having extended an invitation to Stone, Willcox proceeded to the question of who should serve as the fifth member. Willcox thought that it should be an African American: "If you would be willing to take such a position [on the committee], please let me know frankly your judgment and feeling regarding the wisdom of our asking some such colored man as W. E. B. Du Bois or Booker T. Washington to take a place on the committee." Willcox went on

2. "Today in Science History [August 16—Deaths]," www.todayinsci.com/8/8_16.htm. Ripley later joined the faculty at Harvard University and became president of the American Economic Association in 1933 ("American Economic Association: Past Officers," www.vanderbilt.edu/AEA/officerspast.htm). See also William Z. Ripley, "Races in the United States," *Atlantic Monthly* 102 (December 1908): 745–59.

to say that he knew both men "well enough to feel that they would be valuable and suggestive colleagues on such a committee."

Willcox had placed Stone in lofty company when he asked him to serve on a committee with W. E. B. Du Bois or Booker T. Washington. The two men were among the most prominent African Americans in the country. Du Bois was only two years older than Stone, but he had already compiled a remarkable record of first-rate scholarship. Awarded a doctorate in history from Harvard University in 1895—the first African American to achieve that distinction—Du Bois joined the faculty at Atlanta University, where he wrote a series of sociological papers on African Americans. Using black students from Atlanta University, Fisk University, Howard University, and the Meharry Medical College to collect data, Du Bois had studied African American communities throughout the South. By 1900, he had studied more than a dozen black communities in Alabama, Arkansas, Georgia, Kentucky, Tennessee, Texas, Virginia, and the District of Columbia involving thousands of subjects.[3] He had also became the editor of proceedings from an annual conference at Atlanta University entitled *Studies of the Negro Problems*.[4] In 1897, Du Bois had published a remarkably candid assessment of the black race for the American Negro Academy. "Unless we conquer our present vices they will conquer us," he wrote. "We are diseased, we are developing criminal tendencies, and an alarming large percentage of our men and women are sexually impure."[5] Du Bois could hardly be considered to be an apologist for the black race, as was evidenced further by the publication in 1899 of a groundbreaking study of an African American community in an urban setting, *The Philadelphia Negro*. Reviews of this seminal work reflected the diversity of opinions concerning African Americans at the time, but some scholars recognized the book's importance. "A credit to American scholarship," is

3. For example, W. E. Burghardt Du Bois, "The Negro in the Black Belt: Some Social Sketches," *Bulletin of the Department of Labor* ([May 1899]): 401–18; "The Negroes of Farmville, Virginia: A Social Study," *Bulletin of the Department of Labor*, no. 14 (January 1898): 1–38; and *Some Efforts of American Negroes for Their Own Social Betterment* (Atlanta: Atlanta University Press, 1898).

4. For example, W. E. Burghardt Du Bois, ed., *The Negro in Business: Report of a Social Study Made under the Direction of Atlanta University; Together with the Proceedings of the Fourth Conference for the Study of the Negro Problems, Held at Atlanta University, May 30–31, 1899*, Atlanta University Publications no. 4 (Atlanta: [Atlanta University Press], 1899).

5. W. E. Burghardt Du Bois, *The Conservation of Races*, Occasional Papers, no. 2 (Washington, DC: American Negro Academy, 1897), 14. This monograph as well as all of the others cited in this paragraph can be found in TSC.

how a reviewer described *The Philadelphia Negro* in the *Yale Review*, "and a distinct and valuable addition to the world's stock of knowledge concerning an important and obscure theme."[6] Such were Du Bois' qualifications when Willcox was considering him for membership on his committee in February 1901.

Booker T. Washington did not have the scholarly credentials of W. E. B. Du Bois, but he was better known and enjoyed a higher standing among white southerners, who regarded him as a black realist who understood the need for African Americans to maintain a subservient role in the hierarchy of southern race relations.[7] Washington's pragmatic position had been spelled out in a speech at Atlanta's Cotton States and International Exposition in 1895. "The wisest among my race understand that the agitation of questions of social equality is the extremist folly," Washington had observed, "and that progress in the enjoyment of all the privileges that will come to us must be the result of severe and constant struggle rather than of artificial forcing."[8]

Washington's most notable accomplishment to date was the Tuskegee Normal and Industrial Institute in Alabama, which he had founded in 1881. The school was based on the Hampton model and emphasized the acquisition of vocational skills for African Americans. Southern whites liked his approach because it promised to keep black workers "down on the farm" or in the trades. But Washington's vision was also attractive to African Americans who wanted to escape the life of sharecropping as unskilled workers. Washington was also a persuasive speaker who was able to generate considerable interest in his educational programs among wealthy philanthropists in the North. By 1900, Tuskegee Institute was "the best supported black educational institution in the country."[9]

Although Washington and Du Bois adopted separate strategies in their approach to the race problem, their differences were not widely appreciated in 1901, when Willcox was forming his committee, and the open rift between the two prominent black leaders was still two years away.[10]

6. "Brief Recommendations," *Yale Review* 10 (May 1900): 110–11, quoted in David Levering Lewis, *W. E. B. Du Bois: Biography of a Race, 1868–1919* (New York: Holt, 1993), 210.

7. Cresswell, *Rednecks, Redeemers, and Race*, 69.

8. "Speech before the Atlanta Cotton States and International Exposition," www.teachingamericanhistory.org/library.

9. "Booker T. Washington," docsouth.unc.edu/Washington/bio/html.

10. Lewis, *W. E. B. Du Bois*, 297–342. For a contemporary (1908) account of the differences between Washington and Du Bois, see Ray Stannard Baker, *Following the Color Line: An*

Stone's written response to Willcox's invitation to join his committee can not be found in the Willcox Papers, but it is certain that Stone accepted, probably with enthusiasm. Nor is it known which of the two men Stone endorsed as the fifth member. What we do know is that the formation of a "special committee on the economic condition of the American Negro" was announced at the Fourteenth Annual Meeting of the American Economic Association in Washington, D.C., the following December. The members of the committee consisted of Willcox, Newcomb, Ripley, Stone, and Du Bois.[11]

In the meantime, Stone continued his full-time work as a cotton planter and part-time job as an editor, although the opportunity to become part of a national project under the auspices of the American Economic Association may have prompted him to reconsider his editorial work with the *Greenville Times*. Whether by coincidence or not, Stone wrote a series of editorials on the race problem during March 1901, almost as if he was tying up loose ends before relinquishing his editorship. The first of the four editorials endorsed Governor Andrew Longino's efforts to suppress lynching in Mississippi but noted "the utter futility of any and all discussions and effort and legislation which has not the unqualified support of public sentiment," a theme that Stone would reiterate throughout his life and return to with emphasis forty years later during the early days of the modern civil rights movement.[12]

The second editorial appeared a week later in the next issue. The subject was the question of whether the poll tax should be mandatory. Stone did not think that it should because doing so might encourage compliance among African Americans and thus increase the prospects of black suffrage. "While the darkey may have dropped into the habit of staying away from the polls," he wrote, "we esteem it the part of wisdom to assist him by every possible means in making the habit perpetually hereditary."[13] Stone would restate this notion six years later in a paper delivered before the American Sociological Society in Madison, Wisconsin. "In the mass," he would say, "the Southern Negro has not bothered himself about the ballot

---

Account of Negro Citizenship in the American Democracy (New York: Doubleday, Page, 1908; reprint, Williamstown, MA: Corner House, 1973), 219–26.

11. "The Fourteenth Annual Meeting," *Publications of the American Economic Association*, 3rd ser., 3 (February 1902): 47–48.

12. "Lynching as a Habit," *Greenville Times*, March 2, 1901. Stone's most forceful criticism of federal initiatives can be found in "A Mississippian's View of Civil Rights, States Rights and the Reconstruction Background."

13. "The Poll Tax," *Greenville Times*, March 2, 1901.

for more than twenty years—not since his so-called political leaders left him alone."[14]

Stone's next editorial was a response to an editorial that had appeared in the *New Orleans Daily Picayune,* which argued that African Americans in general should not be held responsible for crimes committed by a few. Stone disagreed. "Any man whose knowledge of the negro is from personal contact and observations," as was Stone's, "and not gleaned from the books of long-distance philanthropists, knows that there is an almost universal tendency among them to offer shelter and protection to those of their people who are fugitives from the law."[15] Seven years later, Stone would cite this alleged racial trait as evidence of a primitive black solidarity. "Race consciousness for them [Negroes] is still in its first and lowest stage," Stone wrote in a chapter for his book, "that of the instinctive drawing together—which, for example, guarantees the fleeing criminal a haven with his race, if only his pursuers are white."[16]

The final editorial in the series on the race problem focused on Reconstruction and quoted portions of an article that had appeared in *Atlantic Monthly* by Daniel H. Chamberlain, a Union veteran who had served as governor of South Carolina from 1874 to 1876. Chamberlain had been an advocate for black suffrage when he was elected governor but soon soured on the concept. Defeated in his bid for reelection by Wade Hampton, Chamberlain returned to the North and repudiated his former support for Reconstruction policies. Stone liked what he read of Chamberlain's views. "Governor Chamberlain goes to the root of the whole damnable abomination," Stone observed.[17]

Reconstruction was very much on Stone's mind during the spring of 1901. Members of the Mississippi Historical Society assembled in Meridian for their annual meeting on April 18 and 19, and Stone submitted a paper

14. Stone's talk was published in the *American Journal of Sociology* under the title "Is Race Friction between Blacks and Whites in the United States Growing and Inevitable?" (13 [March 1908]: 676–97) and was reprinted the same year as a chapter in *SARP* (211–41). The quotation comes from page 237 of the book.

15. "'Questions of Crime, Not of Race,'" *Greenville Times,* March 16, 1901. Stone was probably thinking of the sequestering of Robert Charles by a black family in New Orleans less than eight months before.

16. Alfred Holt Stone, "Mr. Roosevelt, the South, and the Negro," in *SARP,* 335.

17. "A Voice from the Past," *Greenville Times,* March 30, 1901. Chamberlain continued writing about his disillusionment with African Americans in the South, and Stone added one of his pamphlets, *Ex-Governor Chamberlain on the Negro Problem* (Charleston, SC: News and Courier, 1904), to his collection.

for publication.[18] The title of the paper was "Mississippi's Constitution and Statutes in Reference to Freedmen, and Their Alleged Relation to the Reconstruction Acts and War Amendments," and it repeated his favorite themes in regard to slavery, Reconstruction, and white paternalism. In addition, it explained that the moral and physical decline of African Americans after emancipation was due to inherited characteristics.

> The war had been brought to a close, and the complete and final manumission of the negro been generally recognized only about six months when they [white southern legislators] entered upon the duty of framing "expedient and proper" legislation for a people which had been in bondage for two centuries and a half. With this people—their habits, when unrestrained, their natural bent of mind, their capacity for labor, under proper supervision, their ingrained tendency to idleness and shiftlessness—these men were familiar through years of intimate personal contact and observation. They knew the negro to be docile, tractable, and obedient to command; they knew him to be abundantly able, physically, to provide from the soil a return for his labor and upon the investment of its owner; they knew, also, that his own wants were easily satisfied, and that, when left to his own devices, he would labor, with neither thought nor care for future needs, only long enough to meagerly supply them; they knew his roving tendency; they knew that his agreement to perform a given work was valueless without some means of its enforcement; they knew that, with the negro, as with other races, idleness begets crime, and that the negro was by nature prone to idleness.[19]

Stone's experiment with black labor was entering its third year when he wrote this paper. Whether it succeeded or failed would be determined, in Stone's mind, by the accuracy of his assessment.

Stone wrote one last editorial on the race problem for the *Greenville Times* in May 1901. It concerned the increasing number of immigrants to the United States from Italy. Although Stone did not advocate their settlement in the South, he did acknowledge their potential as laborers. "If the thief and the thug be kept out[,] the Italian immigrant is as good as any we receive if measured by standards of thrift, economy, willingness to work and adaptability to new employments," he wrote. "As a manual laborer the negro is not

18. "Report of the Proceedings of the Fourth Annual Meeting, April 18 and 19, 1901," in Franklin L. Riley, ed., *Publications of the Mississippi Historical Society* (Harrisburg, PA: Harrisburg Publishing, 1902), 4: 10.

19. Stone, "Mississippi's Constitution and Statutes in Reference to Freedmen," 164.

his equal,—for he has staying power just as great as the latter, with none of his propensity for shirking."[20] It would be four years before Stone publicly advocated the replacement of black labor in the cotton fields with Italian workers,[21] but his close friend in Greenville, LeRoy Percy, had already taken steps in that direction by leasing Sunnyside Plantation just across the river in Arkansas and importing northern Italians through labor agents in New York and New Orleans.[22]

Stone was eager to start work on his study of the economic condition of African Americans and informed Willcox that he was planning to be in Washington that summer. It was to be the first of Stone's many trips to the nation's capital to collect material for his studies. "I am anxious to have a talk with you before your departure [for the fall semester at Cornell]," Stone wrote.[23] Doubtless, Stone wanted to discuss his first venture into the reified atmosphere of academia on the national level, for Stone had agreed to read a paper before the American Economic Association in December.[24]

Willcox's position at the Census Bureau gave him access to a huge database, which he was more than willing to share with his new protégé. Stone was very appreciative of Willcox's offer and asked for books, tables, and data for his talk in December.[25] Stone was going to speak on the subject he knew best, the Mississippi Delta, and he wanted to draw comparisons between black productivity in that region with other parts of the state.

The title of the paper Stone read before the American Economic Association on December 28, 1901, was "The Negro in the Yazoo-Mississippi Delta." Although he focused on the Delta, Stone was convinced that the answer to the larger issue of "the Negro problem" could be found there. In fact, Stone argued that the Delta, with its excellent race relations, demonstrated what could be accomplished elsewhere. "All that I have said of general conditions in the Delta," he concluded, "applies, in greater or less degree, to all the

20. "Italian Immigration," *Greenville Times*, May 11, 1901.

21. Stone wrote two articles on the subject: "The Italian Cotton Grower: The Negro's Problem" and "Italian Cotton-Growers in Arkansas."

22. Wyatt-Brown, "Leroy Percy and Sunnyside," 78.

23. Stone to Willcox from Greenville, April 20, 1901, WFWP.

24. Willcox to Stone, September 7, 1901, WFWP.

25. Stone to Willcox from Greenville, October 26, 1901, WFWP. The letter was the product of Stone's first effort to master the typewriter, and there are numerous corrections and strikeovers. "You have probably observed that I am a novice at handling a typewriter," Stone wrote at the end of the letter, "but I seem particularly anxious to impressthe [*sic*] fact upon you in conclusion."

29,790 square miles of the alluvial valley of the Mississippi. The future of this territory will inevitably be linked with the future of the American Negro." As far as the race problem was concerned, Stone was convinced that African Americans were better off in the Delta than anywhere else in the United States, and the reason for this state of affairs was, at least in Stone's mind, obvious. "It is not claimed that there are no instances of injustice to the Negro [in the Delta]. Not at all," he observed. "But I do claim that nowhere else is his general treatment fairer—nowhere is his remedy more certain. This is but a corollary to the proposition that nowhere in the same extent of territory will be found a greater or more constant demand for his labour. Nowhere does he find a better market for his service, nowhere is he freer to change his local habitation."[26]

The discussant for Stone's paper was James LeGrand Powers, a statistician working with the Census Bureau who later gained fame for his development of an improved key-punch machine that competed successfully with Herman Hollerith's version.[27] Powers complimented Stone "for his very lucid statement of the social and economic situation in the black belt of Mississippi" and acknowledged the depth of Stone's knowledge of the subject. "He [Stone] is a resident of the section under discussion. He has lived and labored with and among the people of whom he has spoken so intelligently." Most importantly, Powers thought of Stone as a role model for the black race.

> The negro is an imitative being. Like all people on a low plane of civilization he needs the influence of example. This is the most important single factor in his elevation, and hence from the moral as well as industrial point of view we are to regret his segregation in the delta counties, or in our city slums where he sees and comes in contact but little with the best side of the white man's civilization, with its patient, continuous labor, its personal frugality, its practice of the virtues. But when there is such segregation, again the only hope of any great advance for the negro is not in the material resources of the section where he makes his home, but in the

26. "Negro in the Yazoo-Mississippi Delta," 123–24.

27. "Early Office Museum: Antique Data Processing Machines," www.officemuseum.com/data_processing_machines.htm. The company producing Hollerith's machine, Computing Tabulating Recording Company, changed its name in 1924 to International Business Machines Corporation (IBM). The company sponsoring Powers's machine merged in 1927 with Remington Typewriter, Rand Kardex, and Dalton Adding Machine to become Remington Rand, Incorporated. See also Martin Campbell-Kelly, "Punched-Card Machinery," in *Computing before Computers*, ed. William Aspray (Ames: Iowa State University Press, 1990), 134–35.

self-sacrificing devotion to his real interests which we find exhibited by Mr. Stone and the best white leaders of the south.[28]

Stone was well-prepared and had capitalized on several things he had going for him in his talk. First, he was articulate. Not only was Stone an effective speaker and able writer, he was good with numbers as well. Although the Scholastic Aptitude Test did not exist when Stone went to college, he would have scored high on both the verbal and quantitative sections had he taken it. Second, he was self-effacing. "In a discussion of these questions in their broader aspects, though I have devoted some years to their consideration," he stated in his opening remarks, "I can claim no peculiar knowledge—no superior wisdom."[29] The lack of bombast and arrogance in Stone's address must have gained him respect among the scholars in his audience and set him apart from other speakers from the Deep South, who, like James K. Vardaman, scared people with the vehemence of their opinions, even if many people in the North silently agreed. Finally, Stone knew his subject, which he highlighted at the beginning of his talk. "A lifetime spent in the 'blackest' of the South's 'black belts,'" as Stone described himself, "a sharer in the association between the two races in the life of the plantation—the most constant and intimate association that is possible between them; a thorough acquaintance with the conditions surrounding the Negro in a section wherein I firmly believe will be discovered the region of his greatest material possibilities; these constitute my only equipment in venturing upon this discussion."[30]

In many ways, Alfred Holt Stone was an ideal spokesperson to defend the institutionalized racism of northern scholars.[31] He was smart, convivial, knowledgeable, and, perhaps most importantly, a southerner. Stone put a good face on Jim Crow, using personal experience and economic data to gloss over the exploitation of black labor in the Delta. Just as he had done in his reflections on Mississippi elections of 1875 and 1876 and the understanding clause in the Mississippi Constitution of 1890, Stone presented his views in

28. [James] LeGrand Powers, "The Negro in the Yazoo-Mississippi Delta: Discussion," *Publications of the American Economic Association*, 3rd ser., 3 (February 1902): 273, 276–77.

29. Stone, "Negro in the Delta," 81.

30. Ibid., 82.

31. The *New York Times* ran a favorable review of Stone's talk ("Economics Are Discussed") in the December 29, 1901, issue. An anonymous review of Stone's talk that appeared in the *Chicago Tribune* four months later cited the same statistics but described the Delta as an anachronism that reflected poorly on Mississippi and its white planters ("An Interesting Anachronism," *Chicago Tribune*, April 27, 1902).

a way that maximized the fair-mindedness of the white power structure and minimized the vulnerability of African Americans.

Using a slang expression from the South, one might say that Alf Stone "cleaned up real good." In fact, he cleaned up so well that he gained admission to the most prestigious gentlemen's club in Washington. Founded in 1878 by the famous explorer and one-armed Union army veteran John Wesley Powell, the Cosmos Club was a social hub for scientists, educators, businessmen, doctors, lawyers, writers, statesmen, and scholars who wanted a congenial place to gather for good food and fellowship. (Herbert Baxter Adams, the father of scientific history in the United States, was a member.)[32] The names of men seeking membership—for it was an all-male organization—were screened by an admissions committee and then sent to the members, 80 percent of whom had to approve. In 1901, the club was located on the east side of Lafayette Square near the White House. The club's main attractions for the members were its dining facilities, called the "Grill Room," and its library, which was well stocked with periodicals and books. By 1900, the Cosmos Club had become the home of Washington's intellectual elite.[33]

By the time Stone read his paper before the American Economic Association, it was clear that he would be spending a good bit of time in Washington to pursue his studies. It is likely that his visit during the summer of 1901 had convinced Stone that the nation's capital with its resources, especially the Library of Congress, was where he needed to be if he wanted to pursue his study of the race problem. That being the case, it would be important to have a place in Washington where he could relax and read after a long day's work in the Library of Congress, and Stone asked Willcox to nominate him for membership in the Cosmos Club.

Once again, Willcox was eager to help. "In accordance with what I understood to be your wish I have posted your name for non-resident membership in the Cosmos Club," he wrote, "and I presume that in the course of two or three months it will be acted upon. Professor Henry C. Adams signed the nomination with me." Willcox also sent Stone under a separate cover the

32. Highham, *History*, 12.

33. Wilcomb E. Washburn, *The Cosmos Club of Washington: A Centennial History, 1878–1978* (Washington, DC: Cosmos Club, 1978), xiv, 15–16, 26–31, 95–99, 109–18, 200–202, 250–51. The Cosmos Club welcomed scholars and scientists (the National Geographic Society was founded there in 1888) but excluded African Americans until the late 1960s. Although the club now accepts women as members, the issue was still being debated in 1978, when its centennial history was published. The Cosmos Club's Web site can be viewed at www.cosmos-club.org.

club's annual publication, which stated, among other things, that the club was composed of "persons interested in science, literature or art." "It is customary to have at least two letters in support of the candidate by those who signed the proposal," Willcox explained. "Such a letter should state the person's college or university experience and also give a list of publications. If you can give me additional information on these points it will help me in preparing the letter."[34]

Stone replied to Willcox's letter as soon as he returned to Greenville from his trip to address the American Economic Association. "I am much obliged to you for posting my name for membership, and to Professor Adams for signing the nomination with you," Stone wrote. "In regard to the matter of qualifying me for membership you have a more difficult task than probably you had anticipated," he added candidly. Stone then went on to explain why he did not have a college degree although he spent three years at the University of Mississippi. However, he did say that he had read law and returned to the university to take a law course, from which he graduated.[35] As far as publications were concerned, Stone wrote that they consisted "in the main of newspaper articles [editorials?] on questions in some way related to the line of work undertaken by me some years since, the study of the negro."[36] He also noted his papers for the Mississippi Historical Society and, of course, the recent paper "The Negro in the Yazoo-Mississippi Delta." Stone concluded by stating, "For a year I was editor of the Greenville Times,—giving up the place a few months since, in order to be able to devote all my leisure time to the work above referred to."[37]

Stone wanted to be a member of the Cosmos Club, but he was honest about his lack of credentials, and attaining that distinction was not his highest priority. "If out of the above you can construct a full fledged member of the Cosmos Club—very good," he wrote, "and I shall appreciate your kindly interest. But I beg of you that you give yourself no trouble in regard to the matter, and I assure you that should I fail of election I shall understand it

34. Willcox to Stone, December 4, 1902 [1901] (misdated). Henry C[arter] Adams was an economist and statistician at the University of Michigan ("Henry Carter Adams, 1851–1921," cepa.newschool.edu/het/profiles/adams.htm). Adams was also a member of the American Economic Association.

35. Stone added, "I practiced for five years." If he did, it is unclear what he practiced, for there is no evidence of his having done any legal work.

36. Stone must have been referring to his editorials in the *Greenville Times* because no other material authored by Stone in periodicals prior to 1902 has been found.

37. Stone to Willcox from Greenville, January 10, 1902, WFWP.

perfectly." Scholarly accomplishment, not social advancement, was his primary goal. "Did I ever tell you," he continued, "that I was trying to shape all my efforts toward the final accomplishment of what I have for some years intended to make my life work, a history of the negro race?" According to Stone, at present there was no adequate history, although "there is a superabundance of books bearing the title." His approach was to be "radically different." "It is my purpose to include a study of all the important African families," he wrote, "and it is this which has led me to the study of African ethnology. Of course such a work would necessitate study in Africa, Central and South America, and the West Indies, and I have no hope of a realization of my design within twelve or fifteen years. It may be that I am never to finish the undertaking, for I shall never publish the work until I have completed it in accordance with my present purpose—that of making it the authority upon the subject of which it treats."[38]

Stone's goal was laudable but naive. What he proposed would have been unrealistic for even the most experienced and knowledgeable scholar. However, one can not blame Stone for thinking big, just as a major professor will not belittle an ambitious doctoral student for choosing an impossible topic for his or her dissertation. Thinking big can lead to other things. In Stone's case, his expansive view of what he intended to accomplish led him to assemble a big—a very big—collection of material on people of African descent. Furthermore, the thinking he expressed in this letter explains why he collected material on the black race in Africa, South America, the West Indies, as well as North America. It also helps explain, at least in part, why he gathered material from different perspectives. The fact that he never did write a definitive history of the black race does not detract from what he accomplished in collecting material for the effort.

As far as the Cosmos Club was concerned, Stone need not have worried about his credentials. Willcox notified Stone in a letter dated February 10, 1902, that the Admissions Committee had voted favorably on his name as a nonresident member and that the members would be voting on his application that very night.[39] Stone cleared that hurdle too, and the Cosmos Club became Stone's base of operations in Washington for the next seven years.[40]

38. Ibid.

39. Willcox to Stone, February 10, 1902, WFWP.

40. Stone's membership lapsed on December 31, 1926 (personal communication [telephone], September 14, 2005, Ms. Joanne Pierre, manager of membership and administrative affairs, Cosmos Club).

8 I AM NOT A NEGROPHOBIST

Stone was preaching to the choir. Two of three audiences he had addressed concerning the race problem to this point, members of the Mississippi Historical Society and readers of the *Greenville Times*, were predictably sympathetic to his point of view. But the warm reception he received from the predominantly northern audience at the annual meeting of the American Economic Association may have surprised him.[1] These well-read intellectuals, most of whom had probably never ventured south of the Mason-Dixon Line, seemed to like what Stone had to say when he explained why strict segregation coupled with white paternalism was the only practical solution to the host of problems resulting from two distinct races living in close proximity.[2]

Stone need not have been surprised, if he was, for many members of the American Economic Association were scientific racists. Their racism had been implied when they published Hoffman's *Race Traits and Tendencies of the American Negro* in 1896, but it was confirmed in May 1902 with the association's publication of Joseph A. Tillinghast's *The Negro in America and Africa*.

1. As of February 1, 1902, there were 965 members or subscribers in the American Economic Association. Of these, only 25 lived in one of the former states of the Confederacy. There was one member in Mississippi (Stone) and none in South Carolina. Texas had the most with 5. The number jumped to 79 if you included the border states (Kentucky, Maryland, and Missouri), and 155 (about 16 percent of the total membership) if the District of Columbia and West Virginia were added. Massachusetts alone had 102 members, New York 183 (American Economic Association, "Geographical Distribution of Members and Subscribers," *Publication of the American Economic Association*, 3rd ser., 3 [February 1902]: 40–41).

2. Stone's point of view in this regard had been developed at length by another Mississippian in 1887 (Fulkerson, *The Negro; As He Was; As He Is; As He Will Be*, 33–34, 89–108, 116–17). It is possible that Fulkerson's work influenced Stone, who was a student at the time.

Walter F. Willcox had been Tillinghast's major professor at Cornell, and he wrote the preface to Tillinghast's monograph.[3] In it, Willcox took the opportunity to absolve slavery for its putative role in the corruption of African Americans. "The merit of the book, in my judgment," Willcox stated, "is to be found . . . in the fact that it brings together two lines of investigation which have hitherto been kept asunder."

> The rapidity with which an uncivilized people may be lifted, or may lift themselves, to the plane of an advanced civilization is still undetermined. To realize that many characteristics of the American negro are part of his inheritance from Africa, and were bred into the race there through long generations, may perhaps strengthen the patience and forbearance of those who seek to expedite his progress. To realize that many faults often attributed to the debasing effects of American slavery, are faults which he shares with his African ancestors and contemporaries, may suggest a juster and more impartial view of the merits and demerits of the economic system which crumbled as a result of the Civil War.[4]

Tillinghast's monograph was divided into three parts: "The Negro in West Africa," "The Negro under American Slavery," and "The Negro as a Free Citizen." Slowly, if not tediously, Tillinghast traced the social, moral, and economic development of the race from its origins in Africa to its present status in America at the turn of the century. The pattern, according to Tillinghast, was clear: a savage beginning, elevation during slavery, and regression to savagism following emancipation. "We can hardly avoid the conclusion that the heavy task laid upon the American Negro, after liberation from slavery, has proved too much for him," he concluded, "and that this people, considered as a whole, is slowly but surely tending to revert. Seized and transplanted unwillingly, forced sharply into new and severely exacting habits of life, held for a time in this condition of strain, and then suddenly released, the Negro finds it surpassingly difficult to suppress the hereditary instincts that do not harmonize with American social organization. He is finding that two or three centuries are all too brief a period in which to compass almost the entire range of human development."[5]

Tillinghast's monograph reflected a widely held belief in the United States at the time that modern science based on Darwinian principles could objec-

3. Aldrich, "Progressive Economists and Scientific Racism," 2.

4. Joseph Alexander Tillinghast, "The Negro in Africa and America," special issue, *Publications of the American Economic Association,* 3rd ser., 3 (May 1902): 6–330.

5. Ibid., 226.

tively account for social problems.[6] As Fredrickson has noted in *The Black Image in the White Mind*, "The conclusions of 'experts' like Hoffman, Willcox, and Tillinghast were widely accepted as the fruit of disinterested scholarly inquiry—which in itself reveals how ready white Americans were by the end of the century to believe the worst about Negro character and other prospects."[7] Mark Aldrich made the same point with a specific reference to Willcox in his article "Progressive Economists and Scientific Racism."

> Believing himself and his data objective, Willcox imagined that he could maintain a middle position between white racists and such black leaders as the eminent sociologist W. E. B. Du Bois. His statistical truths, he believed, would eventually provide the common ground on which both sides could agree. In fact, however, the questions on which Willcox gathered his facts, and his tentative explanations of them, were often motivated and shaped by a deeply held conviction of the inferiority of Negroes. Consequently, his ostensibly objective statistical studies proved to be a valuable source of ammunition to extreme white supremacists. Willcox's career, therefore, reveals an instance in which economists' scientific racist studies helped justify and rationalize the oppression of black Americans.[8]

The prevalence of these racist assumptions among scholars in northern academic circles helped explain the burgeoning relationship between Willcox and Stone. Each had something that the other wanted. Willcox's reputation provided Stone with scientific credibility to bolster the planter's racial biases. Stone, on the other hand, helped Willcox put a human face on his barren statistics.[9] Stone would never be able to acquire the credentials of the respected scholars who assembled to discuss their esoteric theories, but neither could they ever hope to gain the observations that resulted from daily contact over many years with the subjects of their study.[10]

6. The standard work on this topic is Carl N. Degler, *In Search of Human Nature: The Decline and Revival of Darwinism in American Social Thought* (New York: Oxford University Press, 1991). Fredrickson also has a chapter entitled "Darwinism and Race in Conflict" in *Black Image in the White Mind*, 228–53.

7. Fredrickson, *Black Image in the White Mind*, 256.

8. Aldrich, "Progressive Economists and Scientific Racism," 3.

9. For example, see Stone, "Negro in the Yazoo-Mississippi Delta," 96.

10. Willcox occasionally used black anecdotes, notably his detailed description of Sam Hose's lynching in his paper of black criminology, which was later reprinted as a chapter in Stone's *Studies in the American Race Problem* (Walter F. Willcox, "Negro Criminality," 459–63).

Taking their mutual interests into account, it must be added that Stone's moderation in regard to his racism—moderate given the tenor of the times—was the key to his collaboration with Willcox. Extremist views would have disrupted the façade of objectivism, which was a crucial aspect of the scientific racism characteristic of this period.[11] The desire to maintain the appearance of objectivism explains why Willcox collaborated eagerly with Stone, but avoided another well-known authority on the race problem from the Mississippi Delta, James Kimble Vardaman.

Vardaman was nine years older than Stone. In 1896, the same year that Stone and Mary Bailey Ireys were married, Vardaman had begun publishing a newspaper, the *Greenwood Commonwealth,* in Greenwood, Mississippi, another important community in the Delta fifty-five miles due east of Greenville. The *Greenwood Commonwealth* soon became known for its vindictiveness toward African Americans, and the word "nigger" appeared frequently in its columns. Many white people in Mississippi, especially those who were less educated or living in poverty, appeared to like what they read in the *Greenwood Commonwealth,* and Vardaman's popularity soared as he became more radical in his pronouncements concerning the inferiority of the black race.[12] An example of the extremes of which Vardaman was capable can be seen in the editorial he wrote in October 1902 after an alleged rapist had been burned to death by a white mob in Corinth, Mississippi.

> Much has been said and written about the people of Corinth burning the brute who killed Mrs. Cary Whitfield. I am sorry they burned him. It would have been better to have buried him alive, or shot him, or hanged him in the jail. I think they did right to kill the brute, but it would have been better had the crowd been denied admission. It does not help a man morally to look upon a thing of that kind. It is rather hardening. But I sometimes think that one could look upon a scene of that kind and suffer no more moral deterioration than he would by looking upon the burning of an Orangoutang [*sic*] that had stolen a baby or a viper that had stung an

However, as Willcox stated in a footnote, all of his information about the incident had come from columns of the *Atlanta Constitution.*

11. Smith, *Old Creed for the New South,* 107–8.

12. William F. Holmes, *The White Chief: James Kimble Vardaman* (Baton Rouge: Louisiana State University Press, 1970), 5, 58–59, 88–89. Cresswell has noted that Vardaman did not become a virulent racist until he decided to seek the Democratic nomination for governor (*Rednecks, Redeemers, and Race,* 193–94).

> unsuspecting child to death. He ceases to be regarded as a human being, and is only looked upon as a two legged monster. But then, it is not elevating to even look upon the burning of a big monkey. However when one of these devils commit such deeds as this nigger did, somebody must kill him and I am in favor of doing it promptly. In this case I only regret that the brute did not have ten thousand lives to pay for his atrocious deed. An eternity in hell will not be adequate punishment for it.[13]

The lynching that Vardaman was referring to in his editorial had been set in motion on Friday, August 29, 1902, when Cary Whitfield returned home from work about six thirty that evening. Finding the house locked, Whitfield assumed that his wife had gone to the pasture to fetch the cows and set off after her. Not finding her there, he forced entry to the house through a bedroom window and discovered his wife, Minnie, lying dead at the foot of the bed. Word of the murder arrived in Corinth, about a mile north of Whitefield's farm, around eight o'clock that evening.[14]

More than one hundred citizens as well as a number of city and county officers descended on the Whitfield place and tried to track the murderer using bloodhounds. Although the dogs soon lost the scent, the officers were able to arrest four suspects, three of whom were white and one black, and two of whom had been seen at or near the Whitfield place that day. So concerned were the citizens of Alcorn County with the murder of Mrs. Whitfield that they hired a detective from St. Louis, Missouri, to help them catch the culprit. The investigation bogged down, however, when all four of the suspects were able to produce credible alibis, and the detective returned to St. Louis empty-handed.

Three weeks later, a black laborer named William Gibson took his common-law wife to the doctor because he thought that she had been poisoned. No, the doctor told him, she was "in a beastly state of intoxication," and Gibson took her home. The next morning, Gibson beat his wife for her indiscretion, and later that day she told a friend that her husband had told her that he had killed Mrs. Whitfield. Word of the account eventually got back to the white community, and authorities arrested Gibson on Saturday evening, September 27, a month after Mrs. Whitfield had been murdered.

13. "The Corinth Conflagration," *Greenwood Commonwealth*, October 10, 1902.

14. The information in this paragraph and succeeding paragraphs comes from a lengthy article about the lynching entitled "There Was a Strange Face in Hell," which appeared in the *Corinth Herald*, October 2, 1902.

It was reported that Gibson had made a full confession after he was lodged in the county jail. He claimed that he was after money that he supposed to be in the house but insisted that he had not sexually assaulted the victim. The confession was read to a group the citizens assembled at the courthouse, and they were assured that "the person of Mrs. Whitfield had not been debauched." Given those circumstances, those in attendance decided unanimously that Gibson should be hanged by the neck until dead the following afternoon at four o'clock.

Later that evening, the pastor of the local, black Baptist church was allowed to visit the condemned man, and Gibson was said to have offered a more extensive confession as a result of this meeting. Admitting to a string of crimes he had committed under the alias of Tom Clark, Gibson also confessed to having "ravished Mrs. Whitefield before murdering her." The supplemental confession was made public the next morning, and the local newspaper described what happened next.

> When it became generally known that the hellish crime of assault had been committed, all thoughts of hanging as instantly vanished as mist before the rising sun, and in a few minutes eager hands were busily employed preparing for a holocaust. A split railroad switch was secured, together with several wagon loads of pine boxes and dry timbers, and a search begun for a suitable site. After considerable searching an ideal place was located about half a mile northeast of the city, adjoining the negro cemetery on the north. A low depression, flanked on either side by a gradual ascent, formed a natural amphitheater, and there the iron rails were set firmly in the ground, three iron cross-bars bolted thereto, a large platform twelve feet in height [*sic*] built around the iron frame, the space beneath the platform was filled with highly inflammable material previously gathered, about fifty feet square inclosed [*sic*] with barb wire, and the hour of execution patiently awaited.

Word of the execution had been sent out by telephone and telegraph the previous evening, and the roads to Corinth were jammed Sunday morning "with people eager to reach the city at the earliest possible moment." The Corinth paper estimated that as many as eight thousand men, women, and children gathered on the slopes around the pyre to witness the event. Gibson was brought from the jail at two o'clock and chained to the iron cross bars at the neck, waist, and feet. According to the newspaper, Gibson again confessed to his crimes before the murder victim's brother, J. H. Hening, applied a match at 3:35 to the dry fagots at his feet. The victim's husband, Cary Whitfield, lit the pyre from below. The paper left no detail of his death to the reader's imagination.

> As the fire slowly ate its way toward the chained murderer every face in the vast audience was fixed upon him. Nearer and nearer approached the flames, and in a few seconds reached his feet, but he preserved a perfect composure and gave no sign. An instant later the clothing about his lower limbs ignited, his hands opened and closed convulsively, his lips moved as if in prayer, a nervous twitching of the hips was discernable, and then, as a lion leaping upon its prey, the fiery flames leaped heavenly and Tom Clark called piteously to Jesus for mercy. His suffering was awful, but of short duration. For possibly a minute his cries cut the atmosphere in twain and echoed and reverberated o'er the rolling fields, and then all was quiet. Only the crackling of the flames floated on the balmy air. In another minute the heaving of his chest could be plainly seen, steam and foam escaped through his clenched teeth, large pieces of flesh slipped from his arms and legs, while the angry flames at times hid the fast disappearing body from view, then leaped aside as if bidding the populace to again look upon the gruesome sight.

The Corinth newspaper's report of Gibson's lynching is important, not for its investigative rigor, which it lacked, but for what it tells us of the public attitude of white citizens in regard to African Americans in Corinth, Mississippi, at the turn of the century. "Sunday, September 28, 1902, will long be remembered by the eight thousand or more people assembled in Corinth," read the opening paragraph of the story, "and in years to come old, white haired men and women will tell the grand children gathered around their knees of the terrible, but well merited punishment meted out to Tom Clark on that occasion; thereby instilling into their youthful minds the fundamental source of human happiness and peace on earth—the sacredness of womanhood and the sanctity of the home—and thus from generation to generation will the story be handed down to posterity."

These words of generational pride indicated the degree to which things had changed since the days of the night riders during Reconstruction. Back then, the Ku Klux Klan had done its work at night, had worn masks, and had burned crosses, but not people. In 1902, the event was accomplished in broad daylight at an announced time with no attempt to disguise the participants. In fact, every effort was made to publicize the incineration. Lynching had become a tool of social control, a means of intimidating African Americans.[15] The message was clear; if you get out of line, this could happen to you.[16]

Vardaman accepted lynching as necessary for restraining black savagery and capitalized on white prejudice against African Americans when he ran for governor in 1903. Theodore Roosevelt's invitation for Booker T. Washington to dine at the White House during the campaign provided Vardaman with plenty of political ammunition. Roosevelt was "that wild, untamed, self asserted, bronco busting negro dining man who sits in the chair of Washington, Jefferson, and Wm. McKinley," Vardaman cried. The voters in Mississippi wanted to hear more, so Vardaman gave them what they wanted. "Let Teddy take 'coons' to the White House," he exclaimed, "I should not care if the walls of the ancient edifice should become so saturated with the effluvia from the rancid carcasses that a Chinch bug would have to crawl upon the dome to avoid asphyxiation."[17]

There were three candidates in the Democratic primary for governor, and Vardaman came out on top when the votes were counted on August 6, 1903. However, Vardaman had not received a majority, which meant that he had to face the runner-up, Judge Frank Critz, in a second primary three weeks later. The third-place finisher, Edmund Noel, threw his support to Critz, and it looked as if Vardaman would be defeated.[18] But the White Chief, as Vardaman was known, made race the key issue of the campaign. "My election will mean and will be taken by the aspiring, trouble-breeding, ambitious negroes as a condemnation of the white people of Mississippi[,] of Roosevelt's criminal policy of social and political equality," he wrote in an open letter addressed to the "White Democrats of Mississippi." "It [my election] will have a most salutary restraining influence upon them. My defeat will, on the

15. For an extended discussion of the role of lynching in Mississippi during this period, see the chapter "Judge Lynch's Court," in McMillen's *Dark Journey*, 224–53.

16. The Corinth newspaper published what was purported to be a letter that William Gibson wrote to his mother as he awaited death. "Tell my cousins, my brother and my sister good bye, for I am gone," it read in part. "Tell them all out by your settlement to let this be a warning not to do anyone any harm, and never do wrong. . . . Teach all the children how to do. Tell Cousin Joe to raise his child [unintelligible] and never allow them to run around with a rowdy crowd and worthless negroes" (*Corinth Herald*, October 2, 1902). Fredrickson termed lynching "an ultimate sociological method of racial control and repression" (*Black Image in the White Mind*, 272), and Smith has argued that lynching was one of several instruments used to achieve white hegemony in the South (*Old Creed for the New South*, 285–86).

17. Both of Vardaman's quotations come from Holmes, *White Chief*, 105.

18. Cresswell, *Rednecks, Redeemers, and Race*, 197.

other hand, encourage these same negroes to aspire to the unattainable and trouble, discord and demoralization will follow. That is the real issue."[19]

Judge Critz warned the voters that Vardaman's racist views were dangerous and did not bode well for the state. However, Vardaman's race-baiting must have worked because he outpolled Critz 54 percent to 46 percent when all was said and done.[20] Winning the Democratic primary assured the victor success in November now that the Republican Party in Mississippi had ceased to exist, and Vardaman was inaugurated as governor on January 9, 1904.

Vardaman was a smart man and well read. Perhaps not surprisingly, he had latched onto some statistics on black criminality that Walter F. Willcox had presented in a paper before the American Social Science Association in 1899, and Vardaman used Willcox's data in his inaugural address, although he did not mention Willcox by name. The statistics were said to prove that African Americans were the most criminally prone race in the country, even when compared to recent immigrants from the slums of Europe, who were generally considered to be the dregs of society; that black criminality had been increasing steadily since emancipation; and that, most remarkably, African Americans who had obtained an education tended to commit more crimes than those who remained uneducated.[21]

Alfred Holt Stone was aware of Vardaman's use of Willcox's data because he had been in the audience during the campaign when Vardaman quoted the professor's writings, and Stone had written Willcox to tell him of the incident, but he had downplayed its significance because Vardaman had not linked Willcox's name to his racist conclusions. That situation changed dramatically on February 4, 1904, however, when an article by Vardaman entitled "A Governor Bitterly Opposes Negro Education" appeared in *Leslie's Weekly*.[22]

"I am opposed to the nigger's voting," Vardaman wrote, "it matters not what his advertised moral or mental characteristics may be. I am just as much opposed to Booker Washington's with all his Anglo-Saxon reinforcements, voting as I am to voting by the coconut headed, chocolate colored

19. "The Race Issue Is Paramount," *Greenwood Commonwealth*, August 15, 1903. The same letter appeared under the title "Vardaman to the Public: Regards the Race Question as Paramount" in the *Yazoo City Herald*, August 14, 1903, and probably several other newspapers across the state as well.

20. Holmes, *White Chief*, 103–12; Cresswell, *Rednecks, Redeemers, and Race*, 196–98.

21. "Gov. Vardaman's Inaugural Address," *Greenwood Commonwealth*, January 16, 1904.

22. Stone to Willcox from Washington, DC, February 12, 1904, WFWP.

typical little coon, Andy Dotson, who blacks my shoes every morning. Neither one is fit to perform the supreme functions of citizenship." Vardaman then reiterated the points he had made in his inaugural address, but this time he attributed them to Willcox. "These statistics are entirely free from the suspicion of 'race prejudice,'" he assured his readers, "for they were collected by Professor Willcox of Cornell University, a native of Massachusetts."[23]

Willcox was furious when he learned of the article and immediately dashed off a letter to the governor. "My attention has just been called to your statement of my conclusions regarding negro criminology in the issue of Leslie's Weekly of February 4," Willcox began. "I am at a loss to know from what source you have drawn this quotation. It certainly is one I never wrote or uttered and most of the conclusions I do not believe. I shall be glad if you will either verify your quotation, which I think you can not do, or will explain it was an error on your part."[24]

On the same day, Willcox wrote to Stone. "Have you happened to see Governor Vardaman's article in Leslie' Weekly for February 4?" he asked. "I am just sending a denial to Leslie's Weekly and also have written a letter to Governor Vardaman, a copy of which I enclose. I am particularly sorry to have this gross misrepresentation of my opinions published just before my name is likely to appear in connection with the negro bulletin [a publication from the Census Bureau] and am anxious to do everything I can towards correcting the opinion."[25]

Willcox had good reason to be concerned about his being used as an excuse for Vardaman's aggressive assault on black education when the governor referred to Willcox by name in his message to the Mississippi legislature vetoing an appropriation for the State Normal School for black teachers in Holly Springs. Vardaman's veto message with its reference to Willcox was picked up by the northern press and reprinted with comments critical of the governor's educational policies.[26]

As it turned out, Stone had seen the piece in *Leslie's Weekly.* "I can appreciate your annoyance and disgust over Vardaman's article," he replied, "and need not tell you that I shall be glad to assist you in correcting any misapprehension arising from his [Vardaman's] statements. I do not think you need

23. Vardaman, "Governor Bitterly Opposes Negro Education," 104.

24. Willcox to Vardaman, February 10, 1904, WFWP.

25. Willcox to Stone, February 10, 1904, WFWP.

26. For example, see "Vetoed by Governor Vardaman: Because, He Says, Book Learning Is a Detriment to the Negro—No Money for Colored Teachers," *Cincinnati Enquirer,* March 15, 1904.

fear anything on this score, however, for the temper and tone of that article will themselves discredit the statements of the author."[27]

Willcox received Vardaman's reply in a handwritten letter dated February 24, 1904. The governor stated matter-of-factly that his conclusions in regard to black criminality had been taken from an article by Colonel Robert Bingham, the headmaster of a private school for white children in Asheville, North Carolina.[28] Willcox got hold of Bingham's article, and sure enough, the points Vardaman had cited were there, but Bingham had attributed them to Willcox! As it turned out, Bingham acknowledged two sources—Willcox and George T. Winston from Texas.[29] A closer reading of Bingham's article revealed that Bingham had paraphrased Winston, not Willcox. Nevertheless, Bingham had stated in the article that "The conclusions of these two distinguished gentlemen, differing so widely in birth, rearing and environment, are practically the same."[30] Apparently, Vardaman had chosen to identify Willcox as his source, rather than the Texan, in order to capitalize on Willcox's northern, academic credentials.[31]

Willcox's disavowal of the quotation Vardaman had used appeared in the March 17, 1904, issue of *Leslie's Weekly*.[32] Willcox's rebuttal contained a point-by-point examination of Vardaman's seven points. For someone who had been incensed by Vardaman's misuse of his good name, Willcox was remarkably restrained in his effort to set the record straight. Although he rejected Vardaman's claim that educated African Americans were, as a class, more criminal than illiterate members of their race, he basically agreed with much of what Vardaman had to say, terming the governor's conclusions as "in fair accord with the evidence and substantially accurate."[33] Black leaders were alarmed at Willcox's equivocal response. They understood, as apparently

27. Stone to Willcox, February 12, 1904, WFWP.

28. Robert Bingham, *An Ex-Slaveholder's View of the Negro Question in the South* [n.p., 1900; reprinted from the European edition of *Harper's Monthly*], 16 pages. Vardaman's letter is in the WFWP.

29. Winston was president of the University of Texas when he presented these ideas in an address before the National Prison Association in December 1897.

30. Bingham, *Ex-Slaveholder's View*, 11–12. Willcox wrote Vardaman and pointed out his error (Willcox to Vardaman, February 29, 1904, WFWP). If Vardaman responded, his letter is not in the WFWP.

31. In his article, Vardaman mentioned Winston's name in passing. However, the wording just prior to and immediately following the quotation leaves the impression that Vardaman was quoting Willcox.

32. Walter F. Willcox, "Negro Education Not a Source of Crime," *Leslie's Weekly*, March 17, 1904, 252.

33. Willcox agreed with four of Vardaman's conclusions and stated that there were insuffi-

Willcox did not, that his implied endorsement of Vardaman's beliefs strengthened the latter's choke hold over African Americans in Mississippi.[34]

For his part, Stone had kept his distance from Vardaman's views during the campaign and after the election. With all of this encouragement of lynching and race-baiting, Vardaman was giving racism a bad name. Possibly for that reason, Stone studiously avoided writing about lynching. Of the fifty papers, editorials, book reviews, and chapters he wrote about the race problem during his career as a racial theorist, Stone dealt with the subject of lynching at length only three times.[35] The first time was the editorial he wrote in the *Greenville Times* on March 2, 1901, in which he endorsed Governor Longino's attempt to suppress lynching. Stone's concern was that although lynching had become "regarded throughout the south as a justifiable mode of punishment for one crime [rape]," its use had become so widespread that it became the punishment of choice for almost any crime.[36]

The subject of lynching came up a second time in Stone's address to the American Economic Association in December 1901. Early in his talk, Stone reported that lynching rarely occurred in his part of the Delta because strict segregation in Washington County eliminated the need for it to be employed. "The violation by a Negro of the person of a white woman," he remarked, "is with us an unknown crime."[37] Later in his address he returned to the subject again. "The one thing which in the South, directly and indirectly, has been the source of the gravest trouble between the races," he stated, "has been the crime of rape. That it should lead to lynching was inevitable; it was equally inevitable that in time the same mode of punishment would be extended to less grave offences."[38]

It would be six years before Stone took up the subject of lynching once more, and, with this occurrence, for the last time. The occasion was his re-

cient data in regard to a fifth. Of the remaining two, Willcox rejected Vardaman's assertion that more education resulted in higher rates of criminality among African Americans and quibbled with the other. Although "the criminality of the negro race is much higher than that of whites," Willcox acknowledged, it was, nevertheless, lower than that of "Indians [Native Americans] and Mongolians."

34. Aldrich, "Progressive Economists and Scientific Racism," 10–11.

35. Stone seldom wrote about lynching in his letters to Willcox, and when he did, it was usually to reassure Willcox that lynching rarely occurred in Washington County, Mississippi.

36. "Lynching as a Habit," *Greenville Times*, March 2, 1901.

37. Stone, "Negro in the Yazoo-Mississippi Delta," 86.

38. Ibid., 92. Stone went on to quote a professor from North Carolina who, at a recent meeting of the American Academy of Political and Social Science, had offered a gruesome and overdrawn description of black lust. Stone argued that such a characterization was not generally true in the South, especially in those areas in which rigid racial controls existed.

view of two papers for the Southern Historical Association, both dealing directly with the topic. The first was by William H. Fleming (not to be confused with Walter Lynwood Fleming) and was entitled "Slavery and the Race Problem in the South." The second paper, which carried the rather unwieldy title "The Voice of the Third Generation: A Discussion of the Race Question for the Benefit of Those Who Believe that the United States Is a White Man's Country and Should Be Governed by White Men," was written by Henry Peck Fry.[39] Stone had problems with both papers. He faulted Fleming for stating that the crime of black rape "began" only after emancipation, but he saved his severest comments for Fry. First, Stone pointed out that Fry's claim that "There is no record of a negro's having committed an assault on white women before the passage of the fifteenth amendment" was simply not true.[40] Nor could Stone countenance Fry's assertion that "crimes committed by the negro upon the person of white women" were responsible for "ninety-nine per cent. of the lynchings which have taken place in the country." Clearly, Stone had little use for such hyperbole and rejected the arguments of the third generation's self-proclaimed prophet.[41] "The voice of this generation must be that of soberness, of dispassionate appeal to reason, or deliberate statement of unembellished fact, if calm and judicial presentation of every aspect of the case,—if it would be listened to as well as heard," Stone asserted at the end of the review.[42] Stone probably thought, not without merit, that he was describing himself.

39. William H. Fleming, *Slavery and the Race Problem in the South, with Special Reference to the State of Georgia: Address of Hon. Wm. H. Fleming, before the Alumni Society of the State University, Athens, June 19, 1906* (Boston: Estes, [1906?]), 66 pages; and Henry Peck Fry, *The Voice of the Third Generation: A Discussion of the Race Question for the Benefit of Those Who Believe That the United States is a White Man's Country and Should Be Governed by White Men* (Chattanooga, TN: self-published, 1906), 32 pages. Both pamphlets are in TSC.

40. Stone used James Elbert Cutler's book on lynching to substantiate his statements (*Lynch-Law: An Investigation into the History of Lynching in the United States* [New York: Longmans, Green, 1905; reprint, New York: Negro Universities Press, 1969]). He later described Cutler's book as "the only study of the subject of lynching worth the reading" (Stone, "Foundations of Our Differences," 47), which is another example of Stone's use of northern professors to bolster southern arguments.

41. Fry defined the third generation in 1906 as young Americans "ranging in years from eighteen to twenty-seven or eight. Born at a period removed from the horrors of [the Civil] war; unfamiliar with any of its real hardship, yet endowed with courage of his ancestors" (*Voice of the Third Generation*, 5). According to this definition, Stone would have belonged to the second generation.

42. Alfred Holt Stone, "More Race Problem Discussion," *Publications of the Southern History Association* 11 (January 1907): 24, 30–31, 37.

Despite his personal moderation and appeal for evenhandedness, Stone discovered that writing about a topic as controversial as race could get the writer into serious trouble. The controversy began with an article Stone wrote for the May 1903 issue of the *Atlantic Monthly* entitled "The Mulatto Factor in the Race Problem."[43]

Persons of mixed-race ancestry presented something of a dilemma for white southerners. A light complexion for someone with African features was indisputable evidence that a black person had had sex with a white person. During the period of slavery, a child of mixed-race ancestry usually meant that a white slave owner had slept with a black slave, most likely his own.[44] This reality was an embarrassment to white southerners who prided themselves in maintaining a distance between themselves and black slaves, an inferior species of human beings.[45]

The seriousness with which religious leaders in the South at the time of the Civil War took the problem of mixed-race ancestry was illustrated by the Presbyterian Church's decision in the Confederate States to issue a "pastoral letter on the subject of religious instruction of the colored people." The letter emphasized the slave owners' obligation to treat their slaves humanely, and central to that treatment was honoring the institution of marriage among slaves. The reason for this admonition was a concern over the prevalence of sexual intercourse between white males and black females.[46]

> Once more: let the marriage and domestic relations between the slaves be established by law; let chastity and wedded faith be recognized as virtues between blacks as well as white females; and let the law defend them in the maintenance of virtue; and then our female slaves will be, in a great degree, shielded against the contaminating influence of beastly white men. The evil alluded to is enormous and dangerous, both in its influence upon the bond and the free. It degenerates the white man, socially and intrinsically, and in a way that can never be remedied. And, what is worse,

43. Alfred Holt Stone, "The Mulatto Factor in the Race Problem," *Atlantic Monthly* 91 (May 1903): 658–62; reprinted in *SARP*, 425–39.

44. Williamson, *Crucible of Race*, 41, 307–8; see also Smith, *Old Creed for the New South*, 186.

45. See Fredrickson, *Black Image in the White Mind*, 73–76, for a discussion of the polygenetic theory of racial differences.

46. [James A. Lyon], "Slavery, and the Duties Growing Out of the Relation," *Southern Presbyterian Review* 16 (July 1863): 1. A copy of this article is in TSC, and Stone has marked this passage with emphasis.

it destroys confidence, on the part of the slave, in the virtue, integrity, and moral power to elevate and Christianize the slave.[47]

The issue of racial amalgamation was hotly debated on both sides of the Mason-Dixon Line after Lincoln issued his Emancipation Proclamation.[48] In 1864, a pamphlet purported to be written by abolitionists supporting Lincoln's reelection appeared during the presidential campaign. The pamphlet, *Miscegenation: The Theory of the Blending of the Races, Applied to the American White Man and Negro,* advocated the merging of the two races and created an uproar.[49] Although it was later discovered that the pamphlet was hoax, an attempt to create a backlash against the Republican Party,[50] it had given the English language a new word and elevated fears that black men would soon be sleeping with white women.[51]

These fears increased exponentially among white southerners during Reconstruction when African Americans gained a host of civil rights.[52] Although these gains were short-lived and permanently reversed by the ascendancy of white-supremacist governments in all of the former states of the Confederacy by 1880, the concern that race amalgamation was just around the corner continued to fester, pricked sporadically by progressive developments in the North, such as the repeal of antimiscegenation laws.[53]

47. [Lyon], "Slavery, and Duties Growing Out of the Relation," 31.

48. See, for example, Theodore Tilton, *The Negro: A Speech by Theodore Tilton, at Cooper Institute, New York, May 12, 1863, at the Anniversary of the American Anti-Slavery Society,* 2nd ed. (New York: n.p. for the Anti-Slavery Society, 1863), 16 pages. A copy of this article is in TSC, and Stone has marked several passages.

49. [David Goodman Croly], *Miscegenation: The Theory of the Blending of the Races, Applied to the American White Man and Negro* (New York: H. Dexter, Hamilton, & Co., 1864), 72 pages. A glossary of "New Words Used in this Book" appears on page ii. The first word in the list is "miscegenation," which was formed by combining two Latin words, *miscere* (to mix) and *genus* (race). TSC has two copies of this inflammatory pamphlet.

50. Sidney Kaplan, "The Miscegenation Issue in the Election of 1864," *Journal of Negro History* 34 [July 1949]: 274–343; see also Fredrickson, *Black Image in the White Mind,* 171–73.

51. For an example of a reaction to the miscegenation pamphlet, see L. Seaman, *What Miscegenation Is! And What We Are to Expect Now that Mr. Lincoln is Re-Elected* (New York: Waller and Willetts, [1865?]), 8 pages. There is a copy of the pamphlet in TSC.

52. See, for example, M[atthew] F. Stephenson, *Negro Equality: Or the Effects of the Fifteenth Amendment to the Constitution of the United States, on the Morals of the People. Proofs of the Falsity of the Doctrine Are from the Bible, from Physiology, Anatomy, Chemistry, and Their Cognate Sciences* (Augusta, GA: F. M. Singer and Co., 1870), 40 pages. TSC has a copy of this pamphlet, which contains a reference to race "antipathy," a term that would become central to Stone's arguments in *Studies in the American Race Problem* and other writings.

53. See, for example, George Henry, *An Address to the Hon. General Assembly of the State of*

Hoping to discourage interracial intercourse, white supremacists described progeny resulting from these liaisons in the most debasing terms.[54] For example, the president of the Tri-State Medical Association, which encompassed Mississippi, Arkansas, and Tennessee, warned attendees at their annual meeting in Memphis in 1905 against the consequences of white men mingling sexually with an "inferior and corrupt race." "Statistics show that the mulatto, of whatever shade in color," he argued, "is morally and physically the inferior of the pure black, that he is a more dangerous criminal, is comparatively worthless as a laborer, and more subject to tuberculous affections."[55]

Occasionally, the logic underlying the prohibition against interracial intercourse was so convoluted that cause and consequence became the same, as when Mrs. L. H. Harris from Rockmart, Georgia, presumably white, wrote to the editor of the *Independent* to claim that the black rapist was "nearly always a mulatto" with "enough white blood in him to replace native humility and cowardice with Caucasian audacity." She went on to describe "the savage nature and murderous instincts of the wild beast and the cunning lust of a fiend." Because of his presence, the South had become "a smoldering volcano, the dark of its quivering nights lighted here and there with the incendiary's torch or pierced through by the cry of some outraged woman."[56]

The problem with these stereotypic depictions of persons with black and white ancestors as dangerous, worthless, and unhealthy was their absurdity.

Rhode Island, of 1880, and to the Editor of the Providence Journal, On the Repeal of the Law Prohibiting the Intermarriage of Whites and Blacks (Providence, RI: J. A. and R. A. Reid, 1881), 12 pages. A brief preface on the back side of the title page indicates that the law was repealed on March 17, 1881. TSC has a copy of this pamphlet.

54. Fredrickson, *Black Image in the White Mind*, 321

55. H. L. Sutherland, "The Destiny of the American Negro," *Memphis Medical Monthly* (December 1905): 1–10. TSC has a copy of this article. For more on the hypothesized link between mixed-race ancestry and criminality, see Fulkerson, *The Negro; As He Was; As He Is; As He Will Be*, 57.

56. Mrs. L. H. Harris, "A Southern Woman's View," *Independent* 51 (May 18, 1899): 1354–55. The editor of the *Independent* followed Mrs. Harris's vituperation with a brief piece entitled "The Negro and Crime" by W. E. B. Du Bois (*Independent* 51 [May 18, 1899]: 1355–57), in which Du Bois blames segregation for the inflammation of passions and prejudice for both races. "It [segregation] makes it possible for the masses of whites to misinterpret the aims and aspiration of the negroes, to mistake self-reliance for insolence, and commendation of lynch-law for sympathy with crime. It makes it possible for the negroes to believe that the best people of the South hate and despise them, and express their antipathy in proscribing them, taunting them and crucifying them" (1357).

Many of the most accomplished, educated, and talented members of the African American community were of mixed-race ancestry, with Booker T. Washington and W. E. B. Du Bois being the most obvious examples. Furthermore, the pages of northern periodicals that targeted African Americans were full of pictures of African Americans with light complexions who were noteworthy for their accomplishments in business, literature, science, and the arts. Stone was aware of the discrepancy between documented accomplishments of mulattoes and southern folklore describing them as even more inferior than men and women of pure African ancestry. Consequently, Stone cast about and came up with a theory that departed from traditional racial ideology.[57]

Stone first advanced his theory in two of his editorials for the *Greenville Times*. The first was a piece in which he listed the literary achievements of several black authors and noted that "Mr. [William Dean] Howells, particularly, is delighted with this 'proof' of ability of the higher order, this confirmatory evidence that the negro is merely an 'undeveloped,' not an 'inferior' race." Howells's implied endorsement of a social as opposed to genetic cause for racial characteristics would not do. Stone noted that all of the people Howells praised were mulattoes, and rather than attempt to perpetuate the fiction that they were inferior, he acknowledged their accomplishments but attributed their success to the white blood that ran through their veins. Therefore, asserting that mulatto achievements provided evidence that African Americans could attain a level of development on par with the white race was, as the heading of the editorial stated, "a lame conclusion."[58]

Stone returned to the topic a month later. This time he turned his attention to two black magazines, the *New York Age* and the *Colored American Magazine*, which were edited by T. Thomas Fortune and Edward E. Cooper, respectively.[59] "The editors of these sheets," he wrote, "the cuts of the 'distinguished members of the race' which they present, the contributors to their columns, are all strictly mulatto." These men and women with light complexions, he argued, were worlds apart from the black men and women who worked for him at Dunleith. "There is one fact which must impress it-

57. Stone's first editorial for the *Greenville Times* was a complimentary description of Booker T. Washington's emphasis on vocational education and renunciation of political involvement ("Booker T. Washington," *Greenville Times*, September 22, 1900).

58. "A Lame Conclusion," *Greenville Times*, November 3, 1900.

59. There is an almost complete run of the *Colored American Magazine* from 1900 (volume 1) to 1907 (volume 13) in TSC. The near-completeness of the run suggests that Stone was a subscriber and received the issues sequentially as they were issued.

self upon the most casual reader of these papers," he noted, "and that is that the pure negro has but an insignificant part in their conduct or life; indeed he may be said to have absolutely none, except when they go into paroxysms over the lynching of some rapist."[60]

The notion that the admixture of black and white blood would result in offspring that might inherit the strengths of the white race was not new.[61] For example, a Presbyterian minister with antislavery credentials had made this point in 1862 on the eve of emancipation. "The energy, the fire, and activity, the ingenuity and perseverance of the Anglo-Saxon, joined to the plastic docility of the African, is a strange combination, yet one which may be seen every day, and which when made free and permitted to exert its unrestrained power, will be of unmeasured value," he wrote. "The mulatto makes a very bad slave, Anglo-Saxon blood being never intended to run in the veins of a voluntary bondman," the minister added, "but will be a noble freeman."[62] What Stone had done was to borrow an abolitionist's tenet and redefine it in such a way as to distance mulattoes from the black race.

Stone returned to the question of whether mulattoes were truly representative of African Americans in his article for the *Atlantic Monthly*.[63] He began by restating Tillinghast's argument that African Americans had been in a steady decline, morally and physically, since emancipation. "We have too long been guilty of the folly of trying to legislate the Negro into a white man," he wrote, "and a pyramid of failures has apparently not yet convinced us of the futility of the undertaking."

> We have ignored the scientific truth of the ethnic differences among the human family, and have blindly disregarded the fact that the Ne-

60. "Negro Journalism," *Greenville Times*, December 1, 1900.

61. Fredrickson states that the notion that mulattoes were superior to pure-blooded persons of African descent appeared in antislavery literature during the 1850s (*Black Image in the White Mind*, 121).

62. William Aikman, *The Future of the Colored Race in America: Being an Article in the Presbyterian Quarterly Review, of July, 1862* (New York: Anson D. F. Randolph, 1862): 28. TSC has a copy of this article, in which the word "mulatto" is written in pencil in Stone's handwriting at the top of the page, and this passage is marked with emphasis.

63. Willcox may have encouraged Stone to write about this issue by asking Stone for his thoughts on the alleged sterility of mulattoes in a letter dated December 4, 1901, about three weeks before Stone's address at the American Economic Association. Stone responded to Willcox's query briefly in a letter dated January 10, 1902, and at length more than a year later in a letter dated February 13, 1903, both in WFWP. Stone did not address the issue of sterility, which he probably considered to be a moot point, but he did elaborate on several points that were to appear two months later in the *Atlantic Monthly*.

> gro, in common with all other races, possesses certain persistent, distinguishing characteristics. Foolishly attempting to evade the stubborn fact that the Negro in Africa is to-day just what we know him to have been since he first appeared on that continent, we have sought in slavery an excuse for the natural and inevitable resemblance between the native and transplanted branches of the family, and have proceeded toward the American Negro as though heredity could be overridden by constitutions and laws.[64]

Having established that African Americans were innately inferior, Stone picked up where he had left off in his editorials two years before by cautioning the reader not to confuse mulattoes with "the real negro." "One of the greatest needs in the equipment of those who discuss the negroes from a distance is a better knowledge of the real negro," he wrote, "and nothing would so promote this knowledge as a recognition of the fact that in crediting his race with the achievements of its mulatto element they but becloud the question."[65] This time, however, Stone added a new element to his argument. Mulattoes were the primary cause for racial unrest.[66]

"When free from white or mulatto influence the negro is of a contented, happy disposition," Stone argued. "They are docile, tractable, and unambitious—with but few wants, and those easily satisfied. They incline to idleness, and though having a tendency to the commission of petty crimes, are not malicious, and rarely cherish hatred."[67] The problem, according to Stone, was that influential mulattoes, persons who were not representative of the black race, incited African Americans to discontent through their speeches and writings, often using the pages of northern publications that targeted black audiences. "If we will but study the true sources of the agitation over 'negro disfranchisement,' 'negro cars,' the deprivation of 'the negro's rights,' etc.," he wrote, "it will be found that in it all the negro takes but an insignificant if any part. . . . It is the voice of the mulatto, or that of the white politician, that is heard. If the statutes of those states which have been charged with discriminating against the negro were not in any wise enforceable against the mulatto, I strongly suspect that America's race

64. Stone, "The Mulatto Factor in the Race Problem," 426. Pagination for this article comes from the chapter in *Studies in the American Race Problem.*

65. Ibid., 426.

66. Stone had foreshadowed this position in the second of his two editorials on mulattoes in the *Greenville Times* when he described the *New York Age* and the *Colored American Magazine* as "merely vehicles for the expression of underlying antagonisms."

67. Stone, "The Mulatto Factor in the Race Problem," 431

problem would speedily resolve itself into exceedingly small and simple proportions."[68]

Stone's article in the *Atlantic Monthly* reached a much wider audience than he was accustomed to, and not everyone who read it agreed with his theories as enthusiastically as white folk back home in the Delta. The first critical notice of Stone's article appeared in the *Nation* on May 7, 1903. "The so-called achievements of the [black] race, he [Stone] holds, are achievements of mullatoes [*sic*], and due entirely to the white blood in the mulatto," the editors of the *Nation* noted. "This is an old fallacy," they continued, "contradicted by the experience of all teachers and observers of freedmen since the war." This observation was a matter of opinion, which Stone could accept, but it was what the editors of the *Nation* wrote next that bothered him the most. "If the tone of this paper fairly represents the attitude of mind of the entire Committee of the American Economic Association, appointed to investigate the condition of the American negro, a Committee to which Mr. Holt [*sic*] belongs, it is safe to predict that the solution of the problem has little to expect, at least in any direct way, from the action of the Committee."[69]

This allegation was serious, and Stone was quick to issue a disclaimer. "In justice to that committee, I am prompted to make the statement that no member of it was consulted by me in connection with the opinions expressed in the article in question," he wrote to the editors of the *Nation* on May 9, "and that I have no reason to believe that any member of it holds views similar to my own." Stone then went on to point out that he was the only white southerner on a committee that also included W. E. B. Du Bois. He ended his letter by adding that "I had no knowledge of the fact that any statement of my association with that committee was to be made in connection with the appearance of the article noticed by you."[70] Stone was learning firsthand that controversy over a person's views in one area could undermine that person's credibility in another.

Willcox also read Stone's article in the *Atlantic Monthly* and offered some tactful suggestions in a letter to Stone. It is clear that Willcox felt that Stone

68. Ibid., 432–33.

69. *Nation*, May 7, 1903, 376–77.

70. Alfred Holt Stone, "A Disclaimer," *Nation*, May 14, 1903, 393. A heading for the article in the *Atlantic Monthly* described the author as follows: "Mr. Alfred Stone, of Greenville, Miss., has made valuable studies of the negro in the Yazoo-Mississippi Delta, and is a member of the Committee of the American Economic Association appointed to investigate the condition of the American negro."

had gone beyond his data in drawing his conclusions as far as the heredity of racial characteristics were concerned.

> I have read with interest your article in the Atlantic and shall welcome an opportunity to talk it over with you. I find myself in substantial agreement with your views, but would hardly be willing to express that agreement publicly because of the enormous difficulty of the question and the inconclusiveness, as it seems to me, of the evidence. My own tendency is to believe that a large part, perhaps the largest part of the difference between the two races is due to racial heredity. But I believe that to offer this hypothesis in explanation of any observed difference between the races is unsatisfactory, since it amounts to little more than giving the differences another name. I feel bound to explain the differences as far as possible by the obvious social and economic causes, which we can to some degree analyze and estimate. I am disposed to admit that after this process is completed there will be a large unexplained residuum which we might then with some confidence attribute to race inheritance. But to do that at present seems to me unsatisfactory. It is at this point I find myself most radically at variance with many southern whites whose views in the main I agree with. They think the evidence warrants the conclusion that the differences between the two races are almost, if not quite, ineradicable. I do not feel warranted by the present evidence in forming a judgment regarding the persistance [*sic*] of those differences.[71]

Willcox's mild rebuke was well-stated, but it does seem somewhat hypocritical, coming as it did from the person who wrote the preface to Tillinghast's monograph. Had not Willcox stated in the preface that slavery was not to blame for the characteristics Tillinghast had observed among African Americans years later, and was not slavery a major social and economic factor? Whatever the case, Stone passed up the opportunity to extend this debate in writing and responded to Willcox by expressing a desire to discuss the matter "briefly" when Willcox came to Washington later that month.[72]

Apparently, Stone failed to take Willcox's caution about going beyond one's data seriously, for he overstepped his bounds again before the month was out. Raymond Patterson, a feature writer for the *Washington Post*, was

71. Willcox to Stone in Washington, DC, May 23, 1903, WFWP.

72. Stone to Willcox from Washington, DC, June 9, 1903, WFWP. Language in the first sentence of Stone's letter of June 9 makes it clear that it was written after Willcox's letter of May 23. Consequently, it appears that Stone did not attempt to explain his position and possibly did not take Willcox's criticism seriously.

working on a series of articles about African Americans.[73] In his first article, Patterson featured an interview with Booker T. Washington. Among the many topics they discussed, Washington highlighted accomplishments of several black men, and the question of pure versus mixed-race ancestry came up. "I wish I had the time to answer the argument in the Atlantic Monthly, in which Mr. Stone, of Mississippi, who is a most excellent gentleman, makes the charge that the mulattoes, or men of mixed race, are chiefly responsible for stirring up the negroes and inciting them to secure their political rights," Washington remarked to Patterson during the interview. Washington disagreed with Stone's assertion and pointed out that almost all of the African Americans who had been elected to Congress from the South during Reconstruction had been "pure black." Differences between these men and mulattoes were insignificant, Washington suggested. If the mulatto was more persistent, it was "because the mixed blood has been thrown a little more with the whites [that] he has acquired more self-confidence, and is therefore less liable to give in when strongly opposed, and more liable to be persistent in any undertaking than the negro who has not had so much benefit from the white man's example." Washington also discounted the assertion that there were differences between mulattoes and men of pure-African ancestry in terms of their physical characteristics or health. In short, any differences that were observed between the two races stemmed from social, not genetic, factors.[74]

Like any good journalist, Patterson wanted to interview Stone to get his point of view. After all, Stone's article in the *Atlantic Monthly* was the topic of much discussion. Fortunately for Patterson, Stone was in Washington, and Patterson found Stone in an alcove off the reading room in the Library of Congress.[75]

Stone basically repeated for Patterson what he had said in his article, but Patterson wanted more, and Stone gave it to him. He told Patterson that he had personally become a victim of the attempts by mulatto editors of northern publications to incite black workers in the South. Describing

73. The series ran from June 25 to September 7, 1903, and consisted of thirteen articles. These articles can be retrieved online by searching the *Washington Post* archives at www.washingtonpost.com/wp-srv/searches/mainsrch.htm. There is no charge for the search, but there is a modest fee associated with retrieving the full text of each article.

74. "Study of the Negro," *Washington Post*, June 25, 1903.

75. Raymond Patterson, "Mulatto and Negro: Their Final Separation as Solution of Race Problem," *Washington Post*, July 2, 1903. Quotations in the following two paragraphs come from this article.

his rental versus crop-sharing experiment at Dunleith in some detail, Stone complained that seventeen families who had benefited from his largesse had pulled up stakes and moved to a neighboring plantation as soon as they "found themselves with a mule and a little property paid for." Stone believed that, at least in part, "dissention and unrest among the negroes" stirred up by mulatto agitators in the North was the reason for their departure.

Stone was on a roll, "warming up to his subject," as Patterson recalled. The mulatto was on the verge of separating his interest from men of pure, black ancestry, Stone predicted. "From my personal talks with men of the mixed blood I am confident that they all feel the degradation involved in being put into the same class with the black field hand," he asserted. "I know that many mulattoes feel sure about this, and while they continue to assert the oneness of the negro race for obvious reasons of policy, in private conversation they invariably lay stress upon their blood relation to the white man and constantly call attention to their white ancestry."

Patterson's interview with Stone appeared in the July 2, 1903, issue of the *Washington Post*, and Stone immediately saw the implications of what Patterson claimed he had said. Attempting to control the potential for damage to his reputation, Stone immediately wrote to Booker T. Washington. "I shall not attempt to go into the particulars, nor enumerate the instances wherein I have been misquoted, or my meaning utterly misinterpreted," he wrote. "I shall only say that I have never heard, nor claimed to hear, any mulatto 'boast of his white ancestry,' nor give any indication of feeling 'degraded' by any 'negro blood' he might possess." Stone went on to explain that he and Patterson "had a lengthy conversation on the general subject of the so-called race question," and that Patterson had not taken any notes during the meeting. Although Stone admitted that Patterson accurately quoted large portions of what Stone had to say, Stone suggested that the reporter had become confused "in attempting to translate my ideas into his own words."[76]

Stone also wrote to W. E. B. Du Bois in an attempt to blunt the impact of his interview with Patterson. "There appears in the Post this morning an 'interview' which does me an injustice, in that it makes me appear unjust to you, and also puts into my mouth words which I do not use, and represents me as espousing sentiments which I do not entertain." Stone then explained the circumstances surrounding his conversation with Patterson using lan-

76. Alfred Holt Stone to Booker T. Washington from Washington, DC, July 2, 1903, in Louis R. Harlan and John W. Blassingame, *The Booker T. Washington Papers* (Urbana: University of Illinois Press, 1972), 7: 192. This letter is available online through the University of Illinois Press's Web site at www.historycooperative.org/btw/index.html.

guage similar to that he used in his letter to Washington. But his letter to Du Bois was longer, more personal. Clearly, Stone wanted to maintain a friendly relationship with Du Bois, although he did not know the man personally.[77] "I have had the misfortune of being misunderstood before," Stone concluded in his letter to Du Bois, "but I have always felt that I retained the respect of men like yourself, whose good esteem I value, and would not lightly forfeit. But a few interviews such as this, and I would not be entitled to it."[78]

Du Bois replied to Stone's letter on July 14. Although Du Bois' letter has not survived, apparently he made some observations about Stone's predicament with the candor for which Du Bois was famous. "Your suggestions as to the misinterpretation of one's thought prompts me to say that I find myself wondering very often these days if it is all worth while,—this discussing and arguing of what America calls the 'race problem,'" Stone wrote in reply. "For my part I am strongly tempted to say no more on the subject,—for at times I think that for many years the honest opinion of an honest student of the situation, if he be a Southern man," Stone wrote, obviously referring to himself, "will be discounted if in the slightest degree embraced by some demagogue of the Tillman class, and the honest man's motives misunderstood."[79]

Stone wanted to make sure that Du Bois appreciated that his theories and motivations in regard to the race problem did not correspond with those of "Pitchfork Ben" Tillman or other southern politicians like him. After discussing his family's paternalistic affection for black people and describing the enlightened management techniques he had employed at Dunleith, Stone

77. Letters between Stone and Du Bois in the Du Bois Papers (reel 3) suggest that their correspondence was cursory to this point. More specifically, the only correspondence in the Du Bois Papers prior to July 1903 is a note from Stone to Du Bois dated March 2, 1903, thanking him for a copy of "The Negro Artisan" and contributing five dollars to the Atlanta University conference. A draft of Du Bois' brief acknowledgment is scribbled in pencil on the back of Stone's note. Stone's letter to Washington was approximately 400 words; the letter to Du Bois exceeded 1,100.

78. Stone to Du Bois from Washington, DC, July 1 [2?], 1903, DBP.

79. Ibid., July 18, 1903, DBP. Stone was referring to Senator Benjamin R. ("Pitchfork Ben") Tillman of South Carolina. TSC has three of Senator Tillman's speeches in Congress, including one delivered in February 1903, five months before Stone wrote this letter (*The Race Problem: Speech of Hon. Benjamin R. Tillman, of South Carolina, in the Senate of the United States, February 23–24, 1903* [Washington, DC: Allied Printing, 1903], 32 pages). In this speech, Tillman used the controversy over the forced resignation of a black postmistress in Indianola, Mississippi, as a metaphor for a larger issue—the struggle by African Americans for racial equality. Tillman also emphasized that the black rape of white women was the primary reason for a deterioration of race relations in the South.

returned to the problem of his damaged reputation. "In May there appeared in the Atlantic some reflection of mine upon an abstract phase,—so I considered it—of the 'race problem,'" he began.[80]

> For the expression of those ideas I have been abused in newspaper criticisms, letters to papers, and in personal letters to myself.—and my reason for telling you of my relations with, and feeling toward the negro at home, is because it has been the chief burden of all these criticisms that I was a "negrophobist," "embittered by race prejudice," a "negro hater," &c. . . .
>
> If I am to be placed in the Tillman class, and numbered among the "negro haters"—of whom we have none in my section of Mississippi—my honest opinions attributed to law makers, it is quite evident that *I* can accomplish no good by any discussion I engage in.[81]

Despite his efforts at damage control, his article on mulattoes in the *Atlantic Monthly* continued to haunt him. Thomas Nelson Baker, an African American who was both a former slave and Congregationalist minister in Pittsfield, Massachusetts, took exception to a reference to Stone in an anonymous review in *Outlook* of an article Baker had written defending segregation in southern schools and the disfranchisement of African Americans.

The reference was innocuous enough, "He [Baker] agrees with Mr. Alfred H. Stone, who has written on the subject in the 'Atlantic Monthly,' that it is the mulatto, with his dual racial nature, who occasions the most difficult elements in the race problem," but Baker thought that he had been misquoted.[82] "I do not agree with Mr. Stone that the most difficult element of the race problem is the mulatto," Baker wrote in the June 27, 1903, issue of *Outlook*. "From Mr. Stone's point of view, 'the most difficult element' in this problem is the most ambitious; from my point of view, the most difficult element is the element that Mr. Stone calls 'the real negro,' that is, the shiftless and ambitionless negro. The ambitious and thrifty of the race are not all mulattoes—nor are all the shiftless and ambitionless full-blooded blacks. It is remarkable what poor thinking the superior Caucasian can do when domi-

80. Stone to Du Bois, July 18, 1903, DBP.

81. Emphasis in original.

82. "A Negro's View of the Race Problem," *Outlook*, June 13, 1903, 392. Baker was born a slave in Virginia and went on to a distinguished career as a Congregationalist minister in Massachusetts. He earned a Ph.D. from Yale in 1903, making him the first former slave to earn a doctorate (See "No Slave to Fortune," on the Northfield Mount Hermon School Web site, www.nmhschool.org/magazine/2003_fall/no_slave.asp).

nated by prejudice and preconceived opinions."[83] Baker also did not like being lumped into the same category as a southern racist. "Please allow me to say this to your readers, before whom you have set Mr. Stone and myself together," Baker concluded. "You have put me where I do not belong. Please let me change my seat."

Stone must have been irritated when he read Baker's letter. "I have no fault to find with those who take issue with me," he wrote in a letter of protest to *Outlook* that appeared six weeks later,[84] "but I do object to having entirely incorrect motives ascribed to me, and false conclusions drawn from my words. My position is simply that, in its mental and moral equipment, if I may so express it, the Negro race is the inferior to the white; that most of the men who have come to the front as 'race leaders,' and who are pointed to as the highest intellectual types of the race, are, in fact, not really Negroes—under any exact definition of the word, as descriptive of the race; that these mulattoes, from their position of leadership, wield a large influence over the more than nine millions of our population classed as Negroes—and that it behooves them to see to it that influences should be for good rather than for evil." Stone was not finished. "Mr. Baker speaks of the 'poor thinking' the white man is capable of 'when dominated by prejudices and preconceived opinions. It is this, and the term 'negrophobist,' used by some of my critics, to which I take exception."[85]

Although Stone was clearly sailing on troubled waters, he did his best to steer a middle course. By today's standards, most of what he said about African Americans would be seen as abhorrent. But Stone does not live today, and the depth and cruelty of the racism that flourished during this period would almost be unbelievable, if it were not for the record of specific events, like the burning of William Gibson, to remind us of that reality.

Despite his overreliance on heredity as an explanation for the racial differences he observed, and despite his tendency to draw conclusions that were not warranted by his data, it appears that Stone did not intend to harm the men and women who worked on his plantation. Although Stone understood that some people in positions of power, such as Vardaman and Tillman, used pronouncements by scientific racists to justify their oppressive measures, Stone thought that he was appealing to people's better natures in his

83. Thomas Nelson Baker, "An Explanation," *Outlook*, June 27, 1903.

84. Stone reprinted his letter in *Studies in the American Race Problem*, 436–38. This quotation and the next come from page 437 of that text.

85. Ibid., 438.

effort to find a solution to the problems frustrating both races at the time, a point that he emphasized at the end of his letter to *Outlook* in August 1903.

> The people of this country, North and South, white and black, have misunderstood and abused each other long enough over this "race question" to have learned that nothing is to be gained by using harsh terms and indulging in recrimination. Neither section, neither race, can lay claim to the exclusive possession of all the wisdom on the subject, nor justly charge the other with all the prejudice and ignorance. The situation demands a large measure of sympathy and charity from all sides, and the according of honest motives to honest thought, even though it lead to conclusions at variance with one's own. He who cannot criticize without the questioning of motive—who cannot differ without indulging in reproach or abuse—cannot justly claim much of that kindly feeling which should characterize any discussion of the race question to give it one of its highest elements of value.[86]

86. Ibid. This quotation is a continuation of the paragraph in which Stone denied that he was a negrophobist.

9 DESTRUCTIVE PROPENSITY

In the late spring of 1903, Alfred Holt Stone traveled to Washington, D.C., to collect material for his work on the committee for the American Economic Association.[1] He was also interested in pursuing a topic of personal interest, the Thirteenth, Fourteenth, and Fifteenth Amendments—the "War Amendments," as they were known. In addition, he was in the early stages of his work on a history of the black race.[2] The trip in 1903 was to be the first of a string of annual visits to Washington that extended through 1909.

Washington was the obvious place for a researcher to start collecting economic and demographic data. Not only did governmental agencies serve as repositories for this type of information, the Library of Congress was there. In addition, booksellers who specialized in hard-to-find books on a wide variety of topics were concentrated in the District.[3] For someone like Stone, fresh from the Mississippi Delta, Washington was an oasis of information, not to mention that it was a growing city of broad streets and boulevards with good restaurants and friendly people. It may not have been New York when it came to the arts or the theater, but Stone had gone to Washington to work, not play.

1. Stone to Willcox, February 13, 1903, WFWP. Patterson interviewed Stone for his article in the *Washington Post* during this visit. The committee presented its report at the Seventeenth Annual Meeting of the American Economic Association held in Chicago in December 1904 (Walter F. Willcox, W. E. B. Du Bois, H. T. Newcomb, W. Z. Ripley, and A. H. Stone, "The Economic Position of the Negro," *Publications of the American Economic Association* 3rd ser., 6 [February 1905]: 216–21).

2. His research agenda for this period comes from the editor's introduction to Stone's article for the October 1903, issue of the *Southern Workman*, which will be discussed below.

3. Examples of book dealers who provided Stone with material for his collection include Scott's Boston Cheap Book Store, 1103 F Street N.W., and "The Literary Shop," John H. Wills, specializing in Old and Rare Books and Magazines, 506 11th St. N.W.

Stone's successful bid in early 1902 to gain membership in the Cosmos Club was an indication that he was expecting to spend some time in D.C. In March of that year, Stone had asked Willcox to recommend a place where he might find a comfortable room for an extended stay.[4] Stone may have intended to begin his research in Washington that summer but found it necessary to travel in Canada with his wife instead. Mary's health "made it necessary for me to take her out of this climate," Stone had informed Willcox from Greenville.[5]

Stone was able to make the trip to D.C. the next year. Although he had hoped to be in Washington by March 1903, it appears that he did not reach the city until early May, the same month that his article on mulattoes came out in the *Atlantic Monthly*. His room in a boardinghouse on A Street in southeast Washington was ideally located, for the east entrance to the Library of Congress's new Jefferson Building was directly across the street. In fact, the location was so good that the boardinghouse was torn down in the early 1930s to make room for the Adams Building when the Library of Congress expanded.[6]

Despite Stone's ambitious agenda, he took time to write an article for which additional research was unnecessary. The article was entitled "The Negro Farmer in the Mississippi Delta," a topic Stone knew well. Stone wrote the article for the *Southern Workman*, a publication of the Hampton Industrial and Agricultural Institute in Virginia, the place where Booker T. Washington had gone to school and the institution that Washington had used as a model for his teacher and vocational school for African Americans in Tuskegee, Alabama.[7] It may be that Stone wrote the article for the *Southern Workman* as a way of mending fences with Washington, whom he had slighted in the interview with Raymond Patterson. Whatever the case, Stone's article contained nothing about mulattoes, slavery, or innate racial characteristics. In fact, it was puff piece for the Mississippi Delta, which Stone described as "the cream-jug of the continent."[8]

4. Willcox to Stone in Greenville, March 31, 1902, WFWP. Stone's letter to Willcox (March 26, 1902) is not in the Willcox Papers because Willcox forwarded it to a friend who was more familiar with the city.

5. Stone to Willcox from Greenville, Mississippi, February 13, 1903, WFWP.

6. The Jefferson Building opened to the public 1897; the Adams Building in 1939 ("Library of Congress," www.cr.nps.gov/nr/travel/wash/dc79.htm).

7. "Booker T. Washington, 1856–1915," docsouth.unc.edu/washington/bio.html.

8. Stone was quoting a member of the Mississippi River Commission from Indiana (Stone, "Negro Farmer in the Mississippi Delta," 458).

In the article, Stone sought to convince African Americans that the Mississippi Delta offered a certain path to financial independence and prosperity for the black man "with nothing but his hands, and a willingness to work." "Right here let me express the opinion that there is not in America to-day—and I cannot emphasize this too strongly—any avenue open to a laboring man, white or black, wherein a reputation for orderliness, thrift, and straight dealing counts for so much as its possession does with a Negro in my section, and I know of nowhere it may be more readily acquired or promptly appreciated." Stone went on to assure the reader that "there is not one single restriction placed upon the Delta Negro in the matter of acquiring land, there being neither discrimination as to price nor prejudice to color." He concluded by emphasizing the existence of friendly relations between the races in the Delta. In fact, he was satisfied "that these relations are nowhere better than in the rural districts of the Delta. I do not mean to say that we have reached the millennium," he added, "that there is an entire absence of vicious Negroes and mean white men, for we have enough of each. But I do say that there exists among our Negroes a large measure of confidence in our white people, and upon the part of the latter an absolute absence of anything approaching 'hatred' toward the former."[9]

There was a certain irony to Stone's endorsement of opportunity for African Americans in the Mississippi Delta because it coincided with his decision to terminate rental tenancy at Dunleith. Stone had begun the experiment in 1899 to determine whether he could establish a stable labor force, "assured tenantry," as he put it. In a paper he published in 1905, he described its "salient features" as "uniformity of the tenant system, all land being rented at a fixed cash rental; the sale of stock, implements, and wagons to tenants upon exceptionally favorable terms; the exercise of proper supervision over the crop; the use of a contract defining in detail the undertakings of each party; the handling and disposition of the gathered crop by the plantation management."[10]

Initially, the experiment seemed to have been a success. "My negroes have all done exceedingly well this year," Stone had written to Willcox at the end of the season in 1900, two years into the project. His only concern was that his black workers had a tendency to squander their profits without saving anything for the future. In spite of talk among the renters about mov-

9. Ibid., 459–60.

10. Stone, "A Plantation Experiment," 127. This paper was initially published in the *Quarterly Journal of Economics* (19 [February 1905]: 270–87) and later appeared as a chapter in *Studies in the American Race Problem* (125–48).

ing on once they had paid off their debts, he expected about two-thirds of his black families to remain as his tenants for the next year.[11] By the end of 1903, however, Stone had decided to terminate all of his rental agreements and hire labor on a sharecropping basis only. In fact, he had been moving in that direction for several years, converting rental holdings to sharecropping leases as his tenants moved on.[12]

Stone said that he abandoned the experiment because of the turnover in his labor force. Each year about a quarter of the families working in the fields at Dunleith left. By the time the spring of 1904 rolled around, only eight of the seventy-nine families on the place had been with Stone in 1899 when the experiment began. The need to replace departing families with new ones was only part of the problem. Every family that left took with them livestock and implements they had bought using Stone's lenient terms. Eventually, Stone decided that he could no longer afford to prepare black families to go to work for someone else. "In short," he wrote, "we are no longer engaged in the altruistic enterprise of converting shiftless and empty-handed Negroes into desirable and well-equipped tenants for the temporary benefit of other planters."[13]

Stone had hoped that tenant holdings would provide him with a stable work force who would stay on with him year after year, once they were integrated into the system. Turnover, however, meant that Stone was constantly replacing the livestock and implements that departing workers took with them. That process was perhaps too costly, time-consuming, and bothersome to suit him. Sharecropping was a simpler arrangement. Wagons, mules, cows, and implements belonged to the plantation and loaned to the worker as part of the crop-sharing agreement. If the worker left, these items were retained by the plantation and were thus readily available for the next family to use.

It was true that sharecropping required more supervision than tenant farming. Yet, Stone still found time to pursue his racial theories. The key to his freedom from supervision was his partner and friend Julian Fort.[14] A report filed by an agent for the Department of Agriculture's Office of Farm Management in 1916 listed Fort's responsibilities as the plantation's

11. Stone to Willcox from Greenville, Mississippi, December 17, 1900, WFWP.

12. Stone, "Plantation Experiment," 129.

13. Ibid., 132. The turnover was even greater following the first and last years of the experiment.

14. Fort served as an usher at Stone's wedding to Mary in 1896 ("Stone-Ireys," *Greenville Democrat,* July 2, 1896).

general manager as keeping track of accounts at the plantation store and the gin, bookkeeping, purchasing, selling merchandise and cotton, and supervising the overseers.[15] Records do not indicate exactly when Fort joined Stone as a partner, but Carl Kelsey, who visited Stone at Dunleith in the spring of 1902, observed that Stone was able to do his research by "leaving the bulk of the management to his partner,"[16] and Stone confirmed that Fort was his partner in 1905.[17] Fort's management enabled Stone to spend large blocs of time in Washington, making it necessary for him to be at Dunleith only in the fall when the cotton crop was being harvested and put on the market.[18]

Stone's disappointment over the failed experiment darkened his opinion in regard to black workers who left his employment after he had been so good to them.[19] We will never know what was in the minds of those black families who pulled up stakes and left Dunleith after they "found themselves with a mule and a little property paid for," as Stone had described their departure to Patterson in the *Washington Post*. It is possible that the African Americans who left Dunleith might have done so to assert their independence, and that idea did occur to Stone. "They left to get rid of the supervision incident to plantation management and removed a short distance to the property of a non-resident," Stone wrote in describing the motivation of one black family that left Dunleith after a good year.[20]

Despite that awareness, Stone did not interpret their departure as a problem involving the management of labor. He believed that their leaving was a reflection of the innate characteristics stemming from the black family's an-

15. [H. A. Turner,] "Dunleith Plantation, Washington County, Mississippi [January 15, 1916]," U.S. Department of Agriculture, Bureau of Agricultural Economics, Office of Farm Management, Reports, Speeches, and Articles Relating to Farm Management, RG 83 (stack 170, row 37, compartment 89, shelf 6, entry 133), NA, College Park, MD.

16. Carl Kelsey, Review of *Studies in the American Race Problem,* by Alfred Holt Stone, *Annals of the American Academy of Political and Social Science* 33 (March 1909): 217. The year of Kelsey's visit comes from Stone to Willcox, November 6, 1902, WFWP.

17. Alfred Holt Stone, "The Economic Future of the Negro: The Factor of White Competition," *Publications of the American Economic Association,* 3rd ser., 7 (February 1906): 243–94. Stone read the paper before the American Economic Association in December 1905 and referred to Fort as his partner in a footnote when the paper was reprinted as a chapter in *Studies in the American Race Problem* (149–208). Stone also acknowledged Fort's assistance in the preface to the book.

18. For example, see Stone to Willcox, October 16, 1905, WFWP.

19. Stone, "Plantation Experiment," 132; see also Stone to Willcox, December 17, 1900, WFWP.

20. Stone, "Plantation Experiment," 133.

cient past on the African savannas, "a migratory instinct," as he termed it.[21] In addition, Stone believed that African Americans suffered from an "easy-going indolence, which seeks freedom to assert itself, and chafes under restraints which measurably restrict its enjoyment." Stone believed that these characteristics made African Americans inherently "a restless people." "Ever seeking change," he wrote, "they sometimes wander far afield, and traverse the boundaries of states in its pursuit. Again, like one lost in a forest, they move but in a narrow circle, yet always in the same vain, aimless quest. They have been wanderers since emancipation gave free play to native instinct, and I do not see how love of home, in the real sense, could characterize a people who in the mass know so little of such an abode."[22]

The question of whether the problems Stone observed were the result of a migratory instinct or plantation management was being tested by another experiment in plantation labor that was underway near Greenville when Stone decided to terminate his. He certainly knew about the other experiment because it was the brainchild of a close friend, LeRoy Percy. Furthermore, Stone would have been deeply interested in Percy's experiment because it had a direct bearing on Stone's racial theories. The experiment in question was the importation of Italian immigrants to grow cotton, and it took place on a plantation with a deceptively happy name, Sunnyside.

The history of the labor experiment at Sunnyside began in 1886, when three thousand acres of fertile alluvial soil in the southeast corner of Arkansas across the river from Greenville came under the control of Austin Corbin, a financier from New York.[23] Initially, Corbin tried to work the land using black labor, but he had trouble keeping workers who resented "foreign" (that is, out-of-state) management. In 1894, Corbin experimented with convicts but found that option unsatisfactory. Turning to an Italian labor agent in New York, Corbin arranged to resettle immigrants from Italy on Sunnyside. The plan was to subdivide the plantation into small lots and sell them on credit to Italian families who would work the soil to pay off the debt. The first

21. Apparently Stone was in complete agreement with Tillinghast's thesis. For example, Stone wrote, "it is ridiculous and unscientific to shut our eyes to the negro in Africa, untouched by American slavery, and charge to that institution whatever is bad or unpromising in the negro of the South" in a review of Carl Kelsey's book *The Negro Farmer* for the Southern History Association during this period (*Publications of the Southern History Association* 8 [May 1904]: 239).

22. Stone, "Plantation Experiment," 145–46.

23. This acreage, now planted in soybeans, is immediately north of U.S. Highway 82 where it crosses the Mississippi River south of Greenville.

boatload of Italians arrived at Sunnyside from New Orleans on November 29, 1895.[24]

The experiment probably would have fared better had not Corbin been killed in a carriage accident just six months after the first Italian workers arrived. Lacking both his drive and vision, family members who inherited the plantation became discouraged as problems mounted. When a second group of Italian immigrants arrived at Sunnyside in January 1897, the original settlers told the new arrivals of the many problems they had encountered. For example, they did not like the climate, the insects, or the water. Furthermore, disease, particularly malaria, had taken its toll. More importantly, the promise of wealth and independence that had attracted them to the Deep South turned out to be illusive. Disgusted by what they had experienced, many of the families abandoned Sunnyside and relocated to more favorable climes. By 1898, the experiment with Italian labor had collapsed.[25]

Three businessmen in Greenville, partners of O. B. Crittenden & Company, saw an opportunity in the situation at Sunnyside and decided to try to make it work using the families who remained on the plantation as the nucleus for a new cohort of Italian workers. The partners were a cotton factor, Orlando B. Crittenden; a merchant, Morris Rosenstock; and Stone's old friend LeRoy Percy.[26] Crittenden and Rosenstock provided capital for equipment and buildings, and Percy managed the operation. The partners leased the plantation and agreed to split any profits with the Corbin heirs fifty-fifty.[27]

One of the first things the new lessees of Sunnyside had to overcome was the negative publicity resulting from the failure of the first experiment. To that end, Percy instituted a public relations campaign designed to convince Italian officials that resettling Italian families on cotton plantations

24. Willard B. Gatewood, "Sunnyside: The Evolution of an Arkansas Plantation, 1840–1945," and Ernesto R. Milani, "Peonage at Sunnyside and the Reaction of the Italian Government," both in *Shadows over Sunnyside: An Arkansas Plantation in Transition, 1830–1945*, ed. Jeannie M. Whayne (Fayetteville: University of Arkansas Press, 1993): 3, 15–18, and 41, respectively.

25. Gatewood, "Sunnyside," 20–22; Milani, "Peonage at Sunnyside," 41–43. See "A Model Italian Colony in Arkansas," *American Monthly Review of Reviews* (September 1906): 361–62, for a contemporary account of the failure.

26. "O. B. Crittenden & Company," in *The 39th Anniversary Edition: Greenville Times* (Greenville, MS: Times Print and Publishing, 1910; reprint, Greenville, MS: Office Supply Co., 1969).

27. Gatewood, "Sunnyside," 22; Milani, "Peonage at Sunnyside," 44.

in the Deep South was a good idea. The partners also hired labor agents in Italy, New York, and New Orleans to recruit immigrants. It was during this public relations campaign that Stone expressed his views about Italian laborers on the editorial page of the *Greenville Times*. "If the thief and the thug be kept out[,] the Italian immigrant is as good as any we receive," he wrote. "As a manual laborer the negro is not his [the Italian's] equal,—for he has staying power just as great as the latter [the negro], with none of his propensity for shirking."[28]

The editorial was Stone's first public statement pitting the white immigrant against the African American in the job market, and he returned to the topic four years later in an article entitled "The Italian Cotton Grower" for the *South Atlantic Quarterly*. Pointing out that emancipation had forced the black worker to "shift for himself," Stone predicted that white labor, particularly white foreign workers, would break the black monopoly on manual labor in the South. Stone was particularly high on the Italian immigrant, who "has proved his ability to more than meet the negro upon his most favored ground [the Mississippi Delta]." Stone believed that "the Italian works more constantly than the negro and, after one or two years' experience, cultivates more intelligently." Stone was convinced that Italian workers would eventually displace black workers in the South. "Indeed, the matter has long since passed the experimental stage," he asserted. "Measured by whatever standard may be applied[,] the Italian has demonstrated his superiority over the negro as an agriculturalist," and he would alleviate the frustration white planters felt as a result of shortcomings they experienced with the black worker. "The destructive propensity of the negro constitutes today a serious problem on many a well ordered plantation," Stone observed. "It seems difficult to escape the conclusion that back of all this lie the characteristics that apparently have always been a curse to the race—whether in Africa, the Southern States, or the West Indies—shiftlessness and improvidence."[29]

Stone's article in the *South Atlantic Quarterly*, coupled with his article on the failed labor experiment at Dunleith, which appeared a month later, formed the basis for a major paper he was scheduled to present at the annual meeting of the American Economic Association in Baltimore in December 1905. The title of the paper was "The Economic Future of the Negro," and Stone was paired in the session with a second presenter, someone to argue an alternative point of view. Stone's rapid ascendancy to the heights of re-

28. "Italian Immigration," *Greenville Times*, May 11, 1901.

29. Stone, "Italian Cotton Grower," 42–44.

spect in northern academic circles was highlighted by the identity of the co-presenter, W. E. B. Du Bois.

On paper, the confrontation would appear to have been a mismatch. Thirty-five years of age, Stone had done respectable work, but the scope of his research was modest and the data from which he drew his observations were limited. In addition to the articles on Italian laborers at Sunnyside and the labor experiment at Dunleith, Stone had published only three other pieces, not counting the two papers he had presented to the Mississippi Historical Society. They included the paper he had read before the American Economic Association in 1901, the paper on mulattoes for the *Atlantic Monthly,* and the paper outlining opportunities for African Americans in the Mississippi Delta in the *Southern Workman.* As Stone had grown professionally, editors had also sought him out to write book reviews, and he had four of these under his belt by the time he faced Du Bois.[30]

Du Bois' résumé had grown in the years since Willcox had asked Stone to consider Du Bois for membership on the committee for the American Economic Association. Most notably, Du Bois, now thirty-seven years old, had continued to edit the series of studies on African Americans sponsored by Atlanta University.[31] The 1902 study titled *The Negro Artisan* was one of his best. "From responses to comprehensive questionnaires sent to 1,300 skilled laborers in the South," as Du Bois' biographer David Levering Lewis described it, "he [Du Bois] developed a wide-angle socioeconomic photograph of labor and race relations at the turn of the century."[32] Du Bois followed this study with one about religion in the black community the next year.[33]

The hallmark of Du Bois' research was his close attention to detail. Evi-

30. Alfred Holt Stone, review of *The Souls of Black Folk,* by W. E. B. Du Bois, *Publications of the Southern History Association* 7 (September 1903): 395–96; review of *The Negro Farmer,* by Carl Kelsey, *Publications of the Southern History Association* 8 (May 1904): 239–40; review of *The United States in Our Own Time: A History from Reconstruction to Expansion, Being an Extension of "The History of the Last Quarter,"* by Benjamin Andrews, *American Historical Review* 9 (July 1904): 843–44; and "Some Recent Race Problem Literature," *Publications of the Southern History Association* 8 (November 1904): 451–61.

31. For a synopsis of these meetings, see W. E. B. Du Bois, *The Atlanta Conferences,* Atlanta University Leaflet, no. 16 (Atlanta: n.p., 1902), 14 pages. There is a copy of this pamphlet in TSC.

32. Lewis, *W. E. B. Du Bois,* 221.

33. W. E. Burghardt Du Bois, ed., *The Negro Church: Report of a Social Study Made under the Direction of Atlanta University; Together with the Proceedings of the Eighth Conference for the Study of the Negro Problems, Held at Atlanta University, May 26th, 1903,* Atlanta University Publications, no. 8 (Atlanta: Atlanta University Press, 1903), 212 pages. There is a copy of this monograph in TSC.

dence of this characteristic could be found in two projects he authored in 1901 concerning African Americans in Georgia. The larger of the two was a 130-page study of land ownership by African Americans in Georgia between 1875 and 1890.[34] The study was illustrated with eleven maps showing the black population in Georgia by county for every census from 1790 to 1890, inclusive. The smaller of the two works was a sociological analysis of African Americans living in Albany, Georgia, illustrated with eighteen photographs.[35]

But not all of Du Bois' work was statistical, for he also turned his attention to social issues affecting his race, as in an article entitled "The Training of Negroes for Social Power," which appeared in a popular magazine, *Outlook*, in 1903. This article was a clarion call for African Americans to assume "responsibility for their own social regeneration" in a society that Du Bois argued should give them a voice in the decision-making process.[36]

Without a doubt, the most important work that Du Bois published during this period was *The Souls of Black Folk*. Released in April 1903, the book was a series of essays that redefined race relations in the United States. "Du Bois' solution to the race problem in America was new, and it was revolutionary," Joel Williamson has noted in regard to *The Souls of Black Folk*. "Booker T. Washington could imagine no culture worth having that was not thoroughly laced with white ideals and directed toward a consummation of whiteness. Du Bois could boldly envision an evolution that began with a self-propelled withdrawal of black people from the single-minded pursuit of white culture and a plunge into pure blackness with faith that the two would harmoniously join further down the stream of time, without a loss of essential blackness."[37]

Not surprisingly, *The Souls of Black Folk* foreshadowed Du Bois' formal rejection of Booker T. Washington's policy of accommodation with the white power structure. In February 1905, ten months before his impending confrontation with Stone, Du Bois had met secretly with twenty-seven African

34. W. E. Burghardt Du Bois, "The Negro Landholder of Georgia," special issue, *Bulletin of the Department of Labor* [no. 35] (July 1901): 647–777. There is a copy of this monograph in TSC.

35. W. E. Burghardt Du Bois, "The Negro as He Really Is," *World's Work* (June 1901): 848–66. There is a copy of this article in TSC.

36. W. E. B. Du Bois, "The Training of Negroes for Social Power," *Outlook*, October 17, 1903, 409–14. There is a copy of this article in TSC.

37. Williamson, *Crucible of Race*, 399–402.

American leaders in the Michigan Street Baptist Church in Buffalo, New York, and guided their adoption of resolutions that led to the founding of the Niagara Movement. The Niagara Movement's first meeting was five months later on the Canadian side of the Niagara Falls, ten years after Washington had delivered his famous speech on accommodation before the Cotton States and International Exposition in Atlanta. Du Bois was named general secretary of the Niagara Movement, and the group adopted a set of principles that asserted the dignity and manhood of African Americans. Among other things, these principles called for black suffrage, civil rights, economic opportunity, education, equal access to public accommodations, and enforcement of the War Amendments (Thirteenth, Fourteenth, and Fifteenth) of the U.S. Constitution. The Declaration of Principles ended with an exhortation for African Americans to vote, to respect the rights of others, to work, to obey the laws, to be clean and orderly, to send their children to school, and to respect themselves, just as they demanded the respect of others.[38]

Stone was familiar with Du Bois' work. In fact, the first book review Stone wrote was of *The Souls of Black Folk*. Although Stone placed the book's "literary excellence . . . in the first rank of the varied and ever-increasing literature of the 'race problem,'" Stone did not like its tone. "Throughout the book is tinctured with bitterness, a bitterness unfortunate even though pardonable and easily understood by those who are acquainted with something of the life of the author." Referring obliquely to the fact the Du Bois was a mulatto, Stone opined, "It is at once a protest and a plea; a protest against the identification of the individual [such as Du Bois] with the mass,—a plea for public and personal consideration unaffected by questions of color or race." This argument was the same one Stone had made in his article about the mulatto factor, namely, that people of mixed-race ancestry were not representative of black people and thus should not claim to speak for that race. Furthermore, Du Bois, born and raised as he was in the North, was thus doubly discredited. Not only was he a mulatto, he was a Yankee too! "Despite the cry of 'negrophobist' already raised in some quarters to anticipate the suggestion," Stone concluded toward the end of the review with a thinly veiled reference to himself, "the fact remains that to one reared among the negroes of the South—to one who is living a life of daily contact and association with the masses of

38. "The Niagara Movement," www.math.buffalo.edu/~sww/ohistory/hwny-niagara-movement.html; Williamson, *Crucible of Race*, 75–78. For Du Bois' account, see W. E. Burghardt Du Bois, *Dusk of Dawn: An Essay toward an Autobiography of a Race Concept* (New York: Harcourt, Brace, 1940), 87–93.

these people—to one who has enjoyed their confidences and listened to their recitals of grievances and wrongs personal and peculiar to themselves,—to this man it is not 'the souls of *black* folk' thus laid bare."[39]

Four months after Stone's review of *The Souls of Black Folk* appeared in print, Du Bois published an article entitled "The Future of the Negro Race in America" in another popular magazine, the *East and the West*. Du Bois readily admitted that many African Americans were guilty of crimes and acts of immorality. Nevertheless, he disagreed with white theorists who attributed these problems to innate characteristics. Du Bois believed that "the perpetuation and perfection of this present serfdom" for African Americans in the South was responsible for the race's "degeneration into hopelessness, immorality, and crime." Du Bois acknowledged that persons of African descent may be inclined to slip into a moral decay unless encouraged to take a higher road by the provision of good role models and economic opportunity. "Among American negroes the tendencies to degeneration, while not in the ascendancy," as Du Bois put it, "have undoubtedly been encouraged and fostered by the history of the last two decades." The history to which he referred was the establishment of what he called a caste system following the collapse of Reconstruction, a system that was maintained by "the strengthening of present proscriptive laws, the further disfranchisement of black men, and the legal recognition of customary caste distinctions."[40]

Willcox got wind of the article and wrote Du Bois to request a copy. Du Bois complied with the request, and Willcox fired off a letter to Du Bois after receiving the piece. "The fundamental difficulty I feel in accepting your position," Willcox wrote, "is that it is impossible for me to judge how far the present condition of the American Negro is due to persistent characteristics of the people and how far it is due to heavy economic and social pressure upon them, resulting from drawing the color line in society, in politics and in industry." The old nature-nurture debate had come to the forefront once again. "You seem inclined to attribute almost all of it to the latter," Willcox continued. "I do not see that the evidence warrants one in holding either opinion with confidence and therefore for the present I am an agnostic on the subject. Nor do I see any way in which convincing evidence on the question can be derived from an analysis of social processes now in progress in

39. Stone, "Review of *The Souls of Black Folk*," 395, 397; emphasis in original.

40. W. E. Burghardt Du Bois, "The Future of the Negro Race in America," *East and the West*, January 1904, 6–7.

this country. If either factor could be isolated from the other we might derive important evidence, but I do not see how it can be."[41]

Stone may have ignored Willcox's advice when Willcox had lectured Stone on the danger of going beyond his data, but Du Bois was not about to let a self-important academic talk down to him. "The fundamental difficulty in your position is that you are trying to spin a solution on the Negro problem out of the inside of your office," Du Bois began in his reply to Willcox. "It can never be done. You have simply no adequate conception of the Negro problem in the south & of the Negro character & capacity. When you have sat as I have ten years in intimate soul contact with all kinds & conditions of black men you will be less agnostic." Du Bois was well aware that Willcox tended to argue that the problem was too complex to be fully understood when it suited him but would endorse simplistic solutions when they coincided with his biases. "How on earth any fair-minded student of the situation could have stood sponsor for a book like Tillinghast's & actually praised it is simply beyond my comprehension," Du Bois observed. "If you insist on writing about & pronouncing judgment on this problem why not study it? Not from a car-window & associated press dispatches as in your pamphlet on crime but get down here & really study it first hand. Is it a sufficient answer to a problem to say that the data are not sufficient when they lie all about us? There is enough easily obtainable data to take you off the fence if you will study it first hand & not thro' prejudiced eyes—my eyes, or those of others."[42] Du Bois had said what he wanted to say. "Pardon this frankness," he wrote in closing, "but your letter invited it."[43]

Willcox sent Stone his copy of Du Bois' article and enclosed copies of his letter to Du Bois and Du Bois' reply.[44] "The letter seems to me interesting as a [*sic*] evidence of increasing bitterness of feeling on Mr. DuBois' part," Willcox confided in to Stone. "It is surprising to me that a man of his training and real ability should treat intellectual descent [*sic*] from his position as a

41. Willcox to Du Bois, March 13, 1904, in Herbert Aptheker, ed., *Selections, 1877–1934*, vol. 1 of *The Correspondence of W. E. B. Du Bois* (Amherst: University of Massachusetts Press, 1973), 74.

42. The pamphlet on crime to which Du Bois referred was Willcox's 1899 address before the American Social Science Association, the same address that Vardaman had quoted in his inaugural address and that Stone reprinted as a chapter in *Studies in the American Race Problem*.

43. Du Bois to Willcox, March 29, 1904, in Aptheker, *Selections, 1877–1934*, 75.

44. A copy of Du Bois' article from the *East and the West* is in TSC. This copy is probably the one that Du Bois sent to Willcox. Two hands have marked it, possibly Willcox's (in ink) and Stone's (in pencil).

personal matter. Perhaps he feels that a person of my antecedents is morally bound to support the cause of the negro and that it is almost treasonable for one of such antecedents to grow luke warm just when the race is in such dire need."[45]

Willcox's letter caught up with Stone in Asheville, North Carolina, where he had traveled with Mary to visit his brother-in-law, who was hospitalized there.[46] "I note with interest DuBois' letter and your reply," Stone wrote.

> I am reminded of a statement made to me some months ago by Karl Kelsey, of the University of Pa., to the effect that DuBois' bitterness rendered his opinions and conclusions particularly valueless. I might not go this far, but to my mind it seems impossible to reconcile such a mental attitude with anything like fairness or breadth of view. The more I study it the more firm is my conviction, so often expressed to you, that in its last analysis the "negro problem" is an economic problem. Against this view, Dr. DuBois would beyond doubt enter an emphatic protest.
>
> Were you to act on DuBois' rather querulous suggestion, to "get down here and really study it at first hand," there is little doubt in my mind as to your conclusion on the problem. You would not have the "intimate soul contact with all kinds and conditions of black men," of which he speaks; it would, rather, be contact of a more practical kind, the observation of the great negro mass, not of a selected class, and your conclusions would not coincide with his. I feel sympathy for DuBois and feel that his attitude is unfortunate,—both for himself and his race.[47]

Each man was thus well aware of the other's position when Du Bois and Stone faced off at the annual meeting of the American Economic Association in Baltimore.[48] From all indications, the confrontation had been eagerly anticipated because the session was scheduled as the only event on the final day of the meeting, and the panel of five discussants reflected a

45. Willcox to Stone, July 27, 1904, WFWP.

46. The brother-in-law may have been Henry T. Ireys Jr., Stone's buddy in the Levee Guard.

47. Stone to Willcox from Asheville, North Carolina, July 30, 1904. The postal service's efficiency at the turn of the century can be deduced from this exchange of correspondence. Willcox mailed the letter from Ithaca, New York, to Stone at the 200 A Street S.E. address in Washington, D.C., on Saturday, July 27, 1904. Stone replied from Asheville, North Carolina, three days later (Tuesday, July 30).

48. See, for example, Willcox to Stone, April 6, 1905, WFWP.

wide range of opinions.[49] The youngest discussant was Charles Lee Raper, a North Carolinian by birth and the precocious dean of the Graduate School at the University of North Carolina; he had received his doctorate from Columbia just three years before.[50] A second white southerner on the panel was a graduate of Johns Hopkins University and a noted philanthropist from Baltimore, Theodore Marburg.[51] Two of the other discussants held faculty positions at northern universities—M. B. Hammond at Ohio State and Henry W. Farnam at Yale. The fifth discussant was a black administrator at the Tuskegee Institute in Alabama, Roscoe C. Bruce, son of Blanche K. Bruce, the first black senator from Mississippi.[52]

Du Bois went first. The tone of his paper was reasoned and matter-of-act. Rather than speak of African Americans as one homogeneous group, Du Bois talked about four different routes by which black workers had attempted to achieve economic prosperity following emancipation: entry into the professions; migration to cities and towns to become artisans or industrial workers; transformation of field hands into independent farmers; and, finally, economic cooperation among African Americans segregated from the surrounding white economy. According to Du Bois, the first three routes had been closed to African Americans because of racial prejudice. As a result, black professionals and artisans were limited to a black clientele, which was one reason their achievements were generally unknown outside the black community. As far as independent farmers among African Americans were concerned, racial prejudice also prevented black farmers from buying productive farm land, despite claims made by white promoters. "Here first we run flat against one of those traditional statements which pass for truth because unchallenged," Du Bois stated, "namely, that it is easy for the southern Negro to buy land. The letter of this statement is true, but the spirit of it is false. There are vast tracts of land in the south that anybody black or white can but for little or nothing for the simple reasons that they are worth little or nothing."[53]

49. "Supplement, Hand Book of the AEA, 1906," *Publications of the American Economic Association,* 3rd ser., 7 (August 1906): 33.

50. "Charles Lee Raper," archives.syr.edu/arch/faculty/rapbio.htm.

51. "The [Johns Hopkins] Gazette *Online,*" September 27, 1999, www.jhu.edu/~gazette/1999/sep2799/27club.html.

52. "Blanche Kelso Bruce, 1841–1898," bioguide.congress.gov/scripts/guidedisplay.pl?index=B000968.

53. W. E. B. Du Bois, "The Economic Future of the Negro," *Publications of the American Economic Association,* 3rd ser, 7 (February 1906): 230–31.

According to Du Bois, white prejudice against African Americans was a major factor in understanding why black workers tended to be inefficient. "How efficient can he [the black laborer] become with intelligence, technical training, and encouragement?" Du Bois asked. Alluding to competition with white immigrants, Du Bois continued: "That the average Negro laborer today is less efficient than the average European laborer is certain. When, however, you take into account the Negro's ignorance, his past industrial training, and the social atmosphere in which he works it is not so easy to say off-hand what his possible worth is."[54]

Du Bois used farm tenancy to emphasize his point. Despite the potential of this arrangement to encourage thrift and to allow the black farmworker to share profits from a successful harvest, "the system of farm tenancy as practiced over the larger part of the south today is direct encouragement to cheating and peonage, a source of debauching labor, and a feeder of crime and vagrancy. It demands for its support a system of mortgage and contract laws and a method of administration which are a disgrace to 20th century civilization, and for every man which the system has helped to independence it has pushed ten back into slavery."[55]

Stone took the floor after Du Bois was finished. He agreed with Du Bois that black workers were less efficient than their white counterparts, but he rejected the notion that prejudice contributed to this state of affairs. "There seem to be two contributing causes to this situation," Stone stated at the beginning of his paper, "inefficiency, unreliability, and lack of thrift upon the part of the Negro, and prejudice upon the part of the white man." Despite his claim that "I shall not attempt to weigh the one against the other," Stone was quick to assert that prejudice was not a factor affecting black economic advancement in the South, and he quoted Booker T. Washington as his authority. "Whatever other sins the South may be called upon to bear," Washington had written in *Up from Slavery*, "when it comes to business pure and simple, it is in the South that the Negro is given a man's chance in the commercial world."[56]

According to Stone, the problem was not white prejudice against black labor but rather the innate inefficiency of black workers. Stone then proceeded to give examples of the poor work habits and unreliability of African Americans, especially when compared to white workers. To that end, Stone devoted

54. Ibid., 225, 231.

55. Ibid., 237–38.

56. Washington, *Up from Slavery*, quoted in Stone, "Economic Future of the Negro," 164.

a third of his paper to the experiment with Italian laborers at Sunnyside. According to Stone, the success of Italian workers at Sunnyside, coupled with the failure of black workers in his experiment at Dunleith, spoke volumes for the economic future of African Americans in the United States. "I do not speak idly when I say that Southern people in constantly increasing numbers are more and more coming to the conclusion that they must at last put forth a determined effort to render themselves independent of the Negro—" Stone warned toward the end of his paper, "to begin in some degree the final supplanting of the latter with the white man."[57]

Du Bois and Stone had given the discussants something to talk about. Roscoe C. Bruce went first. In a rambling presentation that exceeded his time limit, the educator from Tuskegee provided the audience with data and statistics about black labor in the central prairie belt of Alabama, with which he was familiar, and compared black workers there to those in the Mississippi Delta. Eventually, Bruce got around to his point. He agreed with Stone that "by eliminating white field laborers from direct competition with black, the segregation of Negro population tends to lessen race friction and in the long run to offer the negro a wider range of industrial opportunity." Furthermore, "a second compensation for the absence of the white man's example lies in the fact that the mass of the Negroes quickly find themselves ill prepared to compete even on equal terms with the whites."[58] Bruce's comments were, in fact, an advertisement for the educational model promoted by the Tuskegee Institute and the accommodationist policies of Booker T. Washington.

The next two discussants were enthusiastic in their endorsement of Stone's argument. "Mr. Stone has presented to you that fact that the negro as an agricultural laborer is now losing his position," observed Charles Lee Raper, the young professor from North Carolina, "not because of race prejudice but because of his economic inefficiency. He has shown how the imported Italian laborer is displacing the negro on the Southern cotton farms, in spite of the fact that the Southern cotton planter would because of his temperament and traditions prefer to employ the negro. In other words, he [Stone] has shown that the economic failure of the negro agricultural laborer is due to the lack of efficiency."[59] Theodore Marburg, the philanthropist from Baltimore, echoed Raper's praise for Stone's paper. "Mr. Stone has here put

57. Stone, "Economic Future of the Negro," 174–99.

58. "The Economic Future of the Negro: Discussion," *Publications of the American Economic Association*, 3rd ser., 7 (February 1906): 299–300.

59. Ibid., 306–7.

his finger on a fundamental consideration [the inherent inefficiency of African Americans]," Marburg exclaimed, "one around which the question of the economic future of the negro revolves." Complaining that black workers had accumulated little wealth or property since emancipation "so that the bulk of the taxes in the Southern states are paid by the whites," Marburg affirmed Stone's conviction that the encouragement of immigration from Europe to the South was the only antidote to the "improvident habits" of black workers.[60]

It was not until the fourth speaker that a discussant questioned Stone's assumptions. M. B. Hammond, a professor at Ohio State University, noted that African Americans had, in fact, made some progress since emancipation, such as a decline in the rate of illiteracy and an increase in the number of skilled and semiskilled workers as well as a larger number of landowners. Hammond also believed that racial prejudice in the South was a factor. "I believe that there *is* an industrial prejudice against the negro in the South which is one of the greatest hindrances to his progress and prosperity," he stated. Nevertheless, in regard to the question of black inefficiency, Hammond came down on Stone's side. According to Hammond, the inefficiency of black workers hindered their economic advancement in the North, where, he claimed, prejudice was not a factor. Nevertheless, Hammond did recognize that the lack of incentives had created a double-bind that entrapped both the black worker and the white planter. "We all know that it is difficult for us to make progress in any line when nothing is expected of us," he stated. "The white man in the South does little to help his negro tenants to progress because he feels that his efforts would be wasted, while the negro is little inclined to put forth efforts in this direction because he knows that nothing of this sort is expected of him."[61]

To this point, the discussants had directed almost all of their remarks to Stone. It was not until the fifth and last discussant, Henry W. Farnam, a professor from Yale, that Du Bois' argument received attention. Time for the session was running out, and so Farnam's remarks were brief compared to those of the discussants who had gone before him. "Can we confidently say how much of this temperament is due to racial peculiarities, how much to environment and history?" he asked. Referring to Du Bois' detailed account of a successful landowning system he had described in Georgia, Farnam noted that "If such persons can in a short time become self-supporting land-owners

60. Ibid., 310, 311, 313.

61. Ibid., 314, 316, 317; emphasis in original.

by the application of a system which is not in the least paternal, but merely puts such a premium upon thrift as generally exist in normal society, and is based upon strict business principles, we must receive with some caution the statement that the negro race as such is necessarily shiftless and improvident."[62]

The verbal boxing match between Du Bois and Stone was over. If each discussant counted as one round, the rounds would have been scored three for Stone, one for Du Bois, and one a draw.[63] But Du Bois was not defeated, far from it. In the years immediately ahead, he would distance himself even further from Booker T. Washington's accommodationist policies. Four years after the session, Du Bois would help effect an alliance of "militants" from the Niagara Movement with wealthy white liberals in the North to form the National Association for the Advancement of Colored People (NAACP). Du Bois would be named as the new organization's director of publicity and research, and in that capacity, Du Bois would assume editorship of the NAACP's newsletter, the *Crisis*. His long career as a civil rights leader was just beginning.[64]

Stone's beliefs were also unaffected by his confrontation with Du Bois, as evidenced when another article concerning the Italian labor experiment at Sunnyside appeared in the *Review of Reviews* a year later. Stone began the article with the self-assertive confidence characteristic of the earlier piece in the *South Atlantic Quarterly*. "The inability of the negro, thus far, to hold his own in competition with the Northern white man," Stone wrote, "has been demonstrated so often, and in so many ways, that it is no longer a debatable question." Stone then gave a brief history of the Sunnyside experiment under Austin Corbin and his heirs. He blamed the failure of the first group of Italian immigrants on the managers' failure to "convert a heterogeneous collection of butchers, bakers, tailors, shoemakers, etc., fresh from Italy, into a colony of satisfied and efficient cotton growers." In other words, the labor agents for the first experiment had not been diligent enough in selecting

62. Ibid., 322–23.

63. Coverage in the *Washington Post* would suggest that Stone won the match. The heading for the story read "Negro's Economic Future, Little Hope for His Development in Northern Cities, White Competition Drives Him into Menial Occupations—Discussed by American Economic Association" (*Washington Post*, December 30, 1905). The first paragraph of the four-paragraph article listed the speakers (including the discussants), and the next two paragraphs were devoted entirely to Stone's address. The final paragraph named the association's new officers. There was no reference to Du Bois' talk in the article.

64. Lewis, *W. E. B. Du Bois*, 402–7. Du Bois lived until 1963.

workers who were experienced farmworkers. Nevertheless, Stone was convinced that the new group of Italian workers would succeed, and in so doing, they would supplant African Americans as the plantation's work force. Stone based his prediction by comparing the productivity of the new Italian workers with that of black workers who been hired as replacements for the immigrants who fled when the scheme collapsed in 1898 but continued to work on the plantation after the new group of immigrants arrived. Based on cash balances brought forward and the number of livestock owned by individual workers, Stone estimated that by the end of the 1905 growing season, "Only 2.8 per cent. of the Italians failed to share the community prosperity, while among the negroes 44.7 per cent. found themselves in this condition."[65]

Du Bois read Stone's article and felt that he could not let it pass without a response. "I find the same difficulty with your Review of Reviews article that I find with all your work," Du Bois wrote in the opening sentence of his letter to Stone, "that while apparently in personal conversation and in letters you are perfectly fair, yet in your articles you always give what seems to me an unfair impression. You never say a good word for the Negro, you never give excuses for his condition, you simply attack him." Du Bois acknowledged that Stone's attacks were not a virulent as those of some people, whom he did not name, but Du Bois was convinced that Stone's attacks were more harmful because of the planter's reputation as an expert on the race problem. Du Bois was also convinced that Stone knew better. "No matter what your object was in that Review of Review[s] article, the result was the same, that is, it would give people who did not know the situation the impression that the Negro was a lazy, careless and good-for-nothing workman who under the good circumstances put round him, would not do anything near as much as he ought to do, while the real situation is, as you and I know perfectly well, that considering the tremendous discouragement as the history of the past, there is no set of workingmen in the world that has done so much or improved so much in efficiency as the Negro workingmen of the South." The problem with Stone's public pronouncements, according to Du Bois, was the planter's failure to take both the black and the white perspectives into account: "It seems to me that in your writings, you ought to try and give not the impression to a certain person or boost a certain cause, but give the true picture of the South and if you do, you cannot possibly continue to give the one-sided

65. Stone, "Italian Cotton-Growers in Arkansas," 209–12.

account which you have in the past." Along those lines, Du Bois accepted Stone's flattering depiction of conditions at Dunleith, but Du Bois thought that Stone was too narrowly focused on his own experience. "Even if all you say and have written concerning your Mississippi plantation is perfectly true, yet you know that it is not the rule on Mississippi plantations; you know that the condition of Negroes and the treatment of Negroes on those plantations is a shame and what you ought to do is not to put your plantation forward as a specimen but to so [say?] plainly that it is the exception."[66]

Stone was apparently unfazed by the critical tone of Du Bois' letter. "I wish to thank you for your frank criticism," he wrote in reply. "In so far as your objections to my method of treatment appear to me to be well taken, I trust that I may be able to profit by them." But Stone was willing to go only so far. "I am not prepared to go all the way with you in saying that, everything considered, the Negro workingmen of the South show an improvement in efficiency unequalled by that of any other set of workingmen in the world. But that is neither here nor there, for present. In what I have written it has not been my purpose to 'boost' any cause, convey any specific 'impression,' nor to say any word, good or bad, for either the Southern Negro or the Southern white man. I have simply tried to present facts as to local conditions with which I am familiar. Thus far I have had little, almost nothing, to say as to conditions through the South as a whole."[67]

Stone was getting his facts as to local conditions at Sunnyside directly from his friend LeRoy Percy.[68] However, events were under way as he exchanged jabs with Du Bois that would eventually demonstrate a reality that Stone would prefer to ignore—if cotton growers in the Delta treated white workers like they treated their black workers, they would end up with a white labor force that also was shiftless and demoralized.

The labor experiment at Sunnyside had begun to unravel in 1905 when Edmondo Mayor des Planches, the Italian ambassador to the United States, toured the southern states to see how Italian immigrants were adapting to their new homes.[69] Ambassador des Planches liked much of what he saw when he visited Sunnyside, but he did not find the situation exactly as Stone had described it. "The Italian immigrant at Sunnyside is a human produc-

66. Du Bois to Stone, April 13, 1907, DBP.

67. Stone to Du Bois, April 23, 1907, DBP.

68. Percy to Stone, February 5, 1906, PFP.

69. By coincidence, des Planches' trip occurred three months after Stone's first article on "the Italian Cotton Grower" appeared in the *South Atlantic Quarterly*.

tion machine," he reflected. "He is better off than the black man, more perfect than he is, but like the black man, still a machine."[70]

The fact that Ambassador des Planches penned these observations several years after his visit means that they must be interpreted with caution. Nevertheless, it was clear that dissatisfaction among the Italian workers that des Planches may have observed continued to fester. Eventually, their complaints reached the ears of the Italian consul in New Orleans, who ordered Luigi Villari, the vice consul, to visit Sunnyside. Villari made his first trip to Sunnyside in 1906 and returned for a second visit one year later.[71]

Villari found that new immigrants at Sunnyside were saddled with so many debts that many of the families found it difficult, if not impossible, to get ahead. They were being forced to pay for higher-than-expected transportation costs in addition to mules, cotton seed, tools, and living provisions that were available only through the Sunnyside plantation store at exorbitant prices. The situation was aggravated by a 10 percent rate of interest, which was collected in full regardless of whether the debt was settled in a week or in a year. Furthermore, they were forced to sell their cotton crop to the company, which paid below market prices in Arkansas and then sold it across the river in Greenville for a profit. The immigrant families were also stunned by the climatic conditions they encountered in the Arkansas Delta—stifling heat, mosquitoes, fevers, and unhealthy drinking water. If all of this was not bad enough, rains during the harvest season during the fall of 1906 damaged the crop, which made escaping the vicious cycle of indebtedness next to impossible.[72]

Hard labor and rough living conditions may demoralize a work force, but they are not, by themselves, illegal. But peonage is. Peonage is a system in which workers in debt to their employer are forced to work for that employer until their debts are paid.[73] The employer has a right to expect that his bills will be paid, and there are variety of ways he can settle the debt. For example, he can seize property against which he holds a lien, he can bring suit, or he can garnish the wages of an employee who leaves to work on another plantation. But the employer does not have the right to force the employee

70. Milani, "Peonage at Sunnyside," 45. Brandfon gives a more optimistic account of des Planches' tour (*Cotton Kingdom in the New South*, 154–55).

71. Milani, "Peonage at Sunnyside," 45–46.

72. Randolph H. Boehm, "Mary Grace Quackenbos and the Federal Campaign against Peonage," in *Shadows over Sunnyside*, 52–53.

73. *The American Heritage College Dictionary*, 3rd ed. (Boston: Houghton Mifflin, 1993), 1012. See also Cresswell, *Rednecks, Redeemers, and Race*, 44–45.

to continue to work for him under the same conditions that resulted in the employee's indebtedness. That situation is tantamount to slavery, and involuntary servitude had been outlawed by the Thirteenth Amendment.[74]

Villari informed Ambassador des Planches of what he observed, and des Planches decided that something had to be done. On June 4, 1907, the ambassador filed a formal request for an investigation with the U.S. secretary of state, Elihu Root, and requested that Sunnyside be put at the top of the list. Root turned the matter over to the Justice Department, and Attorney General Charles Bonaparte assigned the case to Mary Grace Quackenbos, a special assistant in the Justice Department who had investigated charges of peonage at labor camps in Florida, Alabama, and Tennessee the year before.[75]

Quackenbos was an interesting person for Bonaparte to pick for the assignment. Divorced and the beneficiary of a comfortable inheritance, Quackenbos had devoted her energies to providing legal assistance to immigrants in New York City. She was a "militant, middle-class reformer whose legal training and personal economic independence enabled her to undertake a crusade against corrupt New York labor agents and their southern employers," as one historian has described her.[76] Whatever her motivation, she was familiar with the problems at Sunnyside and other places throughout the South that recruited immigrants from Europe for labor. As they became disillusioned with their treatment in the South, immigrants had made their way back to New York, where they told of the tactics southern employers had used to entrap them financially and then force them to work. Although Quackenbos had identified forty plantations in the Delta region that employed Italian workers, Sunnyside was an obvious target because it had grown to sixteen

74. "Neither slavery nor involuntary servitude, except as a punishment for crime whereof the party shall have been duly convicted, shall exist within the United States, or any place subject to their jurisdiction."

75. Milani, "Peonage at Sunnyside," 41–43; Boehm, "Mary Grace Quackenbos," 54. For an account of Quackenbos's investigation that takes Percy's point of view, see Lewis Baker, *The Percys of Mississippi: Politics and Literature in the New South* (Baton Rouge: Louisiana State University Press, 1983), 30–31.

76. Jerrell H. Shofner, "Mary Grace Quackenbos: A Visitor Florida Did Not Want," *Florida Historical Quarterly* 58 (January 1980): 275. A contemporary account of her appointment to the Justice Department in the *New York Times* described Quackenbos as "young and pretty, with jet black eyes and hair, a vivacious manner, and a glance of keen penetration. Her figure is tall, slender, and girlish. She always wears well-made black gowns, with touches of white at the neck and sleeves. Her hat is the most distinctive part of her attire, for it is rather wide and flat, and from it, in the back, hang short folds of mourning veiling. Mrs. Quackenbos assumed this dress at the time of the death of her parents a few years ago" ("Woman Will Help in War against Trusts," *New York Times*, September 15, 1907).

thousand acres and was then the largest employer of Italian workers in the Delta on either side of the river.[77]

LeRoy Percy was the person Quackenbos had to deal with, and initially he allowed her free access to financial records and the workers.[78] Percy may have thought that her investigation would be stupid or superficial because Quackenbos was a woman. Perhaps he expected her to ride through the plantation in a carriage without digging too deep. If he did, he was mistaken. Within five days, Quackenbos had interviewed 70 of the 150 families at Sunnyside, and she found many things that bothered her greatly. "I have seen the impoverished condition of these people who originally came from the best parts of Northern Italy and are of splendid Italian stock," she wrote to Attorney General Bonaparte after she arrived. "They are now sick and discouraged—mere shadows of human beings who do not even understand their rights."[79]

Initially, Quackenbos hoped that she could negotiate changes in management-labor relations and thus resolve most of the problems she had discovered. At first, she met with some success using this tactic, but Percy and his partners began to smart under her criticisms and interference with their business affairs. Part of the problem, as they saw it, was that many of the things she stuck her nose into were matters that clearly fell outside the jurisdiction of the federal government.[80]

At this point, Quackenbos was still uncertain as to whether peonage existed at Sunnyside. Although she saw evidence of intimidation, she knew that a number of families in debt to the company had been allowed to leave. But there were other cases in which workers who left were brought back to the plantation and threatened with the chain gang if they fled again. The question of peonage was soon answered more directly when two men who had fallen deeply into debt to the company decided to leave their families for employment in the iron mines in Alabama. Crittenden caught wind of their defection, summoned the sheriff, and pulled the two men from the train in Greenville before it left the station.[81]

77. Mary Grace Quackenbos, "Sunny Side Colony, Ark., O. B. Crittenden & Co., Lessees, Cotton Factors," September 28, 1907, Department of Justice Records, RG 60 (vertical file no. 100937, box 3, folder 4), NA-CP.

78. Boehm, "Mary Grace Quackenbos," 57–59; Wyatt-Brown, "Leroy Percy and Sunnyside," 88.

79. Quackenbos to "The Attorney General," August 1, 1907, Department of Justice Records, RG 60 (vertical file no. 100937), NA-CP.

80. Boehm, "Mary Grace Quackenbos," 58.

81. Ibid., 60–61; Wyatt-Brown, "Leroy Percy and Sunnyside," 83, 89.

Percy got word that Quackenbos was moving toward criminal indictments for peonage, and the gloves came off. Percy wrote to Bonaparte to complain of her investigative tactics and bias in favor of the workers. "Mrs. Quackenbos is a very attractive woman and one of unusual intelligence," he wrote to Attorney General Bonaparte.

> She possibly knows some law, enough for this kind of work, and she is wonderfully industrious, honest, I believe, very much in earnest, and very much interested in her work, which is merely helping the poor. She is by nature a partisan of partisans, and judicial quality with her is simply non-existent. She does not come as an impartial investigator, as an unbiased judge, but, as she says, simply as the attorney and advocate of the poor. She takes no interest whatever in the laborer who is prospering. It is the laborer who is unfortunate or sick who appeals to her womanly sympathies, and on whom she concentrates her attention.[82]

It was true that Quackenbos was passionate about the plight of the Italian workers at Sunnyside, and "Unfortunately her sympathies did outrun her authority as a federal prosecutor," as historian Bertrand Wyatt-Brown has observed.[83] But she was not a lightweight, for she backed up her passion with pages of statistics, depositions, testimony, and observations in her first report to the attorney general on September 28, 1907. The report was sixty-six pages long, and Quackenbos laid out her findings in a logical and coherent manner.

Nevertheless, many people in the Delta did not like Quackenbos interfering with their business, and the *Greenville Times* took a potshot at Quackenbos the day after she submitted her report. "Who Is She?" queried the heading of a paragraph printed above the fold on the front page. "Greenville has a lady lawyer who is stirring up the Italian immigrant question from center to circumference. She comes with a formidable retinue of employees; two interpreters, one stenographer and a law clerk. She has already closed up one importing joint and has the planters of that vicinity worked up over the peonage question. In other words, she is just raising——, as is usual with a professional woman."[84]

82. Percy to "Chas. Bonaparte, United States Attorney General," August 19, 1907, Department of Justice Records, RG 60 (vertical file no. 100937), NA-CP.

83. Wyatt-Brown, "Leroy Percy and Sunnyside," 85.

84. "Who Is She?" *Greenville Times*, September 29, 1907, reprinted from the *Indianola Tocsin*. An editorial in the *Vicksburg Herald* referred to Quackenbos as a "socialist agitator" ("A Serious Accusation," *Vicksburg Herald*, September 8, 1907), to which she objected by confronting the editor personally in his office. The editor retracted the slander but warned that the con-

Resistance to Quackenbos's investigation escalated when her legal portfolio was stolen from her hotel room in Greenville and ended up in the hands of the former Mississippi congressman from the Delta, Thomas C. Catchings.[85] Percy also barred Quackenbos from Sunnyside and wrote a friend of his in Washington—someone he had befriended during a bear hunt in the Delta—President Theodore Roosevelt.[86] The main purpose of his letter was to defend his management of the plantation. However, there were two things Percy wanted the president to do. First, he wanted Roosevelt to have Quackenbos recalled. Second, he wanted to sidetrack her report. "I have no desire whatever to have her report suppressed," he assured the president. "I only ask that no publication be made of it and no action taken under it until it has been verified," and verification "might require an investigation from the Department of Commerce and Labor," he added.[87]

Percy's pull was persuasive, and Quackenbos was recalled from her investigation, but the tumult over her allegations had not run its course.[88] Although charges of peonage were quashed, the investigation was a public relations disaster for O. B. Crittenden & Company. In addition, Quackenbos's report had been shared with the Italian ambassador, and the Italian government temporarily halted the flow of new families to southern plantations. This prohibition would be disastrous for Sunnyside, so Percy and southern representatives in Congress mounted a counterattack to discredit Quackenbos and reestablish the South as a good place for European workers to settle.[89]

Among those taking part in the counterattack was Benjamin G. Hum-

tinued interference by representatives from the Justice Department "is sure to cause discontent and distrust between the Italian tenants and the planters" (untitled remarks, *Vicksburg Herald*, October 2, 1907).

85. Boehm, "Mary Grace Quackenbos," 63–65; Wyatt-Brown, "Leroy Percy and Sunnyside," 91.

86. There are several letters from Roosevelt to Percy in the PFP that suggest the existence of a friendly relationship, such as the one dated November 14, 1911 (box 20, third folder), which begins "I am touched at your writing me . . ." See also Wyatt-Brown, "Leroy Percy and Sunnyside," 84, 89.

87. Percy to "President Theodore Roosevelt," November 13, 1907, Department of Justice Records, RG 60 (vertical file no. 100937), NA-CP. The letter was twelve-and-a-half pages long, typed, and double-spaced.

88. Quackenbos appealed the decision to recall her, but was unsuccessful in getting the decision reversed (Quackenbos to "Mr. Attorney General," November 18, 1907, Department of Justice Records, RG 60 [vertical file no. 100937], NA-CP).

89. Wyatt-Brown, "Leroy Percy and Sunnyside," 92–93.

phreys, a congressman from the Mississippi Delta.[90] On March 2, 1908, Humphreys stood on the floor of Congress to defend "the state of Mississippi against the charges of peonage." The core of Humphreys' lengthy defense was the paper Stone had delivered in 1905 at the American Economic Association meeting in the session with Du Bois. Humphreys described Stone as "a scholarly gentleman who has given a great deal of study to such questions and who has had large experience as a cotton planter, and who, in addition to all that, and best of all, is blessed with an abundance of the saving grace of common sense." Humphreys went on to describe Stone as "perhaps the most profound student of the race question in this country to-day. He is invited frequently by the faculties of all the great colleges of the North to address them on this question, and is everywhere recognized as a man of the highest character and a writer and speaker of unusual ability," Humphreys continued. "Having heard much of the success of the Sunny Side Colony, he went over to investigate conditions there, the particular thought which inspired him being to make a comparative study of the negro and the Italian as farmers on the same plantation and under identical conditions."[91]

Having used Stone's expertise to exonerate Percy of mismanagement, Humphreys sought to discredit Quackenbos. "Now, this good lady has spent much of her useful life in the investigation and study of the unhappy conditions of the extremely poor in the great cities of this country," he observed.

> Whether as a result of that study she reached her conclusion or whether the study was the result of a preformed conviction on the subject, the fact is that when she undertook this investigation she was thoroughly saturated with those ideas of the proper relations of the rich and poor, of society to man, of the individual to society, which are the high tenets of socialism. In this mental condition she was regaled with those stories of cruelties and peonage, of oppression of the weak and friendless by the rich and the powerful on this plantation way down in the black belt of the far away and benighted South.[92]

Humphreys was not the only congressman mad at Quackenbos. Representative Frank Clark from Florida introduced a motion for Congress to

90. Humphreys should not be confused with his father, Benjamin G. Humphreys (1808–1882), who was a Confederate general and the first governor of Mississippi after the Civil War ("The Political Graveyard," politicalgraveyard.com/bio/humphreys.html).

91. Humphreys, *In Defense of the State of Mississippi*, 4. There is a copy of the speech in TSC.

92. Ibid., 10.

investigate her role in the investigation. Clark represented a district that Quackenbos had visited in 1906 to investigate charges of peonage in turpentine camps located there. Clark's motion was referred to John Sharp Williams, a congressman from Mississippi, who steered the matter to the Commission on Immigration, which the House had set up a year earlier to study problems associated with immigration throughout the country. Williams's move was clever because it assured that Quackenbos's report would slip out of sight among the paperwork flooding the commission. Had Clark's original motion been adopted, the whole mess would have been exposed through public hearings.[93]

Williams's move turned out to be brilliant when one of the Senate members of the Immigration Commission, Anselm J. McLaurin, also from Mississippi, died unexpectedly. Governor Edmond F. Noel of Mississippi appointed LeRoy Percy to complete McLaurin's term, and faster than you can say "conflict of interest," Percy was on the Immigration Commission. When the commission's report was finally released a year later, Sunnyside was not even mentioned in the section dealing with peonage.[94]

Despite the success in suppressing Quackenbos's report, the flow of Italian workers to the Delta dried up completely, and Italian families who were already there continued to slip away. An overflow from the Mississippi River in 1912 was the final blow, and Crittenden & Company decided not to renew their lease. Sunnyside was converted to a sharecropping arrangement worked by African Americans. The labor experiment that had started twice with high hopes was finally over.[95]

Still, Stone could not let go of his conviction that the "destructive propensity" of African Americans toward shiftlessness and improvidence made them inherently inferior to white workers and that eventually white labor would force black labor off the land. In December 1907, three months before Humphreys' speech on the floor of the House, Stone read a paper at the meeting of the American Historical Association in which he derogated the productivity of black workers before the Civil War. Using data from censuses of the period, Stone argued that slaves had not cultivated as much cotton as they had been given credit for. "The evidence is clear enough," he stated, "that the white laborer's share in that life [southern economic life] was far greater than was once supposed, but investigation has thus far barely

93. Boehm, "Mary Grace Quackenbos," 73.
94. Ibid., 74.
95. Ibid., 75.

touched the surface of the field." Once again, he predicted that European immigrants would continue to settle in the South, and that this influx would eventually displace African Americans.[96]

The failed labor experiment at Sunnyside and the results of his experiment at Dunleith should have demonstrated clearly to Stone that management was a crucial factor in either improving or attenuating the efficiency of any labor force, black or white. It is true that Stone's treatment of his workers at Dunleith may not have been as high-handed as Percy's at Sunnyside, but the lesson to be learned was clear on both plantations.[97] Workers resented not having a say in the decisions that affected their lives. As Wyatt-Brown has put it, "More than anything else, the planter mentality, along with the lack of full vision that went with it, was LeRoy Percy's undoing at Sunnyside."[98]

The failed experiment at Sunnyside also demonstrated that racial prejudice was an important factor. "The identification with nonwhite labor, especially the Negro, robbed the Italian of his status as a white man," historian Robert L. Brandfon observed in his history of the Delta. "This status decline was reinforced by the servility associated with working on the plantation. In the Delta, no self-respecting white man labored on the huge cotton plantations. This was Negro's work. It was the badge of his inferiority. By replacing the Negro in the same type of work and under the same conditions, the Italians assumed the status of Negroes. One blended into the other, and southern thinking made no effort to distinguish between them."[99]

Racial prejudice interacted with authoritarian management practices to create a work environment that was guaranteed to produce an inefficient la-

96. Alfred Holt Stone, "The Future of the Race Problem in America," *International* 3 (October 1908): 147–56. The historian Eugene Genovese agreed with Stone that slave labor was inefficient but for completely different reasons (e.g., the lack of incentives under the slave system). In fact, Genovese dismisses Stone's argument out of hand: "I do not think that the first [Stone's] argument requires refutation; surely, the negative findings of genetics and anthropology are conclusive and well known" ("The Negro Laborer in African and the Slave South," *Phylon* 21 [4th Quarter 1960]: 344).

97. McMillen characterizes Stone as one of "the most progressive landlords" at the turn of the century (*Dark Journey*, 134). Cresswell grouped Stone with planters who "went out of their way to be generous to their tenants, hoping to encourage them to stay." Cresswell described Percy, on the other hand, as one of the planters who wanted to increase their control over black tenants (*Rednecks, Redeemers, and Race*, 42 and 43, respectively).

98. Wyatt-Brown, "Leroy Percy and Sunnyside," 94.

99. Brandfon, *Cotton Kingdom*, 163–64; Wyatt-Brown, "Leroy Percy and Sunnyside," 87. Southern xenophobia also played a role in dissuading immigrants from settling in the South (Rowland T. Berthoff, "Southern Attitudes toward Immigration, 1865–1914," *Journal of Southern History* 17 [August 1951]: 328–60).

bor force. "They [Italian workers] were not inclined to accept the restrictions that went with being a tenant farmer in the Mississippi Delta," historian James C. Cobb has observed. "These restrictions, including credit only at the plantation commissary, and selling cotton only to the owner or his agent, not only conflicted with what appeared to be a tenant's economic best interest, but they carried with them the stigma of subservience and inequality and thus lumped together all tenants, black and white. Such a situation was unacceptable, socially and economically, to the Italians who had moved to Delta plantations, and they made no secret of their resentment are being treated like blacks."[100]

For someone as intelligent as Stone, the role of social factors, as opposed to innate characteristics, could have explained much about the inefficiency of the black workers about whom he complained. In that regard, Du Bois had been more of a realist than Stone during their confrontation at the American Economic Association in 1905. "That the average Negro laborer today is less efficient than the average European laborer is certain," Du Bois had acknowledged. "When, however, you take into account the Negro's ignorance, his past industrial training, and the social atmosphere in which he works it is not so easy to say off-hand what his possible worth is."[101]

100. Cobb, *Most Southern Place on Earth,* 111–12.
101. Du Bois, "Economic Future of the Negro," 225, 231.

Black workers in a cotton field near Greenville, Mississippi, ca. 1897. J. C. Coovert Collection. Courtesy Mississippi Department of Archives and History.

Cartoon by L. C. Gregg that appeared in the *Atlanta Constitution* shortly after President Theodore Roosevelt invited Booker T. Washington to dinner at the White House. A caption under the cartoon read, "An amusing book, 'My Negro Policy!'" Sullivan, *Pre-War America*, 134.

Portrait of Walter F. Willcox by Prince of Washington, D.C., ca. 1910. Walter F. Willcox Collection, #14/10/821, box 6, folder 14 [platinum print]. Courtesy of the Division of Rare and Manuscript Collections, Cornell University Library.

A white cotton grower weighing the day's labor of his black workers. Cotton Picking Photograph Collection. Courtesy Mississippi Department of Archives and History.

Two men poisoning a cotton field in Claiborne County, Mississippi, in 1910 to kill boll weevils. Allen (Leigh Briscoe) Photograph Collection. Courtesy Mississippi Department of Archives and History. Although the boll weevil was resistant to conventional pesticides, calcium arsenate (now known to be a human carcinogen) applied as a powder to cotton shoots was found to be reasonably effective.

W. E. B. Du Bois, 1904. Digital ID: cph 3a29260, Prints and Photographs Division, Library of Congress.

Booker T. Washington, ca. 1890–1910. Digital ID: cph 3c19898, Prints and Photographs Division, Library of Congress.

Postcard of high water at the Dunleith train station on the Greenville, Columbus, and Birmingham Railroad during the 1912 flood. Greenville, Mississippi, Photograph Collection. Courtesy Mississippi Department of Archives and History.

10

FRANK, WITHOUT BEING OFFENSIVE

Alfred Holt Stone spent most of 1904 through 1906 in Washington, D.C. He usually stayed at the boardinghouse just across the street from the Library of Congress. When Mary accompanied him, as she did in the winter of 1904–5 and again in 1905–6, Stone took a room at the Hotel Driscoll, a modern, five-story building at First and B Streets N.W., facing the Capitol and close to Union Station.[1]

Stone's research was paying off, and his views on the race problem were becoming known outside the circle of academics with whom he interacted at the American Economic Association. In February 1905, Willcox invited him to speak at Cornell University in Ithaca, New York.[2] "My thesis is the influence of the numerical distribution of our negro population upon our respective points of view," Stone informed Willcox on May 5, one week before he was scheduled to speak. The point he wanted to make was that people living in the North would develop the same prejudices concerning black people as southerners if they also lived in a place where African Americans accounted for a vast majority of the population. "My only hope is to suggest some line of thought that possibly may enable others to understand, at least a little better than before, why my people feel and act as they do," as Stone put it. "Certainly I hope that I may induce my audience to believe with me that these people think merely as others think who are similarly circumstanced."[3]

Stone's talk was a success, in large part due to his desire to emphasize

1. Information about the duration of Stone's visits and where he stayed comes from various letters to Willcox in WFWP. The location and description of the Hotel Driscoll comes from postcards of the hotel purchased on eBay.

2. Correspondence between Willcox and Stone dated February 21, 22, and March 2, 1905, WFWP.

3. Stone to Willcox from Washington, DC, May 5, 1905.

the similarities between white people in the North and white people in the South rather than focus on their differences. Furthermore, he distanced himself from the racial vitriol northerners were accustomed to hear from southern bigots. "I am not an extremist," Stone assured his audience at the beginning of his talk, "and I long ago made up my mind to keep faith with myself in this, that I would not utter one word upon the perplexing question [the race problem] of which my conscience did not approve as the prompting of a desire to speak the truth for the truth's sake." Patiently, he spelled out why the South had found it necessary to adopt legislative barriers to the integration of the two races. "I can say that the white people of the South believe that where two races, as widely different as are the white and black, live together in large masses," he explained, "public policy requires the observance of certain regulations in the ordering of the social relations between the two." If the ratio of black people to white people in the North was the same as in the South, Stone predicted, the northern states would embrace Jim Crow too.[4]

Stone's address at Cornell elaborated a theme he had articulated six years before in his first paper for the Mississippi Historical Society. Unless you lived in the Deep South, you could not appreciate the steps that were necessary to maintain harmonious relations between the two races. But the defensive tone of the earlier piece was gone when he spoke to his audience at Cornell. What Stone had discovered in the intervening years was that many well-educated white people in the North thought the same way he did when it came to the race problem. His new approach was so effective in mending sectional bridges that Stone used the address in 1908 as the first chapter in *Studies in the American Race Problem.*[5]

Stone's major interest, however, was not in pointing out that northerners were segregationists too. By 1905, he had thrown himself almost completely into his ambitious plan to write a history of the negro race, "my life work," as he had confided to Willcox in 1902.[6] To that end, Stone had begun collecting articles, pamphlets, monographs, and books about people of

4. Stone, "Race Problem Contrasts and Parallels," 4, 25, 32. Mary accompanied Stone on the trip to Cornell, where they stayed with the Willcoxes (Stone to "Dr. & Mrs. Willcox," May 17, 1905, and Willcox to Stone and his wife, April 4, and May 2, 1905, all in WFWP).

5. "Race Problem Contrasts and Parallels," 3–39. Stone gave the same paper again in 1906 when he visited the University of Michigan.

6. See Stone to Willcox from the Hotel Driscoll in Washington, D. C., April 24, 1906. "Completely" is qualified by "almost" because Stone continued to pursue his interest in the War Amendments.

African descent. Because the scope of his project was broad, Stone wanted information about these people wherever they lived, not just in the United States. In addition, he sought a variety of opinions concerning their strengths and weaknesses as well as how governmental and societal systems affected their behavior. Consequently, the items that came within Stone's reach covered a gamut of disciplines—anthropology, history, sociology, economics, and political science. Furthermore, Stone was eclectic in his collecting. In that regard, Stone's lack of formal training in the areas he was investigating probably helped. He apparently did not know enough to know what not to collect.

Stone sorted the items he collected by format or subject area and had them bound, anywhere from twenty to sixty at time, into hardback volumes. None of the articles in the first five volumes was published after June 1905. In fact, it was not until volume 41 that articles or pamphlets with publication dates occurring in 1906 appeared, and the first item published in 1907, issues of the *Colored American Magazine*, did not show up until volume 89. Although Stone collected only two items published in 1909, both were bound in later volumes, one in volume 95 and the other in volume 97. The rough chronological progression of publication dates throughout the 113 volumes in Stone's collection suggests two things. First, Stone had the items bound as they came into his possession as opposed to waiting to do it when he was finished. This possibility is important because it would indicate that Stone's collecting was an ongoing endeavor begun without a distinct termination point in mind. Second, the rough chronology suggests that the most active period for Stone's collecting was the four-year span from 1905 to 1908.

During those years Stone was a busy man. In addition to his writing papers and giving lectures, he was reading and processing a phenomenal amount of material. His collection of articles, pamphlets, and monographs ended up containing 3,005 items, which meant that, if we credit Stone with working full time on his research for at least six months a year over four years, he was handling approximately 125 new items each month, or about 4 every day, including Saturdays and Sundays. Although one may wonder what Mary thought about all of this collecting, we do know that Stone had some help. "I have a young woman in the Library [of Congress] now who gives me an hour or two every day," Stone wrote to Willcox from Washington in April 1906. "She is doing bibliographical work,—not, however, according to the probably accepted scientific methods, but after ideas of my own."[7] Later,

7. Stone to Willcox, April 24, 1906, WFWP.

when Stone's research received modest funding, he was able to hire additional part-time helpers. Still, he was directly involved in reading and organizing the material that came into his collection. At least 110 items occurring randomly throughout the collection bear the pencil marks of his underlining, his emphasis, or his notes in the margin. Clearly, Stone was a prodigious worker. (See appendix B for an explanation of how Stone's notations in his collection were verified.)

The limitation that Stone faced in regard to collecting material for his book on the history of the black race was not his commitment, nor his energy, nor his work habits, but money.[8] The issue of finances came to the fore during the spring of 1906, when Willcox asked Stone to prepare "a report on the history of the negro population of the United States."[9] Although the topic was directly related to Stone's research interests, and although Willcox offered money from his sources to help cover Stone's expenses in gathering the data he would need for the report, finances had become such a problem that Stone felt obligated to say something about them to Willcox. "I feel that our relations have come to be such that I can speak to you with the utmost frankness, and this I shall take the liberty of doing," Stone began. "As I may have intimated to you before, my financial affairs are subject to very considerable fluctuations, depending as they do upon such uncertain factors as the weather, negro labor and price manipulation. The year just closed,—my business year ends April first,—was a *bad* one, despite the high price for cotton. In consequence I am beginning to ask myself whether I should continue my work here, or return home for a year or so."[10]

Stone's letter stunned Willcox and resulted in a flurry of correspondence between the two men.[11] Willcox knew that he had a good thing in Stone, a smart volunteer who was also a self-starter, and he responded immediately. "I am glad that you have spoken so frankly about your present situation," Willcox replied. "You may be sure that I do not misunderstand it. It ~~will~~ would be a serious disappointment to me if you found it necessary or

8. Stone revised the scope of this work downward as time went on. In January 1908, when he wrote the preface to *Studies in the American Race Problem*, Stone's working title for the book was "Race Relations in America" (xii).

9. Willcox to Stone, April 16, 1906, WFWP.

10. Stone to Willcox, April 24, 1906, WFWP; emphasis in original.

11. Matters discussed in this exchange of letters between Washington, D. C., and Ithaca, New York, could have been handled with two or three long-distance telephone calls. Fortunately for historians, long-distance phone service was in its infancy in 1906, and the insights we gain from reading this correspondence were not lost to history over telephone lines. For an account of telephone service during this period, see Herbert N. Casson, *The History of the Telephone* (Chicago: McClurg, 1910).

best to give up your literary work and return to your plantation."[12] Willcox then commiserated with Stone over the unfortunate reality that authors of academic works could expect little in the way of financial compensation for their efforts. The best situation for Stone, as Willcox saw it, was being offered "a position as a teacher of American History in some worthy American University." However, there was not much that Willcox could do to effect that outcome. Nevertheless, Willcox had an idea. Perhaps he could get Stone an appointment on one of the study groups sponsored by the Carnegie Institution. Maybe he could convince the institution to set up a new study group, which the institution referred to as divisions, with Stone as its head. "My own wish would be that you should be invited to become a colleague with the rest of us on equal terms having an independent appropriation."[13]

The Carnegie Institution had been founded by the fabulously wealthy industrialist Andrew Carnegie in 1902. "It is proposed to found in the city of Washington, an institution which . . . shall in the broadest and most liberal manner encourage investigation, research, and discovery [and] show the application of knowledge to the improvement of mankind," Carnegie had announced at its formation. According to the institution's Web site, "His [Carnegie's] intention was for the institution to be home to exceptional individuals—men and women with imagination and extraordinary dedication capable of working at the cutting edge of their fields."[14]

Willcox had immediately recognized the Carnegie Institution's potential as a source of support for his research and had contacted Colonel Carroll Davidson Wright to suggest that the institution should fund work "along the line of anthropological investigations."[15] Colonel Wright was a statistician who, like Willcox, specialized in demographic studies. He was also one of the institution's founding trustees and director of its Department of Economics and Sociology.[16] Willcox's initiative paid off, for he was asked to head the department's first division, a study group for population and immigration.[17]

12. Willcox to Stone, April 26, 1906, WFWP; strike-through in original.

13. Ibid.

14. "Carnegie Institution," carnegieinstitution.org/about.html.

15. Willcox met with Colonel Wright in Washington on March 28, 1902, which he noted in a letter to Stone dated March 31, WFWP. Willcox was more specific in regard to his research agenda in a letter to Wright on April 8, also in WFWP. Willcox stated that he was interested in an anthropological study of "the American Negro, in the effort to get more light upon the degree of admixture of white blood."

16. "Carroll Davidson Wright Biography," www.bookrags.com/biography-carroll-davidson-wright/index.html, and "Articles of Incorporation," http://carnegieinstitution.org/articles_of_incorporation.html.

17. Carnegie Institution of Washington, *Year Book No. 6, 1907* (Washington, DC: Judd and

Willcox wanted to achieve a similar outcome for Stone. "I wrote suggesting this [Stone's becoming a colleague] to Professor Farnam a few days ago, and also because he has the most influence with Colonel Wright." Henry W. Farnam was the Yale professor who had served as a discussant for the paper session with Stone and Du Bois at the American Economic Association meeting four months earlier. Farnam was also a division head in the institution's Department of Economics and Sociology.[18] "He replies in a letter just received," Willcox continued, quoting Farnam directly: "'I wrote to Colonel Wright Saturday [April 22] that it seems to me that we must secure the help of some first class man to direct and partly carry our field work in the south. I doubt whether any of the existing departments are equipped for it.'" Noting that he and Farnam had arrived at the same conclusion independently, Willcox continued to quote Farnam's letter in his response to Stone's letter:

> "I [Farnam] heartily concur in your suggestion in securing Mr. Alfred H. Stone as a full member. He is not only remarkably equipped, but it seems to me an absolutely essential that we should have a southern man in order to overcome a certain feeling which I can hardly call prejudice; perhaps I had better call it a certain defensive attitude of mine which I have often found among southerners when discussing the negro problem with northerners. I have generally found them remarkably frank and reasonable as soon as that first crust is broken. If it is known that this whole subject is directed by a southerner the crust will not exist. . . . I shall write at once to Colonel Wright suggesting Mr. Stone."[19]

Willcox was pleased that Farnam was in favor of Stone's appointment. "I have hesitated to talk frankly to Colonel Wright for fear that he might suspect me of biassing [*sic*] in your favor on account of my friendship for you." Farnam's endorsement would eliminate the appearance of favoritism. As far as discussing the details of Stone's financial position was concerned, Willcox suggested they wait until they heard from Colonel Wright. But Willcox was not leaving anything to chance. On the same day he wrote to tell Stone the good news, Willcox wrote to Colonel Wright, relaying Farnam's high opinion of Stone's qualifications and his suggestion that Stone "take charge of the

Detweiler for the Institution, 1908), 69; as well as the year books for 1906 and 1908. Colonel Wright died in February 1909.

18. Carnegie Institution of Washington, *Year Book No. 5, 1906* (Washington, DC: Judd and Detweiler for the Institution, 1907), 162. Farnam's division studied social legislation.

19. Willcox to Stone, April 26, 1906, WFWP; ellipses in the original.

work of the negro and his economic history in the south." Willcox seconded Farnam's recommendation and added some praise of his own. "I have known Mr. Stone now for more than six years," Willcox wrote, "and corresponded with him frequently about the negro question and the economic history of the south. In both fields I have found him remarkably well informed and very fair and impartial in his attitude."[20]

Stone was delighted with the prospect of establishing a formal relationship with the Carnegie Institution. "I have your letter of the 26th, but shall not attempt to tell you how much I appreciate it. You know, without being told. Of this I feel sure," Stone replied. "Do you think it would be wise, or best, for me to send Col. Wright copies of the few things I have written?" he asked. "Please advise me frankly. It occurs to me to ask the question merely because I am entirely unknown to Col. Wright." As far as Willcox's suggestion that he should aspire to a professorship, Stone was skeptical. "I am very much afraid that you are a hopelessly biassed [*sic*] man,—certainly as regards my ability to fill a chair of American History. If Mr. Carnegie would endow a chair of 'Negrology' in some institution of learning, I would not mind trying to hold it down,—but my American History is too much confined to the reconstruction period." Nevertheless, Stone was optimistic. "After I get out my book on 'race relations,' and have made a trip or so to the West Indies, possibly something may open up in the line of a lecturing tour for me." Stone ended his letter with a postscript. "Possibly you may think it a good idea for me to mail these pamphlets [reprints] to Col. Wright. If so, the sooner the better."[21]

Willcox had good news when he replied to Stone two days later. "I have a letter in this morning's mail from Colonel Wright indicating that he thinks favorably of trying to make some arrangement with you," Willcox wrote, "but not committing definitely to the wisdom of offering you a position as a member of our board. He says that he is planning for a meeting of the subcommittee on the negro in the near future and I infer that he intends to leave the matter open for discussion and decision at that time." Willcox suggested that Stone send copies of the articles he had written directly to him rather than to Colonel Wright so that Willcox could take them with him to the subcommittee meeting. Willcox also informed Stone that he had recently spent a day with Dr. Robert S. Woodward, president of the Carnegie Institution, when Dr. Woodward had delivered an address at Cornell. "I had opportunity

20. Willcox to Wright, April 26, 1906, WFWP.
21. Stone to Willcox from Washington, DC, April 28, 1906, WFWP.

for quite a little talk with him about the possibility of the work on the negro question," Willcox informed Stone. "He thinks that the trustees are pretty firmly determined not to undertake work on so burning a subject and that only by slow degrees can this opposition be overcome. He is frequently in receipt of letters from men in the south saying they have heard rumors that the institution is to undertake work regarding the negro and protesting vehemently against it."[22]

Stone wrote back to Willcox as soon as he received the letter and promised to send copies of his articles to Willcox. He was concerned, however, over the news that opposition from southern men had discouraged the Carnegie Institution from studying the race problem. "I wish it were possible for me to know who these men are," Stone wrote. "The South has had the misfortune to suffer a great deal through the misrepresentation of some of the citizens of that section. It is difficult to measure public sentiment there by the vehement expressions of a few individuals as it is to guage [*sic*] it by similar methods in the Northern part of the country. Did Dr. Woodward tell you who these men were? or what interests they claim to speak for?"[23]

Willcox had some good news when he replied to Stone's letter, and he had some news that was, if not bad, at least discouraging. The good news was that Colonel Wright had called a meeting of the subcommittee for May 10, 1906, a week from the date of Willcox's letter. The discouraging news was that opposition to the study of racial issues came from the institution's trustees who were concerned about public relations. "Indeed, Dr. Woodward raised the question whether the history of Chinese immigration which is now being written for the institution would be likely to draw down criticism upon the institution by its proof of the great exaggerations and misrepresentations concerning the Chinese which have been prevalent on the Pacific Coast. He did not say who the southern persons are, but he did talk at length on the general question and said that it would probably be easier to get competent investigators for such a subject than to get the social approval and backing requisite to getting it authorized."[24]

There matters stood until May 10, when Willcox wrote Stone again, this time with good news. "As I wrote you hurridly [*sic*] from the train our conference at New York passed off to my entire satisfaction," he began. "The other members of our committee seemed to accept without any objection or re-

22. Willcox to Stone, April 30, 1906, WFWP. Wright's letter to Willcox, dated April 28, is also in the WFWP.

23. Stone to Willcox from Washington, DC, May 1, 1906, WFWP.

24. Willcox to Stone, May 3, 1906, WFWP.

luctance my judgment that you were the best person in sight to undertake the work on the economic history of the negro in the United States. They all agreed too, that the matter was of enough importance to make it desirable to create a separate division for that topic." Willcox also reported that, although members of the committee had voted to invite Stone to start work immediately, Colonel Wright wanted to ask Stone to prepare a syllabus listing the topics he planned to study. Colonel Wright also wanted Willcox to impress on Stone that "we are not engaged in a general study of the negro question but upon a study of the economic history of the United States in which undoubtedly the negro has borne and is bearing an important part." Colonel Wright made it clear that he did not want political issues studied "indirectly under the cloak of economic history." Colonel Wright had also asked Willcox whether Stone would be an easy person to work with. "His anxiety on this point was for two reasons apparently," Willcox reported. "First, that the new field as outlined runs rather across our previous classification, so that work in it will have to be constantly adjusted to the work of others and secondly that he is much pleased and I think a little proud of the entire good feeling and cordial cooperation between the members of our Board thus far and is further desirous that no element of a different sort should come in, but he was perfectly satisfied with my emphatic assurance on that point."[25]

Stone appreciated the importance of collegiality, especially when working on a project as complex and sensitive as the role of African Americans in the nation's economic growth. "Your letter enables me to understand Col. Wright's position quite clearly," Stone wrote in his reply, "and also to sympathize with it. I believe I remarked to you that I was struck with the singular harmony of discussion and action manifest at the Board meeting which I attended. I can readily appreciate Colonel Wright's reluctance to risk any disturbing of such relations through the interjection of an unknown individual into the Board."[26] Stone could not have known at the time he wrote this letter just how important the maintenance of harmonious feelings would become as far as his appointment was concerned. Stone also provided Willcox with a rough list of topics he intended to submit to Colonel Wright. It consisted of twenty-seven items covering almost every aspect of black labor during slavery, after emancipation, and currently as competition for European immigrants. The items covered agricultural and industrial pursuits, as well as

25. Ibid., May 10, 1906, WFWP. Apparently, Willcox's note to Stone from the train was not copied and thus did not survive in the WFWP.

26. Stone to Willcox from Washington, DC, May 12, 1906, WFWP.

governmental policies and demographic shifts. The thoroughness of the list was evidence of the hard work Stone had devoted to his project.[27]

Stone continued to work on his syllabus until it was ready for submission. Farnam had become Stone's sponsor, and it was to the professor at Yale that he sent multiple copies of the final draft at the end of May, but not before sending one to Willcox a day earlier so Willcox could review it one more time.[28] Farnam liked what Stone sent him. "Prof. Farnam writes me as follows," Stone wrote to Willcox on May 31: "'I have only had time to look over the syllabus somewhat hastily, but it seems to me that you have succeeded very well in avoiding conflict with other departments of work. I shall examine it more carefully soon.'"[29] There was nothing else Stone could do but wait.

The wait was difficult for Stone, as evidenced by two letters he wrote to Willcox in June, one dated June 15 and the other three days later. Both letters were uncharacteristic. For one thing, the letters were written in longhand instead of being typed.[30] For another, their content was rambling, disjointed, and repetitive. "I am anxious to do a piece of creditable work, and want to make it thorough," Stone wrote on the first page of the first letter. On the fifth page of the same letter, Stone restated his concern in these words: "But I am extremely anxious to produce something which shall be a credit not only to myself, but to you, who are really responsible for my selection." In between, he agonized over whether enough money would be appropriated by the institution to finish the project. The issue seems to have been whether he would get credit for the time he had already spent doing research, in which case he might be awarded a lump sum on the front end that he could use to hire research assistants. The way in which Stone presented his concerns was somewhat confusing, and he ended the letter by debating whether he should write Colonel Wright directly about the matter. "I shall be guided *entirely* by

27. Apparently, Stone sent Willcox a copy of the syllabus prior to the meeting of the subcommittee on May 10, 1906, because he refers to "the rough outline which I sent you" in his letter dated May 12. An undated copy of the syllabus containing the first twenty-three items is in the WFWP, and Stone continued the outline by adding items 24–27 in his letter. Although the two documents may have been sent on different days, they are treated as one document here because it is clear that Stone intended Willcox to treat them as such (Stone to Willcox from Washington, D. C., May 12, 1906, WFWP).

28. Stone to Willcox from Washington, DC, May 28, 1906, WFWP.

29. Ibid., May 31, 1906, WFWP.

30. Stone started using a typewriter for his letters to Willcox in October 1901 (see chap. 7, n. 25) and rarely diverged from that practice except when he was en route (see, for example, Stone to Willcox from Asheville, NC, July 30, 1904, WFWP).

your [Willcox's] advice as to whether to write Col. Wright," he confided. "If you think it advisable to write him I would like the privilege of telling him that I had consulted you,—but do not hesitate to say so if you prefer that I do not use your name. On the other hand, if you think I would better not write him,—or you think there would be any impropriety in doing so,—you must frankly tell me. It may be just a foolish idea,—but, personally, I wish I did not have to say anything about it, even to you."[31]

Willcox responded to Stone's letter with a note that has not survived. However, its existence can be inferred by Stone's acknowledgment in the letter he wrote on June 18. "I have your note of yesterday, and am grateful to you for your willingness to take up that matter with Col. Wright." Once again, Willcox was acting at Stone's go-between, a role he had played in Stone's appointment to the committee for the American Economic Association. Nevertheless, Stone's uneasiness had not been assuaged. "I wish you lived near to Washington,—or I were nearer to Ithaca," he wrote in a final paragraph. "There are several things about which I would like to talk with you,—but they must wait until the opportunity comes."[32]

When Stone wrote these lines he was unaware of the extent to which Willcox had gone to help Stone secure a position at the Carnegie Institution. On June 18, the same day Stone drafted his letter to Willcox, Willcox wrote to Colonel Wright. In his letter, Willcox did two things that opened the door for Stone. First, Willcox enclosed a typescript of Stone's letter dated June 15 with extraneous or repetitive language removed.[33] By carefully editing Stone's letter, Willcox was able to document Stone's financial concerns without exposing Stone's lack of confidence. The second thing that Willcox did to help Stone was to offer a bankroll for Stone's project. If Colonel Wright was not in a position to budget $200 a month for Stone's work, Willcox wrote, he was willing "to transfer to his [Stone's] account $1,000 of the unexpended balance of the appropriation to my division." Willcox's offer was not only generous; it was tangible proof of Willcox's confidence in his friend's competence as a racial theorist.

Willcox's letter to Colonel Wright did the trick. "I think Mr. Stone better go ahead, making such contracts as he deems wise and sending the bills

31. Stone to Willcox from Washington, DC, June 15, 1906, WFWP; emphasis in original.

32. Ibid., June 18, 1906, WFWP.

33. An undated, single-page typescript entitled "Extracts from letter of Mr. A. H. Stone" is in the folder containing Stone's correspondence, WFWP. Willcox's letter to Wright dated June 18, 1906, in which Willcox refers to "accompanying extracts," is filed in the folder containing Wright's correspondence, WFWP.

to me for payment, without references to any accumulated fund, but having in mind your kind offer to transfer $1,000 from your own allotment," Colonel Wright wrote in reply on June 22. Confident that he could find enough money to fund Stone's work as it progressed, Colonel Wright noted, "I am very much pleased with the arrangements with Mr. Stone, and think the work will be done in a satisfactory manner." [34]

Colonel Wright's decision to go ahead with Stone's project was certainly good news, but Stone still had to wait until the institution's fall meeting for his appointment to be made official.[35] The delay was irksome, but it would have been even more so had Stone known that another researcher was actively seeking funding from the Carnegie Institution for a project that essentially duplicated the one that Stone had proposed.

On May 22, five days before Stone submitted his syllabus to Farnam, W. E. B. Du Bois had written a letter of introduction to Andrew Carnegie, in which Du Bois expressed the hope that Carnegie would remember having met the black scholar when he was introduced to Carnegie by Carl Schurz during an event at Carnegie Hall some years before.[36] Du Bois followed up his introductory letter to Carnegie by submitting an unsolicited research proposal to the institution, in which he offered to undertake a comprehensive study of the black race. The scope of his proposal encompassed a broad range of topics, including the Atlantic slave trade, slavery in the United States, postemancipation education, economics, and politics as they affected black people, as well as anthropological, ethnological, and cultural studies comparing black people in Africa with those in the United States.[37]

Apparently, neither man knew of the other's proposal. Purely by coincidence, Stone had written Du Bois on May 19 to request copies of his recent work. Picking up where they had left off in a friendly exchange of letters in 1903, Stone suggested that they get together sometime when Stone was passing through Atlanta on his way to or from Washington.[38] Du Bois was a busy

34. Wright to Willcox, June 22, 1906, WFWP.

35. The Carnegie Institution operated on a calendar for the fiscal year that ended on September 30 (Carnegie Institution, *Yearbook No. 5, 1906*, 158).

36. Lewis, *W. E. B. Du Bois*, 366–67, 659 n. 47. Carl Schurz was, as Ezra Warner has described him, "revolutionary, orator, ambassador, senator, editor, polemicist, and major general of United States volunteers" during the Civil War. Toward the end of his life, Schurz became "a tireless advocate of equal rights for the Negro, the suppression of the spoils system, anti-imperialism, and the preservation of the public domain" (Warner, *Generals in Blue*, 426–28).

37. Lewis, *W. E. B. Du Bois*, 366–67, 659 n. 48; Du Bois, *Dusk of Dawn*, 84.

38. Stone to Du Bois, May 19, 1906, with a follow-up dated June 5, 1906, both in DBP.

man, and several weeks passed by before he responded. Nevertheless, he did, and Stone's response to Du Bois made it obvious that he was eager to collaborate with Du Bois. "If I do not have an opportunity of seeing you in the meantime," Stone wrote on June 23, "I want to trespass on both your time and your good nature next winter, in the way of correspondence."

> There are a number of things about which I want to ask your opinion and advice, in connection with a study which I am trying to make. I am, and have been for quite a while, engaged in a treatise on "race relations."—"Race Relations in America" probably will be the title. I shall not attempt to go into details here, but I am making an earnest effort to get the point of view of the colored man,—to learn what he thinks and how he feels on the perplexing and ever widening problem. I shall try to be frank, without being offensive to either race, but, no matter what the result, I shall treat some phases of the question more plainly than they have yet been in any work with which I am acquainted. I shall be as honest as I know how to be. The book will employ the reflections and observations of a study which I first seriously began in 1890. Its dominant note will be that the "race problem" is more than a sectional or national [problem],—that it is the world problem of racial contact.[39]

Du Bois must have arched his eyebrows when he received Stone's letter, for here was someone, a white southerner at that, proposing to do precisely what Du Bois wanted to do. However, Stone did not mention the Carnegie Institution or even indicate an anticipated source of funding for his project, so Du Bois was probably unaware that they were competing for the same prize when he responded to Stone's letter.[40] "I shall be glad to help you in any way," Du Bois wrote in reply. "I hope you will have the moral courage, which is so fearfully lacking in the white South, to tell the truth & not to distort it against a helpless people."[41]

The exchange of cordial letters between Stone and Du Bois continued

39. Ibid., from Washington, DC, June 23, 1906, DBP.

40. In his biography of Du Bois (*W. E. B. Du Bois*), Lewis implies that Stone's letter of June 23, 1906, alerted Du Bois to Stone's appointment to the Carnegie Institution. However, reference to the Carnegie Institution does not appear in any of Stone's correspondence with Du Bois until after the institution's board meeting in the fall, at which time Stone's appointment was announced. In *Dusk of Dawn,* Du Bois states that the institution informed him of the status of his proposal indirectly through Stone but does not state when or how that communication occurred (84).

41. An undated draft of Du Bois' response appears on the back of Stone's letter dated June 23, 1906, DBP.

into September. Although a planned meeting between the two men in Atlanta to discuss topics of mutual interest failed to take place because of a delay in Stone's travel to Washington from Mississippi, Stone continued to request material from Du Bois, especially Du Bois' latest bibliography.[42] At some point, probably in early October, approval of Stone's appointment as a division head in the Department of Economics and Sociology was formally announced, and the relationship between Stone and Du Bois changed. Stone now had access to money, and Du Bois needed access to funds for his research.

The nature of Stone's appointment was described in the Carnegie Institution's *Year Book* for 1906. "A new division has been added," Colonel Wright announced, "making twelve in all, and this is entitled 'The Negro in slavery and freedom.' The work of this department is being conducted by Alfred Holt Stone, an educated business man from Mississippi, who is a thorough, impartial, and very candid student of the economic development growing out of negro slavery and the work of the negro under conditions of freedom." Colonel Wright noted that Stone had prepared a syllabus, which would be published and distributed in the near future. "Of course, not much has been done under this division on account of its recent constitution," Colonel Wright continued, "but Mr. Stone has outlined a treatment which is reasonably exhaustive, without, however, trespassing on either the political or social aspects of the topics." Once again, Colonel Wright was trying to make sure that political issues concerning the race problem would not find their way into Stone's work. Stone's study was to be purely economic in its scope. "Mr. Stone will make an effort to interpret the salient features of negro life in relation to their economic significance, both to the race and to the country as a whole, the purpose being to correlate the negro's economic history with that of the American people along certain broad lines, as, for example, through the cotton industry and in the creation of national wealth and favorable trade balances as affected by products closely identified with negro labor."[43]

Colonel Wright concluded his report on the Department of Economics and Sociology by emphasizing the role of the Carnegie Institution as a repository for archival data. "As I have stated in previous reports," he noted, "the

42. Stone to Du Bois in an undated fragment written in late July or early August, as well as letters dated August 7, August 30, and September 15, 1906, DBP. Notations in pencil and the draft of a brief response, probably in Du Bois' hand, appear on this correspondence.

43. Carnegie Institution, *Year Book, 1906*, 158, 163.

really important mission of the Department is to secure a great collection of materials which will be available to the historian not only of economic development, but to others."[44] The importance of the institution's role as a collector of material may have signaled Stone to redouble his efforts to acquire articles, pamphlets, and other material on the Negro race.

Once the announcement was made, Stone lost no time contacting other scholars who might be interested in joining him to document the role of African Americans in shaping the nation's economy. As is inevitably the case when someone has access to money, Stone was able to recruit a number of capable researchers and qualified assistants within months of his appointment. By March 1907, he had twelve: six academics, two independent scholars, a bibliographer, and a staff of three librarians.[45] All of the academic/scholar positions, including Stone's, were unsalaried and part-time, although the Carnegie Institution provided money to help cover expenses related to the cost of their research.[46] Two of the academics, Walter L. Fleming and Ulrich B. Phillips, were young lions who would go on to make their mark on the historiography of the period during their exceedingly prolific careers.[47] Two of the other academics were African Americans. One of the two, Richard R. Wright Jr., was the first African American to earn a doctorate in sociology at the University of Pennsylvania; he would later become president of Wilberforce University in Ohio.[48] The other African American on Stone's team was W. E. B. Du Bois.

Stone's collaboration with Du Bois is instructive for several reasons. For one, it illustrates the professional efficiency with which Stone organized and managed the project. Second, it provides evidence of Stone's ability to work with people who held views that were opposed to his. (It will be recalled that Stone's article "Italian Cotton-Growers in Arkansas" appeared during this period [February 1907] and that Du Bois wrote a critical letter to Stone in April

44. Ibid., 163.

45. Carroll D. Wright to "President [R. S.] Woodward," March 27, 1907, CIA.

46. Carnegie Institution, *Year Book, 1906*, 158.

47. Fleming was thirty-three years old and taught at West Virginia University in 1907 ("Walter L. Fleming (Walter Lynwood), 1874–1932," docsouth.unc.edu/neh/fleming/menu.html). Phillips, only thirty in 1907, was on the faculty at the University of Wisconsin (John David Smith, "W. E. B. Du Bois and Ulrich Bonnell Phillips: Symbolic Antagonists of the Progressive Era," in *Slavery, Race, and American History: Historical Conflict, Trends, and Method, 1866–1953*, by Smith [Armonk, NY: M. E. Sharpe, 1999], 24). The TSC has eleven publications authored by Fleming and seven by Phillips that were published during or before 1907.

48. "Savannah State University," www.georgiaencyclopedia.org/nge/Article.jsp?id=h-1428.

as a result.) Finally, Du Bois' cooperation with Stone demonstrates that one of the most militant of black activists of the time was not about to look a gift horse in the mouth.[49]

In a letter dated December 5, 1906, Stone initiated the idea that Du Bois do some work for Stone.[50] Du Bois responded on January 3 with a financial breakdown of what it would cost to conduct a study similar to "The Negro Artisan," one of his best, which Du Bois had published in 1902.[51] Stone was eager to close the deal and accepted Du Bois' proposal without reservation. "I need not say that I am anxious to see you undertake the study in question," Stone wrote to Du Bois from Dunleith. "I have carefully thought over the matter and shall accept your suggestion as to $1000.00[.] There is no question whatever as to this amount being entirely justified by the importance of the work we have in view. With me the sole consideration has been that of conservatively determining upon how much I would be justified in allotting out of the total sum at my disposal. This part, however, we may consider settled. I would be glad to have you advise me as to the manner in which you want the funds made available for your use."[52] Du Bois responded to Stone's letter with a formal acceptance and a schedule of payments. Stone accepted Du Bois' terms, and the first check for $250 was in Du Bois' hands by February 2.[53]

Bringing Du Bois on board was a personal triumph for Stone, and he had a lot to gain from its accomplishment. For one thing, his collaboration with Du Bois put to rest the contention that Stone was a negrophobist. How could a negrophobist work harmoniously and effectively with an African American known for his black pride? Vardaman certainly could not have pulled that off. In addition, Stone was getting a first-rate researcher. Du Bois was not only a sound methodologist when it came to studying social issues, he was very well connected to the black community, North and South. Du Bois could collect data and get information that Stone could never reach. As far as Du Bois being a mulatto was concerned, apparently Stone was willing to forget his as-

49. For the origin of this proverb, see Bartleyby.com at www.bartleby.com/59/3/dontlookagif.html.

50. Stone's letter did not survive. However, its existence can be inferred by Stone's prompt reply to Du Bois dated December 28, 1906, and confirmed in Du Bois' reply dated January 3, 1907, both in DBP.

51. Du Bois to Stone, January 3, 1907, DBP.

52. Stone to Du Bois, January 15, 1907, DBP.

53. Du Bois to Stone, January 19 and February 2, 1907; Stone to Du Bois from Dunleith, January 21, 1907, all in DBP. Stone sent Du Bois additional installments on April 27 and August 31, 1907, also in DBP.

sertion that people of a mixed-race ancestry could never speak with authority about the black masses Stone was familiar with on the plantations in the South.[54]

Du Bois had much to gain from his collaboration with Stone as well. His unsolicited proposal for funds from the Carnegie Institution had been unsuccessful, something that Du Bois had suspected as soon as he learned of Stone's appointment but became a reality in mid-January 1907, when J. Franklin Jameson, director of the Department of Historical Research, notified Du Bois that there was "no opportunity for doing the sort of work you mention in this department of the Carnegie Institution."[55]

If that news was not bad enough, Du Bois' customary source of funding had dried up as well. Trustees of Atlanta University were short of cash, and the ability of the university to support the twelfth in a series of prestigious Atlanta conferences scheduled for late May 1907 was in doubt. Ironically, political issues, which the trustees of the Carnegie Institution were so eager to avoid, may have played a role in the lukewarm response of the Atlanta University trustees to Du Bois' topic for the twelfth conference: African Americans in politics.[56] To his credit, Stone saw an opening and suggested that Du Bois change the topic to economic conditions in return for underwriting the conference. Du Bois bit, and the conference was held as scheduled on the topic "Economic Cooperation among Negro Americans."[57]

Stone's appointment as division head in the Carnegie Institution was announced in the New York newspapers. Usually, announcements of academic events elicit a yawn, but at least one person who read about Stone's appointment was outraged. "I sincerely trust that the report in the New York papers of March 10 [1907] to the effect that the Carnegie Institution has com-

54. Stone observed that 32.7 percent of African Americans in New England were of mixed-race ancestry as compared to only 11.5 percent in Mississippi ("Foundations of Our Differences," 41).

55. Jameson to Du Bois, January 14, 1907, JFJP. Jameson's blunt rejection of Du Bois' proposal was prompted by a letter that Du Bois had addressed to Jameson six days before. Presumably, Du Bois wanted a definite yes or no from the Carnegie Institution in regard to his proposal for a "history of the Negro race" before accepting Stone's offer. Du Bois' letter to "Mr. Franklin Jamison" [*sic*] is also in JFJP.

56. Lewis, *W. E. B. Du Bois*, 368.

57. The conference report was entitled *Economic Cooperation among Negro Americans. Report of a Social Study Made by Atlanta University, under the Patronage of the Carnegie Institution of Washington, D.C., Together with the Proceedings of the Twelfth Conference on the Study of the Negro Problems, held at Atlanta University, on Tuesday, May the 28th, 1907* (Atlanta: [Atlanta University Press], 1907).

missioned Mr. Alfred Holt Stone to make a study of the negro race during slavery and since the war is not correct," Oswald Garrison Villard wrote to Dr. Woodward, the institution's president, on March 20, 1907. Villard was the owner/editor of the *New York Evening Post* and grandson of the famous antislavery advocate William Lloyd Garrison.[58] "As a life-long student of the negro race, to whose development I am giving year by year more of my time and thought, I should protest very vigorously against the acceptance by the public of anything emanating from Mr. Stone's pen," he continued. "He has written much about the negro problem, and always from the prejudiced point of view. An instance of this is his recent article in the Review of Reviews on the question of negro and Italian labor in Mississippi. The fallacies and absurdities in this article were most admirably pointed out by Professor DuBois of Atlanta University in his recent address before the Ethical Culture Society in this city."[59]

Dr. Woodward responded promptly to Villard's letter. He had not seen the report in the newspaper, nor had he met Mr. Stone. However, he could refer Villard to the institution's *Year Book* for 1906, in which Colonel Wright explained what Stone was up to. "I am disposed to think that Col. Wright and his collaborators," Dr. Woodward reassured Villard, "would not choose a man for such important work without due consideration."[60]

If Dr. Woodward had not seen the report in the newspaper when he responded to Villard's letter, he saw it when Henry S. Pritchett, president of the Carnegie Foundation for the Advancement of Teaching in New York, sent the report to him in the mail. "Mr. Alfred Holt Stone, who is there mentioned as conducting the investigation," Pritchett explained, "has already expressed his attitude so plainly that it seemed to me doubtful whether he could attack this problem in 'a thorough, impartial and candid' way, as is intimated in the clipping, and I am sure that to commission such work to him would cause serious irritation among the more progressive of our negroes like Booker Washington and others who believe in a future for the negro."[61]

Dr. Woodward forwarded a copy of Villard's letter and his reply to Colonel

58. Villard's mother was Garrison's daughter ("Oswald Garrison Villard," www.spartacus.schoolnet.co.uk/USAvillard.htm).

59. Villard to Woodward, March 20, 1907, CIA. Villard suggested that someone like the Reverend Quincy Ewing would have been a better choice for this assignment.

60. Woodward to Villard, March 21, 1907, CIA.

61. Pritchett to Woodward, March 22, 1907, CIA. This letter is the second Woodward received from Pritchett about Stone's appointment. The March 22 letter, which Woodward received on March 24, is placed here chronologically because the first letter has been lost.

Wright on March 22. He also indicated that he had received a letter of inquiry from Pritchett. "I have thought that you should know of this matter in order that you may be able to meet whatever criticism may be raised," he advised. "I know nothing of its merits except what is disclosed in your report and in the letter of Mr. Villard. I assume, however, that you and your colleagues of the Department of Economics and Sociology have not entered into an engagement with Mr. Stone without due consideration."[62]

In the meantime, Villard had responded to Dr. Woodward's letter, noting that he had read Colonel Wright's report in the *Year Book* and "can see no objection to the plan as outlined or to the scope of the investigation." Nevertheless, Villard was unconvinced that Stone was the man for the job. "Judged by his previous work Mr. Stone is hardly the man to undertake such an investigation in the unbiased and scientific spirit which has heretofore actuated the work of the Carnegie Institution."[63]

Dr. Woodward had a public relations problem on his hands, and he tried to handle Stone's detractors with tact and firmness. He used firmness with Pritchett, another employee of the Carnegie trust, with whom he could be frank. "Whence the basis for this article came I am unable to say further than that the information did not come from our office. To what extent matters of the kind may be investigated at present is a matter of doubt, since we may easily waste more energy in friction than can be utilized in useful work. I am by no means satisfied, however, that we should be influenced greatly by the irritation of individuals, since if we gave heed to every such irritation we should certainly accomplish nothing."[64]

Dr. Woodward used more tact in his response to Villard's latest letter. "Please accept my thanks for you letter of the 23d inst. I regret to find, however, that you have made up your mind on the question of the competence of Mr. Stone to accomplish the work he has undertaken," Dr. Woodward wrote. "I have not met him as yet, but some friends who appear to be unbiased express opinions flatly in contradiction with your own. I shall try to weight these opinions pro and con in the scientific spirit which you are good enough to say 'has hitherto actuated the work of the Carnegie Institution,'" he assured Villard. "In any event I shall take pains to see that whatever work on the subject in question may be published is carefully reviewed and pruned of any gross errors."[65]

62. Woodward to Wright, March 22, 1907, CIA.
63. Villard to Woodward, March 23, 1907, CIA.
64. Woodward to Pritchett, March 25, 1907, CIA.
65. Woodward to Villard, March 26, 1907, CIA.

Colonel Wright's response to Dr. Woodward's query arrived two days later. Wright began by emphasizing that Stone was not charged with studying the "race problem." Instead, he was collecting data in regard to "the economic influence of the negro in slavery and freedom on the development of the South," Colonel Wright wrote. "He was selected after most careful consideration and thorough recommendation on the part of Dr. Walter F. Willcox, one of our Collaborators. His name was presented but before action was taken careful inquiry was made as to his ability, integrity, etc., and these having been proven satisfactorily he was unanimously selected." Wright then listed the members of Stone's team, taking pains to point out that two of them (R. R. Wright and W. E. B. Du Bois) were "colored." "Looking at this list I see no opportunity to charge Mr. Stone with prejudice for he has employed some of the very best men who are studying the economic attitude of the negro, both white and colored assistants," Wright noted. "It is absolutely essential that the person in charge of this work relative to the economic value of negro labor should be a white man and a southerner. No colored man or northerner could expect to succeed along the lines of our investigation. Mr. Stone recognized the necessity also of having some colored men to assist him. The prejudices of the two races had to be recognized, and we thought we had made the best combination possible. I think we need have no uneasiness about Mr. Stone," Wright concluded. "I have followed his methods and believe in him."[66]

Dr. Woodward responded to Wright's letter almost immediately, noting that he had recently met with Stone and also that he had sat down for a long conference with Willcox. The result of the two meetings was clear. "My conclusion is that you are quite justified in employing Mr. Stone as the most competent man available for this difficult work," Dr. Woodward wrote. "He appears to me to be exceptionally well qualified for his work."[67]

Stone had won. The endorsement of the president of the Carnegie Institution coupled with a vigorous defense of Stone's qualifications by the department head should have been enough. But winning was not enough, because Willcox was still irked by the criticism of his protégé. "The more I have thought about your statement that certain persons have objected to the appointment of Mr. Stone as a collaborator in our economic history of the United States," Willcox wrote to Dr. Woodward on April 2, "the stronger grows my conviction that the persons so writing ought to be asked to specify

66. Wright to Woodward, March 27, 1907, CIA.
67. Woodward to Wright, March 28, 1907, CIA.

the opinions he has expressed which in their judgement render him ill qualified to do sound and impartial work in this difficult field. It seems to me wrong that his reputation should be endangered and no opportunity for answer or explanation afforded."[68]

Unbeknownst to either Willcox or Dr. Woodward, another critic of Stone's appointment had written Colonel Wright with specific reservations about Stone's appointment. "I see in the newspapers and also in one of the publications of the Department of Economics and Sociology of the Carnegie Institution that Mr. Alfred H. Stone is to make an investigation covering the life of the negro in slavery and freedom as far as the economic elements are concerned," Booker T. Washington wrote to Colonel Wright on March 26, 1907.

> I think I have read with some care almost everything that Mr. Stone has printed on this subject, and I think I am safe in saying that without exception there has been but one conclusion to his investigations, and that is in plain words, to damn the Negro. There are very few white men who have been reared in the South who would be capable of not putting prejudice into such an investigation, and I repeat that I feel quite sure judging by Mr. Stone's previous utterances, that he is not the man to make such an investigation, I mean that will demand respect and confidence. He and I have had considerable correspondence on the subject of the progress of the Negro and I know him pretty well. I have tried over and over to get him to come to Tuskegee or go to Hampton or some other institution and see something of the effects of rational education on the life of the Negro, but I have never succeeded in getting him to go where he could see for himself what is being accomplished. I think the man, in the second place, has been soured by failure in cotton raising in Mississippi.[69]

Dr. Woodward and Willcox did not see this letter because Washington had asked Wright to keep their correspondence confidential.[70] Con-

68. Willcox to Woodward, April 2, 1907, CIA.

69. Washington to Wright, March 26, 1907, BTWP (9: 233–34). The editor's notation for this letter in BTWP states that a copy of this letter is in the Oswald Garrison Villard Papers at Harvard, which suggests that Villard may have played a role in Washington's complaint.

70. Washington began the letter to Colonel Wright as follows: "My dear Mr. Wright: I want to write you regarding a matter which for the present I wish you would consider as confidential, and which I feel deeply concerned about." Colonel Wright honored Washington's request but defended Stone in a letter to Washington dated April 3, 1907 (BTWM [reel 286]). Colonel Wright's letter to Washington contains much of the same language he used in his letter to Dr. Woodward. Colonel Wright followed up with a second letter to Washington dated April 15, 1907 (also BTWM [reel 286]), after meeting with Stone. "I assure you I can find no reason why

sequently, Dr. Woodward ignored Willcox's insistence that Stone's critics be held accountable and responded instead with an unequivocal endorsement of Stone's qualifications. "My impression is that you [Willcox] and the members of the department of Economics and Sociology are quite right," Dr. Woodward declared, "in your estimate of the value of Mr. Stone's work."[71]

One can imagine Washington's uneasiness when, about two weeks after he had written his confidential letter to Colonel Wright, he received a letter from Stone asking him to join his team in the study of the economic impact of African Americans as slaves and freedmen.[72] Stone's letter placed Washington in a difficult position. He had just told Colonel Wright that Stone was "not the man to make such an investigation." Yet, Andrew Carnegie was one of Tuskegee's most generous benefactors, and Stone was acting as an agent of the Carnegie Institution.[73] Washington really had no choice. "I have received your letter of April 10 and am writing to say that, I shall be very glad to cooperate in any manner I can with you or others in covering the subject that you have been requested to deal with by the Carnegie Institution," Washington began in his response to Stone. "I must say in this connection, however, in order to be perfectly frank with you, that I have had fears in the past which I think I have not withheld from you, that some things you have spoken and written have been rather unfair to the Negro, growing out of the fact that you had not come in contact with the results of rational education."

> In most I have read from your pen, the conclusion seems to have been reached by you that the Negro was a failure in the labor world. I may be wrong, but this is the impression I have formed. I think you have reached this conclusion very largely owing to the fact that your contact has been with a certain element of colored people in Mississippi. I believe that a broader contact with the whole problem will give you a different perspective. . . .

Mr. Stone is not doing the work with justice to the Negro and sympathy for him," he informed Washington.

71. Woodward to Willcox, April 14, 1907, CIA.

72. Stone to Washington, April 10, 1907, BTWM (reel 284).

73. For an example of Andrew Carnegie's endorsement of the work being done at Tuskegee and similar institutions, see Andrew Carnegie, *The Negro in America: An Address Delivered before the Philosophical Institution of Edinburgh, 16th October 1907* (Philadelphia: E. A. Wright Bank Note Co. for the Committee of Twelve for the Advancement of the Interests of the Negro Race, [1907?]), 32 pages. There is a copy of this article in TSC.

I hope before you are through with your investigation that you will visit Hampton and Tuskegee and similar institutions. I hope you will also go into some of the small communities where the graduates of these institutions have been exerting their influences.

What I am most interested in is facts as to our real progress. I do not want to cover anything over. No one would suffer more than the Negro by a false view being taken simply for the purpose of pleasing the race.

I repeat, if I can serve you, please be kind enough to let me know.[74]

"I have your valued letter of April 19th," Stone replied, "and beg to thank you both for your frank criticisms and for your expression of willingness to co-operate with me."[75] Stone went on to say that he did not think that Washington's criticism was justified. He had not written anything to suggest "that the Negro is in my opinion 'a failure in the labor world.'" The problem was one of misinterpretation, Stone suggested. His observations had dealt with "a specific class of Negro labor,—that living under the plantation system." His focus had been on "local conditions," not African Americans in general. Nevertheless, Stone was looking forward to expanding the scope of his studies with Washington's cooperation. Stone had been to Hampton Institute the week before, where he had made arrangements with Thomas Jesse Jones, a white professor and the institute's chaplain, to work with him on the project.[76] Stone was convinced that "the influence of the most important Negro Industrial Schools," presumably Hampton and Tuskegee, would open doors for their research. "I appreciate what you say in conclusion," he wrote, "and trust that no one will have any occasion to feel that I have been unfair in my treatment of the subject."

Washington began his reply to Stone's letter with "My dear Mr. Stone:—I have your letter of April 26th and am very glad to receive it."[77] After Washington assured Stone that he wanted to be of service, he returned to the subject of fairness. "I do not want any thing over or under stated regarding the

74. Washington to Stone, April 19, 1907, BTWP (9: 263–65). There is also a copy of this letter on microfilm in BTWM (reel 284), but it is almost impossible to read.

75. Stone to Washington, April 10, 1907, BTWM (reel 284).

76. Thomas Jesse Jones was a Welshman who had immigrated to the United States at an early age. He was later to gain prominence as the foremost advocate for teaching social sciences to African Americans so that they could assimilate more easily with Americans with an European ancestry. W. E. B. Du Bois rejected Jones's educational philosophy because he disliked its accommodationist stance (see Donald Johnson, "W. E. B. Du Bois, Thomas Jesse Jones and the Struggle for Social Education," *Journal of Negro History* [Summer 2000]: 71–95).

77. Washington to Stone, May 1, 1907, BTWM (reel 284).

Negro, simply the facts, because in the last analysis, anything, other than the truth, at this time, would be more harmful than helpful." At least on the surface, Washington wanted Stone to know that he would be a team player. "I thought it would be well to write you frankly, as I did, so that we could understand each other," he concluded, "and I find that I am not mistaken. I think we can cooperate and be mutually helpful."

Despite his frank feedback and obvious reservations, Washington had acquiesced to Stone's request, and Stone now had both wings of the black perspective in the United States cooperating with his research initiative. His appointment as a division head in the Carnegie Institution was the key. It established Stone as a recognized authority on the race problem, it cemented his alliance with powerful men, and it provided him with funding for his work. In short, it placed Stone at the pinnacle of his career as a racial theorist.

11

SEVERE AND DISCRIMINATING CRITICISM

Alfred Holt Stone's appointment as division head for the Carnegie Institution brought him attention, funding, and requests for his work. Stone was becoming well known as a racial theorist, and he wanted to take advantage of his standing. "During the past few weeks I have received a number of inquiries for copies of various papers of mine,—none of which I can supply," Stone wrote to Willcox in early December 1906 from his A Street address in Washington. "Since leaving you the other night," he continued, "the idea occurred to me that it might not be a bad scheme to bring these papers together in the shape of a small volume,—provided I could find a publisher. I wish you would tell me frankly what you think of this suggestion. It seems to me that I might at least thus secure what I would very greatly value,—some severe and discriminating criticism."[1]

Stone did not think that pulling together some of his papers for a new book would interfere with the one he was obligated to write for the Carnegie Institution. "This scheme would mean no delay in my other book," he wrote. "The matter is practically all in print, and could be easily assembled. I could have it indexed at small cost, and without drawing upon my own time. The criticism which I might get would be useful to me in my present work," he added.

Stone had the papers he wanted to publish in mind, and he listed ten of them for Willcox. His plan was to reprint the papers exactly as they appeared when he initially offered them to the public. "My purpose would be to give

1. The quotations in this and the next two paragraphs are from Stone to Wilcox, December 3, 1906, WFWP. As far as Stone's desire to secure severe criticism was concerned, W. E. B. Du Bois' sharp letter in response to Stone's article "Italian Cotton-Growers in Arkansas" and Booker T. Washington's admonishment in his letter agreeing to work with Stone were yet to come. Du Bois and Washington's letters were dated April 13 and 19, 1907, respectively.

the date at which each of the papers was written," he wrote, "explaining that each represented my views at that time, or described then existing conditions. For example, I would not re-write any part of the paper on the negro in the Delta,—though I would state that I could not give the same description of conditions there, particularly as regards to race relations."[2] However, Stone was still struggling with the title, and "The Negro and the Race Problem: Miscellaneous Papers" was the best he could come up with.

Willcox liked the idea of the book, but not the title. "I am heartily in favor of your suggestion to bring together and publish at once your papers dealing with the negro question," he replied. "I cannot see how the publication of such a book could affect unfavorably the book you are now working upon, unless indeed you are planning to incorporate in that [the book for the Carnegie Institution] large fragments from your papers already published." As far as the title was concerned, "How would 'Negroes and Whites in the South Since the War—Addresses and Studies' do?"[3]

Encouraged by Willcox's response, Stone thought that it might be a good idea to have Willcox write an introduction for the new book. However, he did not think much of Willcox's suggestion for a title. "What do you think of this?" Stone asked: "'The Mulatto Factor in the Race Problem: And Other Papers.'"[4] Willcox readily agreed to write an introduction, but he passed up the opportunity to suggest another title, probably because he had something else on his mind. "Let me make a counter suggestion [in addition to the idea of writing an introduction], asking you to express your opinion with entire frankness," he wrote. "What would you think of including, say in an appendix, my paper on Negro Criminality and the address I gave at Montgomery?" Willcox asked. "I mention this because I have frequent requests from persons, particularly persons in the south, for one or another, or both of these papers, and I cannot supply them. My other papers on this subject are in easily accessible sources, but these, especially the latter, cannot easily be found, even in ordinary libraries or in the market. If you have any preference for having the book entirely your own writing do not hesitate to say so."[5]

2. True to his word, Stone added a footnote to the paper when it appeared as a chapter in *Studies in the American Race Problem*. "Writing to-day, 1908, it would be necessary to modify this statement somewhat—certainly for some parts of this territory," he noted in reference to the sentence, "Yet there is no more feeling of fear on the white man's part whether for himself or his children, than in the days of slavery," in the text (91).

3. Willcox to Stone, December 5, 1906, WFWP.

4. Stone to Willcox from Washington, DC, December 8, 1906, WFWP.

5. Willcox to Stone, December 10, 1906, WFWP.

Stone's letter in response to Willcox is missing, but it is clear from Willcox's letter to Stone on December 15 that Stone had agreed to include Willcox's two papers in the yet-to-be-titled book. Sensing an opportunity to include more of his work, Willcox then asked, "What do you think of adding also my brief reply to Gov. Vardaman published in Leslie's 'Weekly'?"[6]

Stone did not think it wise to include Willcox's objection to Vardaman's misappropriation of his data. "My idea is to avoid everything at all calculated to give the book a controversial tone," Stone wrote. In fact, he had decided to ban Vardaman's name from the book.[7] But Stone did like the suggestion Willcox had made to replace his address at the Montgomery Conference with an article entitled "The Probable Increase of the Negro Race in the United States," which Willcox had written for the *Quarterly Journal of Economics*.[8] Willcox had toned down his rhetoric in the later piece, although he was still pessimistic in regard to the future for African Americans. "Relatively to the whites in the south, if not absolutely as measured by any conceivable standard," Willcox had written in 1905, "the negro as a race is losing ground, is being confined more and more to the inferior and less remunerative occupations, and is not sharing proportionately to his numbers in the prosperity of the country as a whole or of the section in which he mainly lives."[9]

Stone's new book project was moving forward with remarkable speed, for within two weeks of broaching the subject, Stone had a draft of Willcox's introduction in hand. The introduction began with an account of Willcox's meeting Stone at the conference on the race problem in Montgomery in May 1900. "From no one have I learned so much as from him about the realities of the race problems in the South," Willcox stated, "and I gladly avail myself of this opportunity to bespeak for his contributions to the subject the attention and the study which their importance deserves." Willcox then remarked on the significance of two men, one from Massachusetts and the other from Mississippi, joining hands "in objective and dispassionate studies of the questions which so long have divided those states and the country."[10]

6. Ibid., December 15, 1906, WFWP.

7. Vardaman's name does not appear in the book.

8. Stone to Willcox from Washington, DC, December 16, WFWP.

9. Walter F. Wilcox, "The Probable Increase of the Negro Race in the United States," *Quarterly Journal of Economics* 19 (August 1905): 545–72. The copy of the article that Willcox sent Stone, the one that Stone copyedited for publication, is in TSC. The quotation can be found on the last page.

10. The quotations come from an undated typescript of the preface that Willcox sent Stone on December 18, 1906. The date of typescript can be determined from a letter that Willcox wrote to Stone the next day. Both the typescript and letter are in WFWP.

> The closet student here [Willcox] will be found not widely at variance in his conclusions from the cotton planter and business man who approaches the subject by a different avenue. The statistician[,] knowing little of the problem beyond what he may read in his figures, finds in Mr. Stone's papers a needed complement, ~~but~~ [and] also in the main I ~~am glad to~~ believe a confirmation of his results. Thus, north and south, theory and practice, figures and concrete specific experience are approaching a unity of conviction. The essays contain no plan of action but are confined to the necessary preliminary work of portraying conditions. Agreement must be reached on these before agreement on action can come into sight.[11]

Stone liked Willcox's proposed contributions to the book and decided to eliminate one or two of his chapters in order to keep the bulk of the volume in check. And slash he did, but he was still having trouble with the title. "What do you think of this?" Stone asked: "'Race Problem Contrasts and Parallels; and other Papers.'" Time was getting short, and Stone was eager to tie up loose ends before he left Washington for Dunleith at the end of December. "I see now that it will be impossible to get the material in the hands of the publisher before I leave," he wrote to Willcox on December 22. "I have worked about eighteen hours a day for the last ten days, on this and the Carnegie business, but I cannot get through in time."[12]

Willcox opened Stone's letters when he returned from a short break for Christmas, and there was little he could add to what Stone had already done, except for a suggestion in regard to the title. "About the title, I hardly feel able to advise, and shall be quite willing to agree to your judgment," he wrote to Stone at Dunleith on January 2, 1907. "The only thing that occurs to me is 'Studies in the American Race Problem.' This implies that the adjustment of the relation between negroes and whites is so much more important than any other race problem that it is entitled to be called 'The American Race Problem,' and the collection makes no claim to deal with the problem in its entirety."[13]

Stone liked Willcox's suggestion for the title and continued to work on

11. The draft of the introduction that Willcox sent Stone in December 1906 appeared in print as it was written, with the exception of copyediting as indicated by strike-throughs and brackets. Willcox's use of the term "closet student" to describe himself probably refers to his academic as opposed to actual experience with the race problem.

12. Stone to Willcox from Washington, DC, dated December 20, 22, and 28, 1906, all in WFWP. It will be recalled that Stone was also putting together a research team for his division in the Carnegie Institution during this time.

13. Willcox to Stone, January 2, 1907, WFWP.

the book at Dunleith, particularly on three new chapters he was writing for the volume. One, entitled "Foundations of Our Differences," was an elaboration of the points he had made in the talk he had given at Cornell in 1905. The other two chapters dealt with the role of African Americans in politics: "Mr. Roosevelt, the South, and the Negro in Politics" and "The Negro in Politics."[14] Apparently, these new chapters gave Stone an outlet for thoughts and opinions that Colonel Wright had prohibited in the volume he was writing for the Carnegie Institution.[15]

Stone had a manuscript ready to send to a publisher on March 11. "Yesterday I sent my ms. to Walter H. Page (Doubleday, Page & Co.), and on Monday [March 18] I shall leave for Washington," Stone informed Willcox in a letter.[16] Stone enclosed a copy of the title page, which followed Willcox's suggestion in regard to the title, and a table of contents. Although three of Stone's chapters and one of Willcox's would be deleted and another by Stone added before the book was to be published, the contents of the volume were essentially in place when Stone submitted his manuscript. "I have made the collection somewhat more elaborate than I had intended,—if I may use the word," Stone explained. "I wrote the Roosevelt and Negro in Politics paper here,—and made them much fuller discussions than I had expected. My reason is this: I must devote practically all my time for a year or more to my Carnegie work. I don't think it will be possible for me to finish my study of Race Relations this year,—if I do the work I want to for Col. Wright. It occurred to me that by enlarging the scope of this volume of papers, its publication would give me a breathing spell for the other study."[17]

Stone needed a breathing spell because "the other study" he was doing for the Carnegie Institution was a substantial endeavor: a volume in the institution's series of books on the economic and industrial history of the United

14. "Mr. Roosevelt, the South, and the Negro" and "The Negro in Politics," both in *SARP*, 242–350, 351–421, respectively. Portions of the chapter on Roosevelt was reprinted in 1916 as "The Negro Problem in the United States: The Psychological Influence of Reconstruction," in *Readings in Social Problems*, ed. A. B. Wolfe (Boston: Ginn, 1916), 665–76.

15. See, for example, Stone to Willcox from Dunleith, February 4 and March 12, 1907, WFWP.

16. Whether intentionally or by coincidence, Stone sent his manuscript to the same company that had published Booker T. Washington's *Up from Slavery* in 1901 and Thomas Dixon's two controversial novels, *The Leopard's Spots: A Romance of the White Man's Burdens* and *The Clansman: An Historical Romance of the Ku Klux Klan*, in 1902 and 1905, respectively. *The Clansman* later became the basis for D. W. Griffith's highly successful film *The Birth of a Nation* in 1914.

17. Stone to Willcox from Dunleith, March 12, 1907, WFWP.

States. The idea for a comprehensive history of the United States' development as a major industrial power had been kicking around in the Carnegie Institution for several years but finally got off the ground when Colonel Wright presented the idea to attendees at the American Economic Association's meeting in New Orleans in 1903. The response had been encouraging, and Colonel Wright followed up with a detailed prospectus at the association's meeting in Chicago the next year. The work was to consist of eleven volumes, with each volume corresponding to a division in the institution's Department of Economics and Sociology. The respective heads of each division were responsible for the content of their volumes. Walter F. Willcox was to oversee the writing of the volume on "Population and Immigration." Colonel Wright was to work on a volume concerning "The Labor Movement," and Henry W. Farnam, Stone's sponsor, was responsible for a volume on "Social Legislation, including Provident Institutions, Insurance, Poor Laws, etc."[18] Stone's appointment had necessitated the creation of a twelfth division, and the title of the volume under Stone's direction was to be that of his division, "The Negro in Slavery and Freedom." (Given Stone's penchant for working on several projects at the same time, linking his books to his collection efforts can be a challenge. See appendix C for an explanation.)

The rough draft of a syllabus that Stone had prepared for Colonel Wright was essentially a list of topics he planned to study, such as "the economic effect of emancipation" and "the effect of the 'slave system,' . . . upon the economic life of the Southern States, as compared with the industrial development of other sections of the country."[19] After he was appointed as division head, Stone developed a full-blown syllabus that became an outline for the book he was to write for the institution. In it, Stone divided the economic contribution of African Americans into four periods: 1619–1793, Beginnings of Slavery; 1793–1860, Slavery as an Economic System; 1860–1880, From Slavery to Freedom; and 1880–1908, Era of Industrialism. Collectively, the four periods encompassed a total of thirty topics he planned to cover in the book, such as "the African and domestic slave trade as related to the increase of American slavery" and "the rise and development of the negro industrial school."[20] Individually or in combination, twenty-two of the twenty-seven topics in Stone's rough draft found their way into the syllabus.

18. Carroll D. Wright, "An Economic History of the United States," *Publications of the American Economic Association*, ser. 3, 6 (May 1905): 160–64.

19. Stone to Willcox from Washington, DC, May 12, 1906, with an enclosure, WFWP.

20. Carnegie Institution of Washington, *Year Book, 1907* (Washington, DC: Judd and Detweiler for the Institution), 74–75.

However, fourteen of the thirty topics in the syllabus were new, six of which fell into the 1619–1793 period (Beginnings of Slavery). Stone also added five new topics to the Era of Industrialism (1880–1908), which suggests that Stone wanted to strengthen those periods with which he was not as familiar. Colonel Wright was pleased with Stone's syllabus. "The scheme submitted herewith is intended to outline a reasonably exhaustive treatment of the economic life of the American negro without trespassing upon either its political or social aspects," Wright wrote as a preface to the syllabus when it was printed in the Carnegie Institution's *Year Book* for 1907.[21]

Stone's syllabus provided an outline for the work he intended to accomplish, but now he needed sources for his work. Extensive collections in research libraries today that contain manuscripts and other material concerning African Americans and slavery did not exist in 1907. Stone attempted to overcome the lack of primary sources by developing an elaborate request for material, which he printed under the title *Material Wanted for an Economic History of the Negro.* The pamphlet consisted of sixteen pages, fourteen of which contained 105 queries that provided concrete examples of the type of material he wanted. "The writer is trying to secure original material for use in writing an economic history of the negro," Stone began in a preface to the pamphlet, "both during slavery and since emancipation. Such a history will of necessity also touch the life of the Southern white man." He wanted to correspond with anyone who might have access to this type of material or knew someone who did. "I shall be grateful to any one into whose hands this circular may fall either for information on any point outlined here or for a word of suggestion as to the whereabouts or ownership of documentary material bearing on such points." Stone ended the preface with a personal note: "I may add that I am a Southern man and that my interest in the subject is genuine. I am temporarily residing in Washington in order to avail myself of its library facilities in connection with this work."[22]

Departing from the organization he had used for his syllabus, Stone divided the *Material Wanted* pamphlet into five subject areas: (1) the antebellum period (2) free Negroes (3) the Negro during the Civil War (4) emancipation and Reconstruction, and (5) the Negro since Reconstruction. Each subject area was defined by a subset of queries. For example, under the antebellum period, he asked, "What were the amusements and recreations of

21. Ibid., 73.

22. Alfred H. Stone, *Material Wanted for an Economic History of the Negro* (N.p., n.d.), 16 pages.

slaves, and what restrictions were placed upon them as to the use of firearms?" Under "free Negroes," he inquired, "To what extent did they become slaveholders and planters?" He also wanted to know, "How far were plantations disorganized, and how did the negro work in absence of owner and overseer?" during the Civil War. As far as Reconstruction was concerned, Stone asked, "Aside from political considerations, what part did the Freedmen's Bureau play in the adjustment of the new economic relations between whites and negroes?"[23]

Given its broad scope, the *Material Wanted* pamphlet included topics that had not appeared in either the rough draft or the syllabus. In addition, topics that had been on Stone's list from the beginning were restated in much more specific terms. For example, "The Freedmen's Bureau as a factor in the evolution of negro labor," which appeared as a single item in the syllabus, was given an operational definition in Stone's request for material that took eight questions to explain.

The historian John David Smith credits Stone with having made two valuable contributions to the historiography of slavery by publishing the *Material Wanted* pamphlet. First, Stone had initiated a search for documentary records "before the establishment of the great archival repositories at Chapel Hill [University of North Carolina], Durham [Duke University], and Baton Rouge [Louisiana State University]." Furthermore, the pamphlet provided an exhaustive list of the types of documentary records researchers needed. Stone's second contribution resulted from the sophistication of the queries he formulated in his search for documents. "These [questions] revealed probing mind, one which viewed slavery not as a monolith, but rather as a fluid system of racial and economic control," Smith wrote. "Although rigidly conservative on the race question of his day, Stone saw considerable flexibility in the slavery system of the old South. The queries included in his pamphlet, quite surprisingly, prefigured much later scholarship. Stone in fact posed many of the questions which still elude modern [ca. 1983] historians of slavery. That he asked them at all is a measure of Stone's importance."[24]

Despite the importance of Stone's groundbreaking work, his contribution has been largely ignored, and several questions concerning the undated *Material Wanted* pamphlet remain unanswered. For example, when was the

23. Ibid., 2–16.

24. John David Smith, "Alfred Holt Stone: Mississippi Planter and Archivist/Historian of Slavery," *Journal of Mississippi History* 45 (November 1983): 262, 266; reprinted in *Slavery, Race, and American History*, by Smith, 55–60.

pamphlet published, who paid for it, and what happened to the documentary material that came into his possession? The first two questions are the least important, and, as fate would have it, the easiest to answer. The key to answering both questions can be found in the address Stone used in the preface to the pamphlet: 124 Third Street N. E., Washington, D.C. This address appears only once in the correspondence that has survived between Stone and Willcox, Stone and Du Bois, or Stone and Washington. It is printed on the stationery of a letter Stone wrote to Du Bois on July 3, 1907, telling him of the book he was in the process of writing as part of the Carnegie Institution's economic and industrial history of the United States. "I am trying to locate documentary manuscript and other material bearing particularly on the ante bellum and earlier post bellum aspects of the subject," he wrote, "such as journals, diaries, account books, account sales, cotton picking records, instructions to overseers, etc., etc." Stone was very specific in regard to what he wanted. "The enclosed circular will indicate the field which I am trying to cover."[25] There seems little doubt that the circular he was referring to was the *Material Wanted* pamphlet, and the date of the letter would suggest that it had been printed during the late spring or early summer of 1907.[26]

In Stone's letter to Du Bois, 124 Third Street N.E. is referred to as a "temporary address." However, other letters Stone wrote during this period make it clear that he was not in Washington when he wrote it. Stone was in Madison, Wisconsin.[27] Ulrich B. Phillips's biographer, Merton L. Dillon, notes that Stone and Phillips were good friends and close colleagues who shared common views in regard to African Americans. Dillon states that at one point Stone stood in as Phillips's temporary replacement at the University of Wisconsin.[28] Whatever the reason Stone was in Madison during the summer of 1907, his absence from Washington would have made it necessary

25. Stone to Du Bois, July 3, 1907, DBP. This piece of correspondence was a form letter. Stone used identical language in a letter he wrote to the historian William E. Dodd a week later (dated July 10, 1907, in William E. Dodd Papers, Library of Congress, quoted in Smith, "Alfred Holt Stone," 264).

26. Stone probably submitted his syllabus for inclusion in the Carnegie Institution *Year Book No. 6, 1907* in late September or early October 1906. The *Material Wanted* pamphlet would have required substantial work beyond what Stone had done for the syllabus, which would have delayed its printing until the winter of 1907 at the earliest.

27. Stone to Du Bois, May 23 and August 8, 1907, both in DBP.

28. Merton L. Dillon, *Ulrich Bonnell Phillips: Historian of the Old South* (Baton Rouge: Louisiana State University Press, 1985), 63–64. Merton places Stone in Madison as Phillips's replacement two years later, but Stone's correspondence with Willcox in 1909 makes that chronology unlikely.

for him to have an address where material he was collecting for his various projects could be processed and held until he returned. It appears that the address at 124 Third Street N.E. served that purpose.

The question of who or what was at that address can be answered by reference to a letter in the archives at the Carnegie Institution.[29] On September 10, 1907, R[obert]. H. Johnston wrote to the Carnegie Institution "in connection with some work I am now doing for Mr. A. H. Stone of the Dept. of Economics and Sociology." Johnston asked for one hundred sheets of Carnegie Institution letterhead and a similar number of envelopes. Johnston's stationery identified him as specializing in "cataloging, indexing, translation, compilation and bibliographic work." His address was 124 Third Street N.E.[30]

Johnston was one of the twelve members of Stone's team named in Colonel Wright's letter to Dr. Woodward in March 1907. Without a doubt, Stone hired Johnston as his bibliographer and gave him responsibility to receive and process documents related to Stone's work on the book for the Carnegie Institution. He also paid Johnston to help him in other ways, such as verifying the references in the collection of papers Stone was hoping to publish as a book and preparation of the index.[31] Johnston's letter also suggests that he played a role in soliciting material, probably by sending letters to dealers who specialized in manuscripts and other historical documents. As a permanent resident of Washington, Johnston was a good person for Stone to use for his "temporary" address.[32] Johnston's letter also strengthens the assumption that Stone paid for the *Material Wanted* pamphlet with funds provided by the Carnegie Institution. That much we know in regard to the questions of when the pamphlet was published and who paid for it.

The larger and much more important question of what happened to the

29. The author would like to thank John David Smith for allowing him access to this correspondence. Stone's folder at the Carnegie Institution was misplaced after Professor Smith photocopied and returned it several years ago. Consequently, it was not available for this researcher to review when he asked to see it on a trip to Washington in February 2005.

30. Johnston to "Secretary, Carnegie Institution," September 10, 1907, CIA.

31. Stone thanked Johnston for these contributions in the preface to *Studies in the American Race Problem*, xiii.

32. Johnston eventually left the Library of Congress to oversee the library for the Bureau of Railway Economics in Washington. In that capacity, he became president of the Special Libraries Association for the 1914–15 term ("Special Libraries Association: Fortieth Anniversary, 1909–1949," *Special Libraries* 40 [April 1949]: 126). Johnston continued to work for Stone into the 1920s, sending him articles related to African Americans and cognate subjects (Memo for "Dr. Stone," May 18, 1925, AHSC).

documents that Stone gained access to as a result of his request for material is more difficult to answer. Although Stone was loquacious in his letters when it came to explaining what he wanted to collect, he failed to tell people what he found. The omission is puzzling, especially in regard to his letters to Willcox, because Stone consulted Willcox on so many other things related to his research. Nevertheless, the correspondence that has survived is silent in regard to his collections. Although we know for a fact that Stone amassed a large collection of printed material, it is less clear whether Stone's effort to collect manuscripts and other primary source material for his volume in the Carnegie series was successful.

If Stone did find any manuscripts, he did not use them in the papers he wrote after distributing the pamphlet. In 1909, for example, he wrote a chapter entitled "American Negro Origins" for *The South in the Building of the Nation*. According to the chapter's bibliography, the sources Stone used were standard published works, such as Du Bois' *Souls of Black Folk*, Hoffman's *Race Traits and Tendencies of the American Negro*, and Washington's *Up from Slavery*. Stone also cited a number of other books and census bulletins plus unnamed publications "of the various American learned societies, and articles under the title 'Negro' in American and English encyclopedias."[33] The bibliography makes no reference to manuscripts, although the topic was easily encompassed by queries in the *Material Wanted* pamphlet.[34]

Timing may be the most important factor to consider in this regard. As we shall see in the next chapter, the productive years of Stone's career as a racial theorist ended in 1909. That being the case, Stone had about a year to collect manuscript material, as opposed to the four years he had to work on the collection of printed material that now resides at the Mississippi Department of Archives and History. And one can assume that tracking down one-of-a-kind manuscripts was significantly harder than obtaining copies of

33. Alfred Holt Stone, "The Negro in the South: American Negro Origins," in *The South in the Building of the Nation*, 12 vols. (Richmond, VA: Southern Historical Publication Society, 1909), 10: 183.

34. For example, entry number 42 under "The Ante Bellum Period" in the *Material Wanted* pamphlet states, "I would like to secure even meager information touching various tribal groups of negroes which remained in tact for a number of years as, for example the 'Moors,' in Delaware, and the 'Malagasy,' or 'Mollygasters,' in Virginia" (8–9). On page 169 of "American Negro Origins," Stone wrote, "At home they [slaves from Madagascar] were known as 'Malagasy,' and the writer has found numerous individuals of this strain who had some vague, traditional knowledge of their origin, usually indicated in the persistence of their original designation, under some such corrupt form as 'Mollygaster' or 'Mollyglaster.'" Of course, Stone may have used manuscript material but failed to acknowledge it in the bibliography.

printed material. Whatever the reason, it appears that Stone did not meet with much success during the time he had available to him to collect plantation records and other documents related to slavery.

On the other hand, one can not help but think that some things had to come his way. Stone was too well known by this time, his connections were too extensive, and his work habits were too efficient for him not to have found something. But if he did, what happened to it? One possibility is that he passed whatever he collected along to someone else. In fact, he may have traded manuscripts for printed material. Given his phenomenal success with collecting articles, pamphlets, and monographs, such an arrangement is a good possibility. But if that is true, who got Stone's stuff?

An excellent candidate for this lucky role would be Ulrich B. Phillips. Seven years younger than Stone, Phillips was a whiz kid when it came to research on southern topics, especially those dealing with slavery. Phillips grew up in Georgia, not far from the Alabama line, and as a result he was familiar, as was Stone, with the race problem as viewed from the white perspective. Phillips had earned a doctorate in history from Columbia University in 1902 after completing undergraduate and master's degrees at the University of Georgia. By the time he joined Stone's team in 1907, Phillips had written several articles and a book entitled *Georgia and States Rights,* which had been his dissertation topic at Columbia. Phillips was a tireless worker and prolific writer. By the time of his untimely death from cancer in 1934 at the age of forty-six, "the Georgian had established himself as the foremost student of southern history."[35]

Phillips and Stone struck up a close friendship because they shared common beliefs. For example, both were convinced that black people were inherently inferior and thought that race, not slavery or class, was the major factor affecting southern history. They thought so much alike that Phillips's biographer has stated: "It would be futile to try to assign priority or influence to either man. Each borrowed freely from the other." Phillips and Stone also had many opportunities to discuss topics of mutual interest and share data gleaned from their research. Phillips visited Stone at Dunleith, and sometimes the two men and their families went on vacations together. One indication of the personal and professional cooperation that blossomed between the two men can be seen by Stone's offering to guarantee the cost of Phil-

35. John David Smith, "W. E. B. Du Bois and Ulrich Bonnell Phillips," 24; Dillon, *Ulrich Bonnell Phillips,* 24, 154–63, 172–74.

lips second book when the young academic was still struggling to make ends meet.[36]

If Stone was to pass along manuscripts he obtained as a result of the *Material Wanted* pamphlet, his friend Phillips would have been a likely recipient. Documents related to the antebellum South were Phillips's specialty.[37] In fact, Phillips was a pioneer in the collection and preservation of plantation records, precisely the sort of material Stone was seeking.[38]

There is evidence that at least some of the primary source material Stone collected made its way into Phillips's hands. In 1918, Phillips published his major work, *American Negro Slavery*. Historian John David Smith has characterized the book as "the first systematic analysis of slavery in the entire South," which "thrust Phillips into the role of the unexcelled authority on the peculiar institution. It eclipsed in scope and detail prior books on North American slavery and has influenced virtually all subsequent works on the subject."[39] Footnotes to *American Negro Slavery* indicate that Phillips relied heavily on primary source material for his research, some of which was in the hands of private collectors.[40] It is not that surprising, therefore, to find that Phillips thanked "Alfred H. Stone of Dunleith, Miss." in a footnote for copies of the records he used in the book "as well as for many other valuable items."[41]

What the footnote in Phillips's book does not tell us is what or how much material Stone contributed to the work, and it is likely that we never will know. The main reason we will never know is because Stone failed to write his much-anticipated work on the role of African Americans in the economic development of the United States. But Stone, as usual, had several balls in the air, and one of these fell into place toward the end of 1907, when he was asked to present a paper to the American Sociological Society at their annual meeting in Madison, Wisconsin.

36. Dillon, *Ulrich Bonnell Phillips*, 63, 122. I have searched in vain for correspondence between Stone and Phillips in the collections of the Phillips Papers at Yale University, the University of North Carolina, and the University of Georgia.

37. John David Smith, "Ulrich Bonnell Phillips's *Plantation and Frontier Documents: 1649–1863*: The Historian as Documentary Editor," in *Slavery, Race, and American History*, 154–64.

38. Smith, "'Keep 'Em in a Fire-Proof Vault,' 138.

39. John David Smith, "The Historiographic Rise, Fall, and Resurrection of Ulrich Bonnell Phillips," in *Slavery, Race, and American History*, by Smith, 45.

40. Dillon, *Ulrich Bonnell Phillips*, 105.

41. Ulrich Bonnell Phillips, *American Negro Slavery: A Survey of the Supply, Employment and Control of Negro Labor as Determined by the Plantation Regime* (New York: Appleton, 1918).

The topic Stone was asked to speak about was racial prejudice, an area that he was familiar with but that diverged substantially from the focus of his work for the Carnegie Institution. The title of the paper he gave was framed as a question: "Is Race Friction between Blacks and Whites in the United States Growing and Inevitable?" Stone's attempt to answer this double-jointed question encompassed his boldest venture into the realm of social psychology. As such, his paper on race friction became the cornerstone in Stone's edifice of racial theory.

The American Sociological Society assembled a panel of eight academics to discuss Stone's paper after he presented it. Six of the eight were from major universities and colleges in the Northeast and Midwest.[42] Walter F. Willcox was also on the panel, as was W. E. B. Du Bois. Although it is evident that Stone circulated his paper ahead of time to each of the discussants, only the feedback from Wilcox and Du Bois is known to have survived.

Willcox was lukewarm in his response to what Stone had written. "The paper seems to be a broad and temperate discussion of the difficult and complicated subject," Willcox wrote. "In reply to your question I think it answers the purpose. I must admit however a slight feeling of disappointment when I first read it." Willcox thought that Stone had "beat about the subject rather than to come at once to a close grapple with it." Willcox suggested that Stone shorten his introduction, especially the part in which he defined his terms.[43]

Stone thanked Willcox for his feedback and stated that he would work on the introduction. "As to my general treatment of the subject," Stone replied, "I believe I wrote you that, candidly, I did not like it. I am glad that after thinking it over you feel that it is probably the better line for me to follow,—rather than that of answering the question [about racial prejudice] categorically and marshalling the concrete evidence in support of my position." Apparently, Stone agreed with Wilcox that he might have taken a more hard-line approach.[44]

Du Bois' note to Stone about the paper was brief and to the point. "I have just read your paper which you are to present at the Sociological Society in Wisconsin. I want to thank you for it," Du Bois wrote. "I think it by all means the fairest paper that you have written and strikes a note which I have not

42. The institutions represented by discussants at Stone's talk were Indiana University, Smith College, University of Illinois, Miami (of Ohio) University, Lake Forest College in Chicago, and Syracuse University.

43. Willcox to Stone, December 11, 1907, WFWP.

44. Stone to Willcox from Dunleith, December 18, 1907, WFWP.

seen in your other papers. If this spirit of discussion could be held to in the race problem, I am sure we should accomplish great things."[45]

It is tempting to think that Du Bois liked Stone's paper because Stone had, according to Willcox, pulled his punches. But that explanation is hard to reconcile with someone like W. E. B. Du Bois. Du Bois would not have been fooled by a wolf's argument disguised with sheepish language. It is more likely that Du Bois liked the paper because Stone had been honest.

What Stone wrote in his paper on race friction may have seemed mild to Willcox, but one can only imagine what Willcox and Stone had shared with one another in their private conversations about race. In public, however, Stone had downplayed the rougher edges of his racial theories in an effort to cultivate the image as someone who was moderate and reasonable. Had not Stone denied that racial prejudice was a factor that affected the economic future of African Americans in his address at the American Economic Association meeting two years before? In his paper on race friction, Stone had finally acknowledged that white people in the South were prejudiced against black people, something that everyone knew but sometimes had trouble admitting.

Stone not only admitted that racial prejudice among white southerners existed, he resurrected an old term for it—antipathy.[46] Prejudice may be a predilection, Stone wrote, but antipathy was a much stronger emotion, a "'natural contrariety,' 'incompatibility,' or 'repugnancy of qualities,'" as he defined it. Furthermore, antipathy was innate, a "constitutional feeling" over which individuals have no control.[47] Stone was breaking new ground with these remarks. To this point, he had argued that the "destructive propensities" of black people were inherited. Now, he was claiming that the feelings of white people toward persons of African descent were inherited too.

According to Stone, race friction was the result of too many black people

45. Du Bois to Stone, December 26, 1907, DBP.

46. "Is there antipathy to the Negro in the South?" was one of the questions raised by the Southern Society for the Promotion of the Study of Race Conditions and Problems in the South at its conference in Montgomery, Alabama, in 1900.

47. Alfred Holt Stone, "Is Race Friction between Blacks and Whites in the United States Growing and Inevitable?" *American Journal of Sociology* 13 (March 1908), 676–97; reprinted under the title "Race Friction" in *SARP,* 211–41. Quotations in this chapter are taken from the book because several minor revisions and copyediting changes were made to the article before it was reprinted. Pagination for the article is noted in parentheses following citations of the book version. For example, the quotations in this paragraph come from page 212 in the book (677 in the article).

around a white person, which increased racial antipathy. Stone described the process as "a wholly real feeling of 'pressure,' which comes to the white man almost instinctively in the presence of a mass of people of a different race." Stone went on to argue that white feelings of racial antipathy were aggravated further whenever black people ignored the fact that they were inferior. "Post bellum racial difficulties are largely manifestations growing out of the novel claim to equality made by the Negro after emancipation," he observed. Slavery, according to Stone, was a good thing because it enforced a status of inferiority for black people, thus relieving feelings of racial pressure in those parts of the country in which persons of African descent lived in very large numbers. "Slavery was responsible [for race friction] only in so far as it was responsible for bringing the races into contact," he explained. "The institution, *per se,* was not only not the cause of the problem, but, on the other hand, it actually furnished the basis of contact which as long as it existed minimized the problems which result from racial contact."[48]

From Stone's perspective, the demise of slavery made the establishment of a segregated society necessary. "The proposition is too elementary for discussion," Stone noted, "that the white man when confronted with a sufficient number of Negroes to create in his mind a sense of political unrest and danger, either alters his form of government in order to be rid of the incubus, or destroys the political strength of the negro by force, by evasion, or by direct action." Maintenance of a rigid color line minimized, if it did not actually prevent, race friction. The only reason there was not as much race friction in the North was because of "the economic segregation of the Negro," especially in larger cities. As far as Stone was concerned, the "undeveloped mental state" of black people in the South made it possible for them to accept segregation and live contentedly under those conditions, something that "a more highly organized and sensitive race would have thrown off, or destroyed itself in the effort to do so. . . . In the main, the millions [of Negroes] in the South live at peace with their white neighbors."[49] Furthermore, segregation worked because the subordination of an inferior race (black people) was "an inherited part of his instinctive mental equipment."[50]

The major threat to this satisfactory state of affairs, as Stone saw it, was

48. Stone, "Race Friction," 217 (680), 219–20 (681–82). Stone credited a professor at Harvard with influencing his thinking in regard to slavery's role as a modulator for good race relations (see N[athaniel] S[outhgate] Shaler, "Race Prejudices," *Atlantic Monthly,* October 1886, 510–18). A copy of Professor Shaler's article is in TSC.

49. Stone, "Race Friction," 227–28 (687–88), 235 (692).

50. Ibid., 235–36 (693–94).

the growth of black consciousness. "There is a large and steadily increasing group of men, more or less related to the Negro by blood and wholly identified with him by American social usage, who refuse to accept quietly the white man's attitude toward the race." Once again, Stone was blaming persons of mixed-race ancestry for stirring up trouble, but was he talking about Du Bois? If so, Du Bois probably applauded Stone's honesty when he warned that "The white man is no wiser than the ostrich if he refuses to see the truth that in the possibilities of race friction the Negro's increasing consciousness of race is to play a part scarcely less important than the white man's racial antipathies, prejudices, or whatever we may elect to call them."[51]

The question of whether Stone was referring to Du Bois was answered when Stone quoted Du Bois to illustrate what he was talking about. "In its final analysis the sum and substance of the ultimate demand of those Americans of African descent whose mental attainments and social equipment identify them much more closely with the Anglo-Saxon than with the Negro masses, is definitely and clearly stated in the words of Dr. DuBois":

> "There is left the last alternative—the raising of the Negro in America to full rights and citizenship. And I [Du Bois] mean by this, no half-way measures; I mean full and fair equality. That is, the chance to obtain work regardless of colour, to aspire to position and preferment on the basis of desert alone, to have the right to use public conveniences, to enter public places of amusement on the same terms as other people, and to be received socially by such persons as might wish to receive them. These are not extravagant demands, and yet their granting means the abolition of the colour line. The question is: Can American Negroes hope to attain this result?"[52]

The quotation Stone used came from the article Du Bois had written for the *East and the West* in 1904, the same article that Willcox had criticized for its angry tone. This time around, however, Stone had an answer to Du Bois' question.

> With equal clearness and precision, and with full comprehension of its larger meaning and significance and ultimate possibilities, the American white man answers the question in the language of another eminent American sociologist, Professor Edward A. Ross, in contrasting the attitudes of Anglo-Saxons and Latins toward other races on the continent:

51. Ibid., 240 (696).
52. Ibid., 240–41 (696–97).

"The superiority of a race cannot be preserved without pride of blood and an uncompromising attitude toward lower races. . . . Whatever may be thought of . . . [this] policy, the net result is that North America from the Behring [*sic*] Sea to the Rio Grande is dedicated to the highest type of civilization; while for centuries the rest of our hemisphere will drag the ball and chain of hybridism."[53]

Thus, the issue was joined. Stone had answered the question of whether race friction was inevitable. It was. The question of whether race friction in the United States would increase depended on whether black people continued to insist that they should be treated as equals. Stone was convinced that the only way to lessen the pressure for racial conflict was for African Americans to assume a passive role as the inferior race.

By and large, the discussants liked what Stone had to say. Willcox and five of the six northern academics endorsed Stone's theory of race friction and offered variations on the theme. For example, Willcox wondered if race friction would level off as a formal caste system similar to the one in India took hold, thereby replacing slavery as an effective means of social control.[54] U. G. Weatherly from Indiana University believed that interference by well-meaning friends of the African Americans in the North was the primary cause for an increase in race friction, and two discussants were concerned that race friction would increase in the North as more African Americans moved there.[55]

Only John Spencer Bassett from Smith College raised the possibility that racial antipathy might be a social phenomenon rather than an inherited characteristic, an interesting point for someone to make at the annual meeting of the American Sociological Society. Bassett argued that when the country was founded, the attitude toward African Americans was more tolerant than it was today, and he used Thomas Jefferson's views on slavery as an example. According to Bassett, sometime around 1830, as sectional passions became inflamed, tolerance turned to antipathy and distrust. Bassett wondered if the more tolerant attitudes of the founding fathers could be recaptured as a

53. Ibid., 241 (697).

54. This summary of the discussants' comments comes from "Discussion of the Paper by Alfred Holt Stone, 'Is Race Friction between Blacks and Whites in the United States Growing and Inevitable?'" *American Journal of Sociology* 13 (May 1908): 820–33.

55. Weatherly was a founding member of the American Sociological Society and would later serve as its president in 1923 ("Ulysses Grant Weatherly," www2.asanet.org/governance/weatherly.html).

means for easing race friction. Although Bassett's notion was hardly original, he was clearly in a minority when he expressed these thoughts.[56]

Unfortunately, Du Bois could not afford to attend the meeting, which meant that Stone's response after the discussants' comments did not address issues that Du Bois would raise when his comments were submitted after the fact.[57] Not that it mattered, because Stone had little to say. Noting that all of the discussants' papers except for one "agree substantially with my fundamental propositions," Stone focused on what Bassett had said, especially the part in which Bassett asserted that the traditional Anglo-Saxon attitude toward persons of African descent was one of tolerance. "Here he [Bassett] falls into a very common error," Stone remarked. "The relations of which he speaks are characteristic of many communities in which the number of negroes is small as compared with the white population, combined with a certain primitive social state." In fact, Stone explained, the change in race relations after 1830 that Bassett had noted was the result of an increase in the number of black slaves in the United States. "But I have no quarrel with those who differ from my views," Stone said as he drew to close. "I usually avoid discussing the future. My life work is trying to learn something about the past and present—and endeavoring correctly to interpret what I learn. As I look at the history of race relations the world over, it seems to me almost utopian for us to flatter ourselves that we can escape continued race friction in the United States."[58]

Stone still had not seen Du Bois' comments when he wrote Du Bois five days later from Chicago en route to Greenville. Stone thanked Du Bois for his note of December 26, in which Du Bois had judged Stone's paper to be the fairest he had written so far. "I have been conscious at times of a feeling that you did not understand me, and have wished for an opportunity to have a long, *confidential* chat with you," Stone wrote. "There is much that I would like to say to you,—of my personal attitude & feeling,—of my practical expe-

56. Although born and reared in North Carolina, Bassett had been out of step with white southern values before. In 1903, Bassett got in trouble while on the faculty at Trinity College, now Duke University, when he wrote that Booker T. Washington was second only to Robert E. Lee among southerners born in the last one hundred years. Bassett fled to Smith College because of the stink that indiscretion caused in North Carolina ("John Spencer Basset," www.lib.duke.edu/archives/history/j_s_bassett.html). For a detailed account of Bassett's troubles at Trinity College, see Williamson, *Crucible of Race*, 261–71.

57. Lewis, *W. E. B. Du Bois*, 373; Editor's note preceding Du Bois' comments in "Discussion of the Paper by Alfred Holt Stone," 834.

58. "Response by Mr. Stone" in "Discussion of the Paper by Alfred Holt Stone," 838–40.

riences and observations. You might not,—probably would not, agree with much that I might say,—but I believe that always thereafter you would think me sincere, honest and with no purpose or wish to do aught but justice to and to feel aught but kindly interest in, the Colored American people. I believe also that I could discuss things in the situation which would be new to you,—even with your broad knowledge of the subject." Stone ended his letter with an invitation to Du Bois to visit when he was in Washington.[59]

It would be a while before Stone reached Washington because he had gone to Dunleith to work on the manuscript of *Studies in the American Race Problem,* which had been accepted for publication. To that end, Stone spent most of January and February 1908 reading and correcting proofs.[60] He was far enough along by January 25 to ask Willcox to proof his chapters. Three days later, Stone wrote the preface and sent his corrections to the publisher, Doubleday, Page. In the preface, Stone noted that "For some years I have been working on a study the results of which I had hoped to submit by this time in the concrete form under the title 'Race Relations in America.' The scope of the work, however, has postponed its completion. These papers in the main are by products of investigations in the broader field."[61] Among the "by products" was a new chapter Stone had added, his paper on race friction.[62] Stone received the corrected galleys in early February, and on March 3, Stone sent the final proof to the publisher, just before he returned to Washington. Although Stone continued to make corrections during the spring as the publisher brought them to his attention, Stone's work on the book was finished.[63]

Stone got hold of a copy of Du Bois' comments on Stone's paper at Madison when he returned to Washington. In them, Du Bois raised three questions: (1) "Is the old status of acknowledged superiority and inferiority between the white and black races in American longer possible?" (2) "Are the race differences in this case irreconcilable?" and (3) "Is racial separation practicable?"

Du Bois' answer to the first question was no. Despite what Stone had

59. Stone to Du Bois from Chicago, January 4, 1908, DBP; emphasis in original.

60. Stone to Willcox from Dunleith, January 14, 1908, WFWP.

61. Stone, *SARP,* xii.

62. Stone dropped the long title he had used for his address in Madison and substituted the economical "Race Friction" instead (*SARP,* 211).

63. Letters in WFWP, Stone's folder, dated February 6, March 3, March 8, and May 25, 1908. The date of the preface comes from *SARP,* xiii.

said, the mass of black people were not willing to accept their current status as second-class citizens. In other words, the millions of black folk in South did not want to live in peace with their white neighbors if doing so meant acknowledging their inferiority. As far as the second question was concerned, Du Bois thought that more work had to be done to determine whether the apparent differences between the two races were innate or the result of generations of social abuse.[64] Du Bois had little confidence in research on the topic to date. "The data upon which the mass of men, even intelligent men, are basing their conclusions today, the basis which they are putting back of their treatment of the Negro, is a most ludicrous and harmful conglomeration of myth, falsehood, and desire."[65]

Du Bois' answer to his third question was equally direct. The separation of races was impracticable, he wrote, because the world was shrinking. As a consequence, Anglo Saxons would eventually have to rub shoulders with people of many races, not just persons of African descent. Noting that the European powers were colonizing remote parts of the world, Du Bois asked, "If the world can enter Asia, why cannot Asians enter the world?" Economic factors would ensure that people of different races would end up working together. "Race segregation in the future is going to be impossible," he wrote, "primarily because these races are needed more and more in the world's economy." Du Bois ended his comments on Stone's paper with an admonition. "Rhetoric like that quoted by Mr. Stone is not in itself of particular importance," he wrote, "except when it encourages those Philistines who really

64. Du Bois noted the work of Franz Boas at Columbia University and quoted him as saying "that an unbiased estimate of the anthropological evidence so far brought forward does not permit us to countenance the belief in a racial inferiority, which would unfit an individual of the Negro race to take part in modern civilization." For a discussion of Boas's work on racial differences, especially in regard to African Americans, see Degler, *In Search of Human Nature*, 74–80.

65. The reader will note that sometimes the first letter in the word "Negro" is capitalized, as in Du Bois' comments, and that it is not capitalized at other times, as in Stone's comments. Although that usage appears to be inconsistent, it reflects a policy in effect at the time that consistently honored the writer's (e.g., Du Bois' or Stone's) preference. Stone and Willcox debated whether "Negro" should be capitalized in *Studies in the American Race Problem* and finally decided to go with a capital *N* (Stone to Willcox, March 13, 1908). Willcox spelled out his reasons for spelling "negro" with a lowercase *n* in an exchange of letters with H. B. Frissell, principal of Hampton Normal and Agricultural Institute, in April 1907, although Willcox stated that he was willing to change his position "if this usage [spelling *negro* with a lowercase *n*] is distasteful to a large body of my fellow countrymen." Typescripts of the three letters in the Willcox-Frissell correspondence regarding this issue can be found in appendix D.

believe that Anglo-Saxons owe their pre-eminence in some lines to lynching, lying, and slavery, and the studied insult of their helpless Neighbors. God save us from such social philosophy!"[66]

Stone wrote Du Bois on March 13 and asked Du Bois for a progress report on his work for Stone to submit at a meeting of division heads with Colonel Wright in April. It was obvious that Stone had read Du Bois' comments on his paper, but he seemed more interested in something Du Bois had written in his letter to Stone just prior to the meeting in Madison. What was it that made his paper "fairer than any other I had written?" Stone asked. Eventually, Stone acknowledged Du Bois' comments as a discussant. "I was sorry that you put in your discussion of my paper the statement that I had 'often expressed the cheerful hope that white labor would supplant the Negro in the South,'" Stone complained. "I let it go without a reply [in a previous letter?],—but you are mistaken. I have never expressed any such hope or wish,—cheerful or otherwise,—and you would be puzzled to give a reference for the statement. But we shan't fall out about it," Stone promised. "Only I wish you would let me know some time when you are bound this way."[67]

Du Bois was prompt to respond, promising to have the first half of his report to Stone within days. The rest would be ready by May 1. With that matter out of the way, Du Bois answered Stone's question about the paper on race friction. "With regard to your Madison paper: What I had noticed was more of a disposition than I had ever seen in any of your papers, to consider the point of view of the Negro as a man."

> Always there seems to me, on your part, a lack of that consideration. You speak continually from the point of view of the white world and seem to consider it sufficient, if you have proven that the thing that you believe in is a good thing for a certain race; that we, as a race have the right to be considered in our wishes, and wants and ambitions, is a thing that gets very little expression in your writing. I am perfectly free to say that in this

66. Du Bois' comments, "Discussion of the Paper by Alfred Holt Stone," 834–38.

67. Stone to Du Bois, March 13, 1908, DBP. The context for Du Bois' remark about Stone's "cheerful hope" is as follows: "Race segregation in the future is going to be impossible primarily because these races are needed more and more in the world's economy. Mr. Stone has often expressed the cheerful hope that the Negro would be supplanted by the white man as worker in the South. But the thing does not happen. On the contrary there are today more Negroes working steadily and efficiently than ever before in the world's history. The world is beginning to work for the world" (Du Bois' comments, "Discussion of the Paper by Alfred Holt Stone," 837). This remark is one of only two references to Stone by name that Du Bois made in his comments. The second occurs in the next to last sentence of Du Bois' comments, as quoted above.

> Carnegie report, I fear that what we are going to have is the point of view of the South which never has recognized the Negro as an independent wishing, thinking and feeling man. I hope I am going to be mistaken, and your Madison paper certainly gave some grounds for that hope.
>
> My criticism as to the "cheerful hope" that the white labor would supplant the Negro in the South is based upon such phrases as that in your Baltimore paper [at the American Economic Association meeting in 1905], when you said that "unstable as water" the Negro could not prevail. There are Negroes unstable as water and there are white men, but to put instability as a characteristic of the Negro race is grievously wrong; and it seems to me that both in the phrase and in a number of others, you have expressed not a fact but a wish.[68]

Du Bois ended his letter to Stone with "I trust I may see you in Washington. I shall let you know when I pass through and can stop."

It took Stone almost two months to respond, and when he did, his letter dealt exclusively with the next installment of Du Bois' payment for the work he had done for the Carnegie Institution.[69] Du Bois replied with a brief note on June 1 acknowledging receipt of the check, and there matters stood until October, when Stone wrote again. If there was an exchange of letters in between, they have not survived, and it is unlikely that the two men met for a conversation in Washington. Du Bois' biographer does not mention such an encounter, and Du Bois' meeting with Stone for a private discussion would have warranted mentioning. Apparently, Stone's letter in October was in response to Du Bois' brief communication of June 1. "Your note is at hand," Stone began, "and I owe you an apology all round,—first, for not writing you long ago about the report, second, for not at once replying to this last letter."

> But I shall not try to make excuses. The only valid ones would be that I was away when the reports first came, that I have had two spells of fever since, and have had to make three trips South. Mainly, I have been harassed by business troubles, and have allowed everything else to go.
>
> The reports did come, and were duly distributed. They brought favorable comments from the various members of the board, particularly from Col. Wright,—for all of which, and to you, I am grateful and acknowledge my thanks.

68. Du Bois to Stone, March 30, 1908, DBP.
69. Stone to Du Bois, May 20, 1908, DBP.

> As I wrote you quite a while ago, I did not mean to impose on your grant for these copies,—hence I enclose my personal check to cover their price.[70]

Stone closed "With regards and best wishes," and with those words, correspondence between the white planter and the black activist ended.

Studies in the American Race Problem was released on October 7, 1908, three days before Stone's last letter to Du Bois.[71] Reviews of the attractive 555-page volume were overwhelmingly favorable. A theme that ran through most was the perception that Stone's approach to racial issues was even-handed. "It is impossible here to do justice to the entire frankness, calmness, and honesty with which Mr. Stone has set forth his conclusions on a subject of such vital importance to the country at large," an anonymous reviewer in the *New York Times Book Review* wrote in January 1909.[72] John Spencer Bassett, the discussant at Madison who had objected to Stone's emphasis on heredity, wrote that "This [book] is, probably, the most considerate and reasonable statement of the negro question ever made by a Southerner who holds the views usually held by his people."[73] A reviewer in the *American Journal of Sociology* praised Stone's "earnestness and fairness in seeking and presenting the truth without fear or favor."[74] Carl Kelsey called Stone "a frank and honest student" in his review, while a reviewer in the *Bulletin of the American Geographical Society* acknowledged that "If not free from bias he [Stone] is at least dispassionate and his work is a valuable contribution toward a fair expression of the best southern attitude."[75] Cleland Boyd McAfee echoed this observation when he wrote that "The better sentiment of the Southern people could hardly be more clearly presented than by Mr. Stone."[76]

It was one thing to praise Stone's objectivity, but it was quite another to

70. Ibid., October 10, 1908, DBP.

71. Willcox to Stone, October 14, 1908, WFWP.

72. "Question of Race Diversely Viewed," *New York Times Book Review*, January 9, 1909.

73. John Spencer Basset, review of *Studies in the American Race Problem*, by Alfred Holt Stone, *American Historical Review* 14 (July 1909): 837. This statement is interesting because it came from someone who was born and reared in the South.

74. F. W. Blackmar, review of *Studies in the American Race Problem*, by Alfred Holt Stone, *American Historical Review* 14 (May 1909): 838.

75. Carl Kelsey, review of *Studies in the American Race Problem*, by Alfred Holt Stone, *Annals of the American Academy of Political and Social Science* 33 (March 1909): 217; Mark Jefferson, review of *Studies in the American Race Problem*, by Alfred Holt Stone, *Bulletin of the American Geographical Society* 43, no. 4 (1911): 299.

76. Cleland Boyd McAfee, review of *Studies in the American Race Problem*, by Alfred Holt Stone, *Journal of the Royal African Society* 8 (January 1909): 148.

judge his book as among the best works available on the topic, a topic that had gained a popularity in 1909 that is difficult to appreciate today. But judge it as best, some did. "One of the best studies yet made of the problems growing out of the presence of blacks and whites in our country" is how Kelsey described Stone's book in his review, although he cautioned the reader that "no one need accept Mr. Stone's conclusions."[77] A reviewer in the *American Journal of Sociology* was even more enthusiastic: "For, with due respect to some other valuable publications on the same subject, the book as it appears represents the most valuable contribution yet appearing on the race problem in the United States."[78]

Rave reviews sold books, and that reality must have pleased Stone because he had guaranteed the cost of publication.[79] With reviews like these, it is doubtful that Stone had to pay a penny.[80] The book certainly had an impact, as was illustrated in an exchange of letters in November 1908 between Du Bois and Charles Francis Adams, an amateur historian and businessman who contributed articles to *Century Magazine* that disparaged African Americans.[81] Du Bois had written Adams to complain about one of his articles, actually a speech that Adams had given in Richmond, Virginia. Du Bois thought that Adams's comments were "ill-considered and wrong but [also] distinctly sensational in the worst sense."[82] "One of the most unfortunate things about the Negro problem is that persons who 'do not for a moment profess to be informed on the subject' insist on informing others," Du Bois wrote. "This, for a person who apparently boasts of advanced scientific knowledge is most

77. Kelsey, "Review," 218.

78. Blackmar, "Review," 838. Stone's neighbors in Greenville also actively endorsed his book in the national media (e.g., Orvil A. Williamson, from Greenville, Mississippi, "The North and the Negro Problem," *New York Times*, April 9, 1923).

79. Stone to Willcox from Washington, DC, October 16, 1908, WFWP.

80. Stone did not have to worry. *Studies in the American Race Problem* was adopted by white universities in the South as a text for their courses on the race problem, which were popular during this period (Charles Hillman Brough, "Work of the Commission of Southern Universities on the Race Question," *Annals of the American Academy of Political and Social Science* 49 [September 1913]: 49).

81. For example, see Charles Francis Adams, "Reflex Light from Africa," *Century Magazine*, May 1906, 101–11, in which he argued that the notion African Americans can compete with European Americans intellectually and culturally "is the sheerest of delusions." There is a copy of this article in TSC.

82. Du Bois to Adams, December 15, 1908, in Aptheker, *Selections*, 144. This letter was a continuation of the correspondence that began between the two men in November. The article in question was *"The Solid South" and the Afro-American Race Problem. Speech of Charles Francis Adams at Academy of Music, Richmond, VA., Oct. 24, 1908* (Boston, 1908).

deplorable, and I trust that before publishing further matter on the race problem, you will study it. To that end, I am sending you some literature."[83]

Adams was the grandson of John Quincy Adams, the abolitionist, and great-grandson of John Adams, the president, and he was not about to take Du Bois' criticism lying down. "I infer that your purpose is to administer me a rebuke, . . ." he replied hotly. "I shall read the three publications you have sent me with close attention, . . . Meanwhile, I am forced to say that by all odds, and to an incomparable degree, the most illuminating book in relation to the problem I have come across is that of Mr. Stone, entitled, *Studies in the American Race Problem*, to which, in the form of a quotation from Booker T. Washington, I made allusion in my speech at Richmond. It may be ignorance on my part—and, I presume, it is," Adams concluded, "but I do not know any direction in which to go for safer guides than to Mr. Washington and Mr. Stone."[84]

83. Du Bois to Adams, November 23, 1908, in Aptheker, *Selections*, 142–43.

84. Adams to Du Bois, November 28, 1908, in Aptheker, *Selections*, 143. Adams's favorable assessment of Stone's work may have been influenced by the review Stone wrote of Adams's article ("Reflex Light from Africa"), in which Stone praised Adams for his perspicacity and for making a "distinct contribution" to the literature on the race problem ("More Race Problem Literature," *Publications of the Southern History Association* 10 [July 1906]: 225–27).

12 A SLAVE TO BUSINESS

Not everybody was impressed with *Studies in the American Race Problem.* William Jay Schieffelin, a wealthy New Yorker, the great-great grandson of Chief Justice John Jay, and a generous supporter of the Tuskegee Institute,[1] wrote a critical review of Stone's book in the *Political Science Quarterly.* Schieffelin acknowledged that Stone marshaled his facts and presented them in a logical and impartial manner, but Schieffelin asserted that Stone "is guilty, nevertheless, of glaring inconsistencies." For example, although Stone argued that "the [Negro] problem is a national one," he "shows impatience not only of much of the comment that comes from the North, but of the endeavor of northern men to contribute towards a solution." Along the same lines, Schieffelin pointed out that Stone "warns the reader against generalizing from special instances [for example, the accomplishments of talented African Americans], and then he cites a simple case from which he himself draws pessimistic conclusions [Stone's experience with seventy-five black families at Dunleith]." Schieffelin also criticized Stone for comparing the efficiency of black laborers to Italian immigrants. "The very fact that these immigrants have come to America," Schieffelin observed, "shows them to be more resourceful and enterprising and intent upon success than their fellows who remain at home."[2]

Yet, there were things about Stone's book that Schieffelin liked—the chapter "Race Friction," for example. "In it he presents many interesting facts," Schieffelin wrote; "his diagnosis, in the reviewer's opinion, is correct;

1. "Miniature of Mrs. William Jay Schieffelin (1871–1948)," luceweb.nyhistory.org/luceweb/item_detail.htm?qmkey=16703; "Choate and Twain Plead for Tuskegee," *New York Times,* January 23, 1906.

2. William Jay Schieffelin, review of *Studies in the American Race Problem,* by Alfred Holt Stone, *Political Science Quarterly* 24 (December 1909): 701–2.

but the treatment which he proposes, namely, that the Negro should be let alone, is gravely inadequate." Schieffelin also liked Stone's chapter on race and politics. However, he complained that the chapter entitled "The Mulatto Factor" was too short. "Instead of brooding over 'the horrors of reconstruction,'" which Stone had done with his account of black political participation in the previous chapter, "the author should labor to create a prevailing public opinion against the menace of miscegenation. Why did he not take this opportunity to proclaim the gospel of race integrity and to demand that it be practiced by the men of the South?" Not surprisingly, Schieffelin ended his review of *Studies in the American Race Problem* by endorsing another book. "Since Mr. Stone's articles appeared," Schieffelin wrote, "a very able and temperate book has been published, entitled *The Basis of Ascendancy,* by Edgar Gardner Murphy. Its perusal will lead the reader to a higher and more hopeful view of this great question than is taken by Mr. Stone."[3]

Edgar Gardner Murphy was the rector at St. John's Episcopal Church in Montgomery, Alabama, who organized the three-day conference on the race problem that Stone had attended in 1900. Murphy's tenure as a priest in Montgomery had been cut short, however, because of his involvement in progressive causes, such as the campaign against child labor in Alabama. Murphy was also against the unilateral disfranchisement of African Americans. If illiterate and uneducated citizens should be barred from voting, he argued, those restrictions should be applied to white people and black people equally.[4] But Murphy was not for the integration of races; his progressivism did not go that far.[5] As Murphy's biographer, Hugh C. Bailey, put it, "Even with these advanced ideas, Murphy still held that the Negro could be granted his civil, political, and industrial rights *without* social integration and race amalgamation."[6]

The Basis of Ascendancy was a blueprint for how the South could grow economically and enhance its standing both politically and morally by end-

3. Ibid.

4. Fredrickson, *Black Image in the White Mind,* 284 n.

5. For an example of Murphy's defense of segregationist principles, see Edgar Gardner Murphy, *The White Man and the Negro at the South: An Address Delivered under the Invitation of the American Academy of Political and Social Science, the American Society for the Extension of University Teaching, and the Civic Club of Philadelphia, in the Church of the Holy Trinity, Philadelphia, on the Evening of March 8th, A.D. 1900* (N.p., [1900?]), 55 pages. There is a copy of this pamphlet in TSC.

6. Bailey, *Edgar Gardner Murphy,* 192; emphasis added.

ing its obsession with keeping black people down. According to Murphy, the overly repressive social legislation engendered by the emergence of Jim Crow cut both ways. It was true that segregation kept African Americans at a distance from white folk, but these laws also regimented white society and limited freedoms white citizens should be able to enjoy. "The law which does not protect the weak," Murphy wrote, "will not—and in the end cannot—protect the strong."[7]

Lest Murphy be accused of being a negrophile, he was quick to distance himself from the views of theorists such as Josiah Royce, a professor of philosophy at Harvard, who argued that the two races were more alike than they were different because they shared a common heritage.[8] Although the two races may have sprung from the same stock, Murphy was convinced that people of African descent were, at least at present, inferior to white folk. On the other hand, Murphy rejected the heavy emphasis on the black race's inherited characteristics found in Stone's book. "If the former [Royce] seems to underestimate the meaning of race," Murphy wrote in the preface to his book, "the latter [Stone] seems to me to give to the fact of race a disproportionate significance."[9]

Coming as it did within months of the publication of *Studies in the American Race Problem*, Murphy's book offered stiff competition for Stone's position on the race problem. Both authors were southern men (Murphy was born in Arkansas), and they were almost the same age (Murphy was born in 1869, Stone in 1870). Both Murphy and Stone agreed that a segregated society was necessary in areas where the two races lived side by side in large numbers, and both men thought that persons of African descent were inferior to white people. But they differed sharply in regard to their assessment of the black potential for improvement and gain. "That the negro's powers are not the powers of the white man, and that his present capacities are at many points not equal to the economic competition presented by the stronger race," Murphy wrote, "is due in part to the fact that the negro is a

7. Edgar Gardner Murphy, *The Basis of Ascendancy: A Discussion of Certain Principles of Public Policy Involved in the Development of the Southern States* (New York: Longmans, Green, 1909), 34–35. Washington read and liked Murphy's book, writing Murphy in a letter, "what you have written comes as near being an accurate and, I might even say, scientific statement, as any book I have yet read, on this subject" (Washington to Murphy, undated, BTWM [reel 61]).

8. Josiah Royce, "Race Questions and Prejudices," *International Journal of Ethics* (April 1906): 265–88. There is a copy of this article in TSC.

9. Murphy, *Basis for Ascendancy*, xiii–xiv.

negro: but how far is it also due to the fact that that sense of 'security as to his property and his person,' which is the best school of his economic virtues, has not always been adequately permitted him, and that that 'hope,' without which, as Marshall declares, 'there is no enterprise,' is as yet but partially enjoyed?"[10]

Murphy's identification of social factors as the reason why African Americans were still struggling economically and socially was not original. However, the publication of his book emphasized the great divide between racial theorists who attributed the shiftlessness and improvidence of black people primarily to social factors and those who were convinced that those characteristics were evidence of innate racial traits. Given Murphy's willingness to consider social explanations for the black inadequacies, *The Basis of Ascendancy* was decidedly more optimistic than *Studies in the American Race Problem,* and people noticed. Here was a white southerner who was against race mixing but who wanted to give African Americans a fair shake.

Compared to Murphy, Stone did not appear to be as moderate and fair-minded as many people had initially thought. He may not have been a negrophobist like Vardaman, but Stone was more intransigent when it came to finding solutions to the race problem that would elevate black people in the South than many educated white people in the North found acceptable. Whether for this reason or for some other cause, the Executive Committee of the Carnegie Institution's board of trustees decided in April 1909 that it was time to reevaluate the desirability of continuing Stone's division in the Department of Economics and Sociology, and Murphy had a hand in this turn of events.

Murphy had been on Stone's case since he learned that the Mississippi planter was writing a book about the race problem. In August 1908, about six weeks before *Studies in the American Race Problem* was released, Murphy wrote the editor of *Outlook* to complain about how Stone's book was being marketed. Murphy accused Stone of using his position at the Carnegie Institution to boost the sales of his book, which had nothing to do with Stone's work at the institution. "Mr. Stone, as stated in the advertisement of the volume, has been chosen by the Carnegie Institution to write the economic history of the negro," Murphy wrote. "This volume, however, is a publication from Mr. Stone and not from the Carnegie Institution." It was not so much the potential increase in sales that bothered Murphy; it was the enhanced

10. Ibid., xv. Murphy was referring to a conservative economist, Alfred Marshall, who held a professorship at Cambridge University in England.

credibility Stone's book would gain if readers thought that it had been sponsored by the institution.[11]

Outlook did not publish Murphy's letter to the editor, but Murphy sent a copy to Booker T. Washington along with a "confidential" explanation of what concerned him the most about Stone's book. "It is not scientific in its temper, but it possesses a vigorous, controversial quality that will do much to give it currency and effect," Murphy wrote. "It will, I think, do great harm. I believe that Mr. Stone is a thoroughly honest and conscientious man, and I have high regard for many of his personal qualities"; he added, "his selection [by the Carnegie Institution], however, for the sort of work he is doing is a real calamity."[12]

Washington did not think that there would be much in Stone's book to worry about after it was released. "I do not share this opinion as to the harm that will be done by Mr. Stone's book," Washington wrote to Murphy in a reply dated August 29. "Of course, among a certain element it will have an influence for harm, but human nature, as I observe it, is so constructed that it does not take kindly to a description of a failure. It is hard to get up enthusiasm in connection with a funeral procession. . . . I am not saying, of course, that the Negro race is a failure, Mr. Stone writes largely from that point of view, hence there is no rallying point for the general reader." One thing Washington did agree with Murphy on, however, was the undesirability of Stone's having been appointed as a division head for the Carnegie Institution. "I entered my earnest protest both to the President of Carnegie Board [of Trustees] and to Mr. Carroll D. Wright," Washington wrote. "I told them plainly that it was unfair to the race to have such a man in charge of such important work, who really does not believe that the race can succeed."[13]

In spite of his confidence that Stone's book would not damage race relations, Washington was not about to keep his dissatisfaction with Stone's appointment as head of the division to study "the Negro in Slavery and Freedom" to himself, and he had the opportunity to express his reservations regarding the appointment on February 9, 1909, when he sat down with the great man himself, Andrew Carnegie.[14] Murphy's criticism of Stone allowed

11. Murphy to the editor of the *Outlook*, August 26, 1908, copy in BTWM (reel 61).

12. Murphy to Washington, August 26, 1908, BTWM (reel 61); see also Louis R. Harlan, *Booker T. Washington: The Wizard of Tuskegee, 1901–1915* (New York: Oxford University Press, 1983), 259–60.

13. Washington to Murphy, August 29, 1908, BTWP (9: 614–15), also in BTWM (reel 61).

14. Ibid., February 9, 1909, BTWM (reel 61). "I had a good chance to talk at length with Mr. Carnegie about you and your work to-day" is how Washington began the letter.

Washington to bring up the issue of Stone's credentials without appearing to be self-serving. Washington was simply passing along a different perspective from another white southerner who was also interested in a resolution to the race problem.[15]

The question of whether Stone's division should be continued could not have come at a worse time. Colonel Wright died in late February 1909,[16] and Dr. Woodward invited the division heads to meet on March 20 to discuss the future of the Department of Economics and Sociology. Although the future of Stone's division was not on the agenda, Stone reported to Willcox, who could not attend, what had transpired when the division heads met. "The one clear impression which I carried away from the conference with Pres. Woodward was that the Trustees of the Institution had determined that they would commit themselves to *nothing* with regard to the future of the work of our department," Stone wrote.

> He [Dr. Woodward] said that it was impossible to promise additional funds even for the completion of the work,—no matter how important the latter might be or how urgent the need. He repeatedly emphasized the statement that the Institution had no further funds for this department, and that the Trustees were most anxious to see the present work brought to a close. He also stated that the Trustees would be glad to be relieved of the burden of publishing the results of our work, and would willingly turn the entire undertaking over to any private concern. He said that it was not a question of the value of what we were doing, but of its cost, at a time when the Institution is in pressing need of funds.
>
> I do not want to create the impression that Pres. Woodward seemed inclined to disparage the work or the department. He did not,—but, on the contrary, spoke appreciatively of both. But his one thought seemed to be of the necessity of winding up the work without additional cost to the Institution,—beyond the original appropriation. The future of the department was touched upon, and he said that it had been created for one particular undertaking, and might be terminated with the completion of that work, or might be continued. That must be determined when

15. Murphy's article "Backward or Forward?" had appeared in the January 1909 issue of the *South Atlantic Quarterly* (vol. 8, pp. 19–38), in which he called for a new perspective on the race problem. Although Murphy did not mention Stone by name, the optimistic piece was clearly at odds with the pessimistic tone in *Studies in the American Race Problem*. Apparently, Washington used Murphy's article to bring Stone's work into the discussion.

16. "Carroll Davidson Wright Biography," www.bookrags.com/biography-carroll-davidson-wright/index.html.

> the time came. He only knew that the Trustees would promise nothing now,—nor in any way commit themselves to a continuation of the department.[17]

Stone did not say anything in his letter to Willcox about the issue of whether his division should be terminated, regardless of what happened to the department. It is possible, even likely, that Stone did not know that his credentials had been questioned. Whatever the case, the Executive Committee asked Professor Henry W. Farnam, who had taken over as interim chair of the department after Colonel Wright's death, to look into the matter. A discussant for Stone's paper at the American Economic Association meeting in 1905, Farnam knew Stone's work well, and he took his time before submitting a report. Eventually, Farnam had it ready for the institution's annual meeting in September.

Farnam began his report by noting that Stone's division was not included in the original plan of work for the department. He then provided a brief explanation of how the study group for "The Negro in Slavery and Freedom" had come about. "The important part played by the negro in the economic life of the United States need hardly be dwelt on here," Farnam wrote. "The collaborators were unanimous in feeling that, if we failed to adequately trace the history of the negro in this country, we should omit one of the most important chapters in our economic history." Having reaffirmed the importance of establishing a division to study the topic, Farnam turned his attention to the question of whether Stone was the best man to serve as its head. "There is no doubt that a subject that has played too important a part in the political history of the United States is a difficult one to treat with absolute impartiality," Farnam continued, "and that whoever is called upon to handle it will be criticized in some quarters."

> Almost any northern man, who might study this question, would be apt to be considered in the South ignorant and ill informed, if not biased; almost any southerner might be held in the North to be prejudiced. Looking at the work, as it will be judged by future generations, the board felt that Mr. Stone, by virtue of the careful study which he has given to this subject through many years, and of his recognized fair mindedness, was the most competent man in the country to undertake it, and considered themselves fortunate in being able to secure his co-operation. To eliminate his division from the work of the Department of Economics [and So-

17. Stone to Willcox from Dunleith, July 1, 1909; emphasis in original.

> ciology] at this time would not only omit an extremely important feature of our economic history but would lay us open to the charge of not acting in good faith toward a collaborator, who has already spent a good deal of labor on the work.[18]

Farnam ended his report by emphasizing that the heads of the other divisions had been unanimous in their support of Stone's work. Not only had they welcomed Stone into their ranks, they had reduced their share of the trustees' allocation by adding a new division. They certainly would not have taken that step, Farnam assured Dr. Woodward, "unless they had been convinced of its importance."

Farnam's unequivocal endorsement of Stone's qualifications must have been accepted at face value, because Stone's work received a glowing review in the institution's *Year Book* for 1909. The annual report for the department in 1908 had highlighted the work of Stone's research team,[19] but a year later the focus was entirely on Alfred Holt Stone. His book for the institute's series on the economic history of the United States had grown to three volumes: the first covering the period of slavery from its introduction into the West Indies down to the invention of the cotton gin in 1793, the second covering slavery from 1793 to 1861, and the final volume being the economic history of African Americans after emancipation to the present. The *Year Book* reported that Stone had gathered material for the first volume and was in the process of writing it. Although he also had some material for the other two volumes, their completion would depend on how much time he could spare from his business interests to work on the project.[20]

The reference to Stone's business interests was not perfunctory because Stone's business was in trouble. As the entry for Stone's division in the institution's *Year Book* for 1909 indicated, "Mr. Alfred H. Stone, who has charge of the topic and who has already made gratifying progress with it, found himself obliged, on account of the ravages of the cotton boll weevil, to suspend literary work in the spring of the present year and devote himself for a time to the warfare against this pest, which has already done so much damage to the country at large, and which seriously threatened investments in Missis-

18. Farnam to Woodward, September 17, 1909, CIA.

19. Carnegie Institution, *Year Book No. 7, 1908*, 85. Three of Stone's original team are identified by name in Colonel Wright's report for 1908. However, four new names were also mentioned, including Thomas Jesse Jones of the Hampton Institute, but Du Bois' name does not appear.

20. Carnegie Institution of Washington, *Year Book No. 8, 1909* (Judd and Detweiler for the Institution, 1910), 80.

sippi on the income from which Mr. Stone was replying to pursue his scientific work."[21]

The interruption to his work was not a surprise to Stone because he had seen it coming. In fact, all of the cotton planters in Mississippi had seen it coming. The small, grayish, long-snouted beetle from Mexico had crossed the Rio Grande near Brownsville, Texas, around 1892. Like two other unwelcome invaders from Mexico in the next century, armadillos and fire ants, the boll weevil hopped, skipped, and jumped across Texas, traveling about fifty-five miles on average each year. Sometimes the little insect flew on its own accord, sometimes prevailing winds helped it along, and sometimes it hitchhiked on trucks taking cotton to the gin. It was probably the last-named mode of transportation that introduced the boll weevil to cotton patches around Natchez in 1907.[22] It was only a matter of time before the boll weevil either made its way north from Natchez or crossed the river at Greenville to invade Washington County. By the summer of 1909, it was there.[23]

The boll weevil damaged the crop by feeding on the buds of the cotton plant, called squares, or the bolls, as they were forming. The females caused the most damage, however, when they deposited a single egg through a puncture made by the pest's long snout. Within a few days, the egg hatched, and a small grub began to feed inside the square or boll. Damage resulting from the grub's feeding either caused the square to fall off the plant, thus prohibiting a boll from forming, or prevented the boll from attaining its optimal growth.[24] Unlike chestnut blight or Dutch elm disease, both of which kill the host,[25] the boll weevil reduces the yield of cotton plants by interfering with their normal development. As such, the boll weevil became yet another factor, along with the soil, cultivation, and weather, in determining how much cotton could be harvested.

Stone was acutely aware of this new challenge to the cotton industry in the Delta, and wrote an article about it even before the little insect crossed the river into Washington County. Entitled "Negro Labor and the Boll-Weevil,"

21. Ibid.

22. Blake Layton, *The Boll Weevil in Mississippi: Gone, But Not Forgotten* ([Starkville?]: Mississippi State University Extension Service, n.d.), 1. This publication is available online at msucares.com/pubs/publications/p2294.html.

23. Stone to Willcox from Dunleith, November 18, 1901, WFWP.

24. Layton, *Boll Weevil in Mississippi*, 5.

25. "Dutch Elm Disease," www.ext.nodak.edu/extpubs/plantsci/trees/pp324w.htm; "Revitalization of the Majestic Chestnut: Chestnut Blight Disease," www.apsnet.org/online/feature/chestnut/.

Stone's article appeared in the March 1909 issue of the *Annals of the American Academy of Political and Social Sciences.*[26] In it, Stone pursued a theme that he had followed for several years, namely that African Americans were inefficient workers. What made this article different from his other work was the lack of focus on political or social issues. This article was entirely economic in its scope, although Stone's pessimistic assessment of the potential of persons of African descent to prosper in difficult circumstances was evident. Up to now, Stone argued, black laborers in the South had had it easy. The soil was rich, and the weather was generally favorable. Consequently, slipshod attention to the crop's cultivation and neglect of sound business practices did not matter. But now they did.

> The advent of the boll weevil means that both the negro and the white man must change—or go to the wall. The unbusinesslike methods of the latter in dealing with the former must cease. There must be an end to the granting of absurd credit, the advancing of money on ridiculous demands, just because it is the custom of the community, the offering of unfair or dishonest inducements to labor in order to secure it from another plantation or community. These methods cannot continue after the appearance of a pest which can reduce the crop of a plantation from 1,700 bales in one year to 210 in another. By the same token, the negro must raise vegetables, instead of weeds, in his garden; he must discontinue senseless demands upon his merchant or planter for things he does not need, to be paid for out of a crop which may never be gathered. He must learn to provide for himself, at least between the time when one crop is gathered and another pitched—instead of going to the commissary fifty-two weeks in the year. He must do these things—or another class of labor will be found which will. The negro is no more immune to the operation of elementary laws than is the white man.[27]

Although Stone was critical of the shortcomings of white men as well as black, his prediction for the future was decidedly one-sided. "Under the new order, the white man has in his greater intelligence and keener appreciation of the necessity of mending his ways, an advantage over the negro," Stone wrote. "The masses of the latter are handicapped by a concentration

26. Alfred Holt Stone, "Negro Labor and the Boll-Weevil," *Annals of the American Academy of Political and Social Sciences* 33 (March 1909): 167–74.

27. Ibid. 173.

of thought upon immediate desires, with a contemptuous disregard of either future good or future evil."[28]

Stone practiced what he preached and made plans to move back to Mississippi until the crisis caused by the invasion of the boll weevil was under control. On April 2, two weeks after the meeting of the division heads with Dr. Woodward, Stone publicly announced that his permanent address would be "Dunleith, Mississippi." Dr. Woodward acknowledged Stone's change of address in a letter. "I trust that your labors in combating the boll weevil will not seriously interfere with your capital studies on social and economic questions," Dr. Woodward wrote. "There is great need everywhere now of men in business who are likewise students of contemporary problems." Stone thanked Dr. Woodward for his kind words in a handwritten note and promised "to continue my work,—regardless of the boll weevil."[29]

Continuing his work turned out to be harder than Stone expected. In transferring Stone's books and papers from Washington to Dunleith, the moving company lost two important boxes. "I don't suppose I shall ever be able accurately to measure the extent of the loss to me involved in the going astray of my boxes of papers," Stone wrote to Willcox.[30]

> Whenever I undertake to do my work I find something missing. In handling my things for shipping from Washington I secured the services of the largest and best equipped concern in the city,—the Merchants Transfer and Storage Company. I made my arrangements with the manager in person, and no transaction with any of his subordinates. I explained the nature and value of that portion of the shipment which included my boxes and papers, and paid for the manager's personal supervision. To guarantee myself against any loss I took an entire car, and arranged to have it come through from Washington under unbroken seals. Notwithstanding all these precautions, the contents of the car checked two boxes short on arrival here. After more than eighteen months the railway company is still unable to locate the missing boxes, and I have no idea that they will ever come to light. To complicate matters, the manager of the shipping company died just as he was instituting a search for the boxes in Washington, and no one else in the office seems to know any-

28. Ibid., 169–70.

29. Change of address card postmarked April 2, 1909; Woodward to Stone, April 12, 1909; and Stone to Woodward from Dunleith, April 15, 1909, all in CIA.

30. Stone to Willcox from Dunleith, November 18, 1910, WFWP.

> thing about the matter. The lost matter included three completed chapters of volume one, and partly filled-in skeletons of several other chapters, a nearly complete outline of volume two, and a considerable collection of notes and other manuscript material.[31]

Farnam reported the loss in the Carnegie Institution's *Year Book* and noted that the moving company had given up trying to find the missing boxes after a year's search. "Mr. Stone will have the sympathy of his colleagues in this serious loss, which goes beyond the pecuniary value of the books and notes and includes three chapters of his work practically completed and the skeletons of several others," Farnam commiserated. "Fortunately, he has not lost courage, but is prosecuting the work with such material as he can use on his plantation."[32]

Despite the loss of books and notes, Stone was able to publish another article in the *Annals of the American Academy of Political and Social Sciences* entitled "The Negro and Agricultural Development," but that piece was clearly an offshoot of work that he had already done for the first of the three volumes planned for the Carnegie Institution's series on the economic history of the United States.[33] Although he also agreed to write several book reviews, Stone's correspondence with academic friends and colleagues fell off, and it was not until the cotton crop for 1909 had been harvested and sold that he responded to a letter from Willcox urging Stone to come up to Washington for a visit.[34] "The possibility of a trip to Washington appears remote for me just

31. This loss may explain what happened to some of the manuscripts Stone might have collected as a result of his *Material Wanted* pamphlet. However, Stone's letter does not make it clear whether he is referring to primary source material or drafts of his work.

32. Carnegie Institution, *Year Book No. 11, 1912* (Washington, DC: Judd and Detweiler for the Institution, 1913), 72.

33. Alfred Holt Stone, "The Negro and Agricultural Development," *Annals of the American Academy of Political and Social Sciences* 35 (January 1910): 8–15. The article focused on the introduction of slaves in the British West Indies and the subsequent role of slave labor in the economic development of colonial America, topics that fell within the scope of the first volume.

34. Stone published three reviews in the *American Historical Review* (*AHR*) during this period: review of *The Negro in the New World*, by Harry H. Johnston, *AHR* 16 (April 1910): 650–52; review of *A Documentary History of American Industrial Society*, ed. John R. Commons, Ulrich B. Phillips, Eugene A. Gilmore, Helen L. Sumner, and John B. Andrews, *AHR* 16 (October 1910): 137–39; and review of *The Interest of America in International Conditions*, by A. T. Mahan, *AHR* 16 (April 1911): 652–54. He also contributed a review of *The Southern South* by Albert Bushnell Hart to the *American Political Science Review* (4 [November 1910]: 609–14).

now," Stone replied. "I hope to be free to move about somewhat after a year or two, but for the immediate future I must be a slave to business."[35]

Stone found himself "moving about" sooner than he anticipated, but it was not to continue his research on the economic history of African Americans. Stone and his partner, Julian Fort, hit the road in 1910 for a two-week trip to determine the impact of the boll weevil on cotton production in the Deep South. Traveling by train and car, they covered 1,600 miles through the "swamp regions" of southern and south central Louisiana, along the Mississippi, Atchafalaya, and lower Red rivers, into the alluvial parishes in northwestern Louisiana, and up the Red River into the timbered counties in north and east Texas. Back in Mississippi, they inspected cotton farms in the Natchez district and the region south of the Big Black River. Stone and Fort purposely visited cotton farms located in diverse latitudes and a variety of physiographic regions. At each stop they interviewed from three to eight persons, asking about how the advent of the boll weevil interacted with other variables, such as rainfall, labor conditions, and management practices.[36]

The First National Bank of Greenville published their thirty-four-page report, which offered their careful and systematic analysis of the havoc the little bug from Mexico was wreaking. According to Stone and Fort, the picture was not as bleak as some people thought. The conscientious cotton grower could still make a crop if he did not panic. "We cannot make cotton without labor, and we cannot hold our labor if we pursue the suicidal policy of not only becoming frightened ourselves, but of showing our fright to our negroes," they wrote. "We may depend upon it that our conduct, whatever it may be, will be reflected in that of your labor. And the labor cannot live unless it is fed. If we do as a good many planters in Louisiana did—throw up our hands and tell our negroes that we can no longer take care of them, it will not take them as long to find other homes as it will take us to find other labor. The boll weevil cannot put this country out of business, but we can easily be bankrupted by our own folly."[37]

Stone and Fort were convinced that the best palliative for boll weevil anxiety was the careful cultivation and management of the crop.[38] Stone

35. Stone to Willcox from Dunleith, February 24, 1910, WFWP.

36. Alfred Holt Stone and Julian H. Fort, *The Truth about the Boll Weevil: Being Some Observations on Cotton Growing under Boll Weevil Conditions in Certain Areas of Louisiana, Texas and Mississippi* ([Greenville, MS]: n.p. for the First National Bank, 1910), 34 pages.

37. Stone and Fort, *The Truth about the Boll Weevil*, [29].

38. Stone became a national expert in the cotton-growers' war on the boll weevil. He was

spent 1910 working with his partner to prepare for the onslaught. "I find that I returned to the personal supervision of my affairs none too soon," Stone wrote to Willcox shortly before the report of their inspection tour through the South was published. "The boll weevil reached this country last year, and has now spread pretty nearly over it."

> I am on the edge of the county opposite to point of approach, and the pest stopped its fall migration within seven miles of me. This means that I shall be infested next spring. An agricultural routine of many years is not easy to revolutionize, and the economic changes now necessary are little short of revolutionary. The process demands time, judgment, patience and various other things,—but above all, unremitting attention to details. For the next two years I see small opportunity for leaving home. But I intend to be present at the next meeting of our board [division heads of the Department of Economics and Sociology], if that is in any way possible.[39]

The boll weevil reached Dunleith in 1911, as Stone had predicted, and he stuck close to home. "I was especially sorry, after the receipt of your note," Stone wrote to Willcox in January 1912, "that I could not attend the Washington meetings [of the American Economic Association]."[40] The distance, the expense, and pressing matters at home prevented him from attending. "I wish I could talk over the labor situation with you," he continued. "The American cotton crop for 1911 was the largest on record, meaning, of course, a great decline in price, from recent levels. The Delta crop was the shortest within the memory of any of us. In addition to the financial embarrassment which this combination entailed, the boll weevil had by the fall of 1911 become sufficiently numerous to be taken seriously into account. The short crop in the Delta was not due to the weevil, but our negroes thought so, and we were threatened with an exodus."[41]

Stone attempted to maintain his black labor force with generous incentives. "I determined to try once more an appeal to reason," he reported. "We

convinced that "grit and eternal watchfulness" were the key elements for victory ("Cotton Men Hope to Conquer Weevil," *New York Times*, September 25, 1921).

39. Stone to Willcox from Dunleith, November 18, 1910, WFWP.

40. American Economic Association, "The Twenty-fourth Annual Meeting," *American Economic Review* 2 (March 1912): 1–4. Professor Farnam was president of the association at the meeting.

41. Stone to Willcox from Dunleith, January 6, 1912, WFWP.

told our people that we would balance their 1911 accounts, and and [*sic*] let them begin 1912 unhampered by last year's disaster. We further offered to take care of them during 1912 on exactly the same basis as formerly, and agreed to charge off anything they might owe at the end of this year also,—thus absolutely guaranteeing them against any possible contingency." Despite these incentives, many of his black workers left, and Stone attributed their restlessness to innate characteristics. "In the face of every inducement which we could offer, we lost 50% of our labor. The irritating feature of it all was that these negroes, claiming to be afraid of the boll weevil, went to an adjoining county, only a few miles away, and to a much worse infected locality than this. It is the same old story; a restless spirit and a roving disposition. That is about all I can make out of it."[42]

Stone ended his letter to Willcox with an ominous reference to the weather. "We have had a most unusual winter, for this country," he wrote. "We had more than twenty-two inches of rain in December,—about eighteen above normal. Our roads became actually impassable. At the beginning of January the temperature dropped to 10, which is very low for us. The churned up mud in the roads was frozen hard, and we are now unable to travel, even [on] horseback. I live on the lower part of the plantation, two miles from the office, and my movements have been on foot. I have seen it colder here, once or twice below zero, but I have never seen such rains. It is the combination of the two which has put the country temporarily out of business."[43]

The wet winter was a harbinger of worse things to come. Heavy spring rains in the Ohio Valley had dumped an unusual amount of water in the creeks, streams, and tributaries feeding the Mississippi River. The river rose and the flood tide flowed downstream. Nevertheless, people living along its banks in Mississippi put their faith in the levees that the Mississippi Levee District had constructed to prevent flooding.[44] Unfortunately, the district authority's decision to build levees higher meant that the river did more damage if those levees broke. The wet spring of 1912 was a bad one, coming as it did on the heels of a wet winter that had saturated the ground, and seventeen of eighteen river gauges from Cairo, Illinois, to the mouth of the river recorded new heights.[45]

42. Ibid.

43. Ibid., November 18, 1910, WFWP.

44. Saikku, *This Delta, This Land*, 144–51.

45. John M. Barry, *Rising Tide: The Great Mississippi Flood of 1927 and How It Changed America* (New York: Simon and Schuster, 1997), 158. An article in the *Greenville Daily Democrat*

Stone and the other citizens of Washington County read reports of overflows in Kentucky, Tennessee, and Arkansas as the flood crest moved southward.[46] Finally, the levees began to break upriver from Greenville. "To-night's paper brings me word that the levees in your immediate vicinity are breaking and makes me fear that your plantation is now under water," Willcox wrote to Stone on April 18.[47]

Stone's reply, if he had the opportunity to write one, has not survived, but it is clear that he had other things on his mind. The levee had given way just south of Beulah, Mississippi, at a place about thirty miles north of Greenville where the levee had failed twice before in 1882 and 1884. The problem was not an overflow because engineers using black workers had topped the seventeen-foot-high levee along that stretch of the river with sandbags. The problem was a "sand boil," a leak where water under tremendous pressure from the raging river had worked its way through a weak spot in the base of the levee. By four o'clock Wednesday afternoon, April 17, the boil had developed into a crevasse, and by eight thirty that night the crevasse had become a break. Within fifteen minutes, the crown of the levee had fallen in, and the break in the sandy loam of the levee quickly widened to 150 feet. It would extend to over 1,000 feet within twenty-four hours.[48]

Millions of gallons of water rushed through the break in the levee onto the flat farmland. The torrent extended for more than a mile inland, washing houses off of their foundations and floating them on the flood "like

---

explained how the convergence of several factors resulted in spring floods on the Mississippi River: "A cold early winter, resulting in a frozen soil and the accumulation of a considerable supply of unmelted snow over the Ohio watershed; then in February a storm, almost invariably from the southwest, accompanied by heavy rains and abnormally high temperatures over the Gulf States and the Ohio valley. The ice in the Ohio River gives way, the run-off from the heavy rains is greatly augmented on account of the frozen soil, and the water from the melted snow is added to the total volume" ("Present Flood Not Greatest Possible," *Greenville Daily Democrat*, May 7). The only way the flood of 1912 departed from this scenario was its timing; the thaw and runoff were about a month later than usual. Interestingly, an authority with the U.S. Weather Bureau was quoted in this article as saying that the flood of 1912 was not nearly as bad as it could have been if more river systems feeding the Mississippi River (e.g., the Missouri River) had been involved. His observation would be confirmed fifteen years later.

46. "Hickman, Ky., Engulfed by Break in the Levee" and "Main Levees Hold Firm against Rising River," *Greenville Daily Democrat*, April 2 and 4, 1912.

47. Willcox to Stone, April 18, 1912, WFWP.

48. "Break at Hughes [Plantation]" and "Crevasse of Thousand Feet at Hughes," *Greenville Daily Democrat*, April 18 (two editions), 1912.

chicken-coops." Fourteen black men and women died, one of them an elderly woman who had been marooned at Benoit but died of exhaustion in a skiff after she had been rescued.[49]

Water will always seek low ground, and there were two water systems draining that part of the Delta. Deer Creek flowed from Lake Bolivar directly south of the break and continued southward toward Greenville before a slight rise in the ground caused it to turn eastward toward Stoneville (near Leland), where it ran under a bridge on the Southern Railroad line to Greenville. Farther east, Bayou Phalia flowed from north to south, crossing the Southern line about five miles east of Stoneville. Below Stoneville, the two streams flowed side by side, coming as close as a mile from each other at Wilmot, six miles south of Leland. The two water systems were therefore like a large funnel with the neck at Leland. Dunleith was in the neck of the funnel.[50]

Within fifty-two hours, most of Dunleith was underwater.[51] The only exceptions were slight elevations of land, ancient remnants of high ground that local residents optimistically referred to as "ridges." Fortunately, Stone's house was on one of these ridges not far from a mound built hundreds of years earlier by Native Americans. Obviously, the original inhabitants of this country knew where to go to stay dry.[52]

By Sunday, April 21, the depth of water in the fields at Dunleith had reached four feet, eight inches, a rise of fifteen inches in twenty-four hours. By Monday, most of Dunleith was under more than five feet of water, and the Greenville newspaper reported that the worst flood conditions "are centered in the vicinity of Dunleith, Holly Knowe *[sic]* and Holly Ridge," a triangular-shaped area along Bayou Phalia east of Leland. The only good news was that families from plantations in the area were safe. An estimated 2,500 to 3,000 people, mostly African Americans, were being sheltered on high ground at Dunleith and Hollyknowe. Another 1,000 were living in boxcars on the

49. "Fourteen Negroes Drowned in Vicinity of the Crevasse," *Greenville Daily Democrat*, April 19, 1912.

50. Deer Creek flows south until it enters the Yazoo River north of Vicksburg. The Bayou Phalia flows south until it makes a right-angle turn about twenty miles south of Leland to enter the Big Sunflower River, which eventually flows into the Yazoo River near Satartia. According to a contemporary account, the "bulk of the flood waters" flowed into Bogue Phalia ("Fourteen Negroes Drowned").

51. [Alfred Holt Stone], "The Overflow," *Staple Cotton Review* 5 (May 1927): 2.

52. "Steamboats Will Carry the Mails," *Greenville Daily Democrat*, April 20, 1912. For the topography of this area, see the Holly Ridge, Mississippi, quadrant in the 7.5 minute series of the U.S. Geological Survey.

railroad tracks at Leland. Finally, on Tuesday, April 23, the water stopped rising.[53]

The flood crest had passed, and overflows or breaks in the levees were now occurring south of Vicksburg. The mighty Mississippi was wreaking havoc in Louisiana.[54] Nevertheless, it would be another week and a half before the water at Dunleith and the surrounding area began to ebb. Life in the Delta began to return to normal, but the flood had interrupted cultivation of the cotton crop and left a lot of work to do.[55] The carcasses of dead livestock littered the ground, and a slimy ooze deposited by the floodwater was everywhere. Citizens of the Delta breathed a sigh of relief, not knowing that, incredibly, the same thing would happen again south of Greenville the next year.[56]

It was a hard time for Stone. Understandably, his research on the economic impact of African Americans on the country's growth and development came to a halt, and his correspondence with associates in academic circles ceased. "It is a long time since I have heard from you and I suppose from various references that Professor Farnam has made at our meetings that you have been going through a pretty strenuous time in the effort to save your investment under most trying conditions," Willcox wrote to Stone on April 22, 1914. Apparently, Willcox was referring to the annual meeting of the Carnegie Institution in 1913, at which Farnam had reported that Stone was "in the South in order to look after his property interests." Still, Farnam was optimistic. "The prospects of his [Stone's] resuming active work on his division are considerably better than they were a year ago."[57]

Encouraged by Farnam's optimism, Willcox asked in his letter whether Stone could get away for a meeting of the division heads in New York at the end of May. Stone was delighted to receive Willcox's letter. "You cannot imagine how much pleasure your letter gave me," he replied. By coinci-

53. "Flood Conditions Intensified in Bogue Phalia District," "Large Number of Sufferers Are Still without Any Relief," and "Flood Waters Are Stationary in Upper Part of Bogue District," *Greenville Daily Democrat,* April 21, 22, and 23, 1912, respectively.

54. "Lower Louisiana Levees Are in Critical Condition," *Greenville Daily Democrat,* May 2, 1912.

55. "Lower Section of Louisiana Is Flooded by Two New Breaks," *Greenville Daily Democrat,* May 2, 1912.

56. Barry, *Rising Tide,* 158. For accounts of the 1913 flood, see "All High Water Records Broken" and "Levee Gives Away above Mayersville," *Greenville Daily Democrat,* April 10 and 24, 1913, respectively.

57. Carnegie Institution of Washington, *Year Book No. 12, 1913* (Washington, DC: Judd and Detweiler for the Institution, 1914), 95–96.

dence, Farnam had just written Stone a day or two before, asking if he could come to the May meeting. Stone thought that he would go. "Frankly,—I am going more from a desire to see you than from a sense of duty to attend the meeting," he confided to Willcox. "In fact, in view of the state of my work I do not see that much is to be gained, for my division, by my going. But you may count on my being there." In fact, Stone was going to use the trip north as an opportunity to meet another obligation. In 1913, he had agreed to give a lecture for the Phelps-Stokes Foundation at the University of Virginia in Charlottesville. The foundation sponsored an annual program on race relations in the South, and Stone's being invited was recognition of his prominence in the field. The flood of 1913 had prevented Stone from making the trip, but he thought that he could do so now. Stone also suggested that he could run up to Ithaca before the meeting and ride on the train down to New York with Willcox.[58]

Stone was willing to make the effort to get back in the swing of things, but his thoughts were still shackled by the practical problems he faced trying to keep Dunleith afloat. "My state of mind during the past two years would have made me an interesting study for a psychologist," Stone wrote in his response to Willcox.

> It would be idle for me to say that during all this time I have been "unable to write," or "could not find time to write,"—or to urge any other any other stock explanation for my failure to do what I all the time wanted to do, and the doing of which would have given me genuine pleasure. I did begin a number of letters to you. I began three, I recall, the very week I finally wrote Farnam. The trouble seemed to be that there was so very much that I wanted to say, and it all was so very far removed from the current of my daily thought, that I could not get my mind in the proper trim to say it just at the moment when I might have had the time to put my words on paper. Then too, there were the manifold interruptions,—which hopelessly wrecked my train of thought, even where they did not actually disturb my physical ability to stick to my desk. It was a matter constantly in and on my mind. Daily I was quite sure that I should sit down quietly and have a long chat with you the very next day. But I always seemed too everlastingly tired to write. But I feel much better these days,—and am much more cheerful over things in general.[59]

58. Stone to Willcox from Dunleith, April 30, 1914, WFWP.
59. Ibid.

Stone made it to the meeting, spoke at the University of Virginia, and visited the Willcoxes in Ithaca.[60] His talk in Virginia was a good one: clear, temperate, and logical. Stone covered ground that he was familiar with. Slavery had been "one of the great commercial tools of the world, which could not have developed if the world had not demanded its services." Southerners may have inherited the institution, but they were not responsible for its origin. Europeans, particularly British and Spanish slave traders, had introduced slavery to this continent. The end of slavery had been anticipated, but Reconstruction messed everything up by preventing southerners from instituting social policies that would have allowed the two races to live together in harmony. Northerners did not understand the race problem in the South and had no right to impose their solutions for it. "I know his ["the negro's"] faults, his vices, his weaknesses, the inherited and apparently ineradicable limitation of his character," Stone told the audience. "I appreciate also the difficulties grounded in certain restrictions placed by the federal constitution [the War Amendments] upon the action of the state,—restrictions which make it impossible frankly to treat the negro as a negro." Anticipating a phrase that would become cliché among opponents of the modern civil rights movement fifty years later, Stone remarked that "You cannot legislate a people into either morality or prosperity." His solution was an entirely southern one, in which white men provided appropriate role models that African Americans could imitate. "The negro is an imitative creature," Stone observed, "or we may say that his race is in its imitative stage." According to Stone, becoming a mentor to the black man was the southern white man's primary responsibility to the African Americans.[61]

Despite the delivery of his address in Virginia and participation at the spring meeting of division heads for the Department of Economics and Sociology, Stone did not regain momentum for his research. "For the reasons explained in earlier reports," Farnam wrote in the report that appeared in the Carnegie Institution's *Year Book* for 1914, the study of the Negro problem, under Mr. A. H. Stone, . . . [is] suspended for the present."[62] With this cryptic announcement, Stone's formal association with the Carnegie Institution came to an end.

As if the boll weevil and flood of 1912 were not enough of a burden, in 1917 Stone experienced another setback that came out of the blue. His longtime

60. For confirmation of Stone's trip to Ithaca, see Willcox to Stone, May 26, 1914, WFWP.

61. Alfred Holt Stone, "The Responsibility of the Southern White Man to the Negro," in *Lectures and Addresses on the Negro in the South* (Charlottesville, VA: Michie, [1915?]), 5–18.

62. Carnegie Institution of Washington, *Year Book No. 13, 1914* (Washington, DC: Judd and Detweiler for the Institution, 1915), 105–6.

partner, Julian H. Fort, the man who had run Dunleith while Stone was pursuing his career as a racial theorist in Washington, died. Fort was only forty-three, and the stroke he suffered in the spring came as a complete surprise. Partially paralyzed, Fort held on through the summer and fall, only to expire a week after Thanksgiving. His death could not have come at a worse time.[63]

Although Stone did not give any more talks, write any more articles, or review any more books about racial issues after his address in Virginia, it was not until 1918 that Stone's career as a racial theorist can be said to have definitely ended. The occasion marking this milestone was Stone's agreeing to review Ulrich Bonnell Phillip's groundbreaking book, *American Negro Slavery.*[64] Both Phillips and the book review editor, J. Franklin Jameson, were friends of Stone, and the book was ideal for Stone to review because of his extensive work on the topic for the Carnegie Institution.[65]

Jameson was pleased that Stone had agreed to review Phillips's book and immediately sent him a copy. Jameson set a deadline of August 10, 1918, so that he could put the review in the October issue. Jameson was effusive in his praise for Stone's qualifications. "We [Jameson and a mutual acquaintance that Jameson had referred to in the letter] should both like mighty well to see you Governor of Mississippi," he wrote, "but if you can do more good in other ways, so be it. I wish, however, that some time it might come about that your duties should lie permanently in Washington city instead of in Washington county."[66]

The United States had entered the First World War by then, and Stone was heavily involved with the war effort in Mississippi.[67] Consequently, he

63. "Julian H. Fort Dead," *Greenville Daily-Democrat-Times* (afternoon edition), November 29, 1917.

64. Stone to Jameson from Dunleith, June 28, 1918, JFJP; Phillips, *American Negro Slavery.*

65. John Franklin Jameson was a well-connected and influential historian who later oversaw the centralization of federal records in the National Archives Building in Washington, D.C. He also served as the editor of the *American Historical Review* for thirty-two years, was instrumental in the publication of the *Dictionary of American Biography,* and headed the Carnegie Institution's Department of Historical Research for twenty-five years. In 1907, Jameson was the first professional historian to be elected as the president of the American Historical Association, establishing a precedent that has continued more or less intact until today (Highham, *History,* 20–25; David D. Van Tassel, "John Franklin Jameson," in *Keepers of the Past,* ed. Clifford L. Lord [Chapel Hill: University of North Carolina Press, 1965], 81–82). For an indication of the Jameson-Phillips-Stone connection, see Phillips to Jameson, July 8, 1908, JFJP.

66. Jameson to Stone, July 12, 1918, JFJP.

67. Stone was active as a volunteer with the Red Cross, YMCA, and Jewish War Relief, an organization established in 1914 to disperse humanitarian aid to Jews in European war zones and Palestine (Albert Lucas, "American Jewish Relief in the World War," *Annals of the American Academy of Political and Social Science* 79 [September 1918]: 221–28). He also served on Wash-

did not have time to answer a lengthy letter from Jameson in August that sought his suggestions regarding a project to document African slavery in the United States.[68] Nor did he find time to write the review of Phillips's book. Jameson waited seven weeks before prompting Stone for a response. "I do not wish to press you unduly, but if in the near future I might have your reply to my letter of August 5, I should appreciate it very much," Jameson wrote. "I desire to have your excellent judgment on the subject before proceeding much further with the work in question."[69]

"I am not fighting in this war," Stone replied, "but I am doing lots of running. In four weeks I have not spent four consecutive days at home. I am on my way to Memphis & Atlanta today. I hope to return Wednesday, and shall write you at once. I was much interested in what you wrote, and appreciated your calling on [me]."[70]

The next correspondence Jameson received from Dunleith was not from Stone. It was from his wife, Mary. "Mr. Stone asks me to tell you that he was taken ill the day after his return from his recent trip," she wrote. "He has had a severe attack of influenza. While he has, so far, escaped pneumonia, he is far from well. He wants me to assure you that he will write you, just as soon as he is able to sit up."[71] The obstacles Stone had to overcome to continue his research were daunting: the boll weevil, loss of research material, floods, a world war, the death of his partner, and now influenza. Events seemed to be conspiring to keep him from being a racial theorist.

The review of Phillips's book came out a year later in the October 1919 issue of the *American Historical Review*, but Stone did not write it.[72] The author was Theodore D. Jervey, an amateur historian and attorney from Charleston, South Carolina.[73] Although Stone would live a very full life in the thirty-five remaining years allotted to him, his career as a racial theorist was over.

---

ington County's Council of Defense. Mary Stone also took part in several volunteer projects, including the YWCA and the Woman's Committee for the Council of Defense. For documentation of these activities, see the World War I Scrapbook for Washington County compiled by the Daughters of the American Revolution, Belvidere Chapter, at MDAH.

68. Jameson to Stone, August 5, 1918, JFJP.

69. Ibid., September 25, 1918, JFJP.

70. Stone to Jameson from Dunleith, October 4, 1918, JFJP.

71. Mary Ireys Stone to Jameson, October 18, 1918, JFJP.

72. Theo[dore] D. Jervey, review of *American Negro Slavery: A Survey of the Supply, Employment, and Control of Negro Labor as Determined by the Plantation Regime*, by Ulrich Bonnell Phillips, *American Historical Review* 25 (October 1919): 117–18.

73. Smith, *Old Creed for the New South*, 269.

13

IN PUBLIC DUTY

Alf Stone pursued multiple, overlapping interests during the remainder of his life. One of those was cotton, which he grew at Dunleith. He also became the first vice president of the Staple Cotton Cooperative Association and edited its newsletter from 1923 until his death. Another of Stone's interests was public service, first as a state legislator and later as the state's tax commissioner for four consecutive terms from 1932 until his death. Stone was also interested in history and spent much of his free time pursuing the state's storied past, first as a president of the Mississippi Historical Society and later as a member of the board of trustees for the Mississippi Department of Archives and History (MDAH) from 1933 until his death. The pattern of Stone's achievements during the last thirty-five years of his life is clear. If he latched on to something he was interested in, he stuck with it until he died.

Cotton was the reason Stone gave up his career as a racial theorist, and it was the business of growing cotton that consumed most of his attention after he returned to Dunleith in 1909. The boll weevil was the first challenge he faced, but it would not be the last. An overflow from the Mississippi River in 1912 inundated Dunleith, and flooding the next year created havoc throughout the Delta, although it did not damage Dunleith extensively.[1]

The floodwaters had scarcely receded when the First World War erupted in Europe. Although it would be some time before the impact made itself felt in the United States, Stone assumed an active part in response to the worldwide crisis when it arrived. He was a member of the Washington County

1. One reason the flood of 1913 did less damage was because it occurred in January, before the cotton crop had been planted (Stone, "The Overflow," 3).

Council of Defense and the local chapter of the Red Cross. In addition, he was active in work for Jewish war relief and organized the black workers at Dunleith for a war bond drive. Mary Stone was also active in the Woman's Committee for the Council of Defense and the YWCA.[2]

In 1916, Stone was elected to the Mississippi State Legislature as a representative from Washington County. He served two, four-year terms. His tenure in the legislature gave Stone experience in state governance that would later prove to be useful in many ways. Perhaps the most important experience in that regard came as the result of his being appointed to a joint committee charged with studying the tax equalization of property throughout the state.[3]

Stone's service in the legislature also allowed him to become an advocate for historic preservation. Within one week of assuming office, Stone introduced a bill to determine the feasibility of restoring the Old Capitol. The state's new Capitol building had been dedicated on Jefferson Davis's birthday in 1903, and the Old Capitol on State Street was in a state of ruin. Some people wanted to tear it down, but Stone and other legislators fought for its renovation. They won, and Stone was largely responsible for allowing the Old Capitol to duck the wrecking ball.[4]

Despite his legislative duties in Jackson and interest in historic preservation, Stone's heart was in the Delta. Traditionally, cotton growers in the Mississippi Delta were at the mercy of the cotton brokers, called "factors," who bought their crop and sold it at a profit to buyers for the spinning mills.[5] Cotton growers had little leverage in these transactions because they did not have an avenue by which to sell their cotton directly to the mills. As a result, factors paid the growers low prices, regardless of what the market was doing, and financial necessity usually prevented the growers from holding on to

2. Undated newspaper clippings in Washington County, World War I Scrapbook, Daughters of the American Revolution, Belvidere Chapter, MDAH (Z1899).

3. Mississippi Legislature, 1916–1918, *Joint Report of the Senate and House Committees Appointed at the Session on 1916 to Consider the State's Revenue System and Fiscal Affairs* (Jackson, MS: Tucker Printing, [1918]), 63 pages.

4. A. S. Coody, "Repair of and Changes in the Old Capitol," *Journal of Mississippi History* 11 (April 1949): 89–96; John Ray Skates, *Mississippi's Old Capitol: Biography of a Building* (Jackson: Mississippi Department of Archives and History, 1990), 125–26.

5. Stone was particularly interested in cotton factors, as was evidenced by his paper "The Cotton Factorage System of the Southern States," which he read at the 1914 annual meeting of the American Historical Association in Chicago. It was later published in the *American Historical Review* (20 [April 1915]: 557–65).

their crop until they could negotiate a better deal. What the growers needed was a middle man who would treat them fairly; and a growers' cooperative, in which members would share the profits, was an organization that could accomplish that goal.

In 1919, Oscar E. Bledsoe, a cotton planter near Greenwood, Mississippi, overheard two cotton factors talking on a train to Memphis. They were bragging about how easy it was to manipulate what they paid growers for their crop. The conversation angered Bledsoe, and he determined to do something about it. As soon as he got back from Memphis, he talked with three other men from the Delta who knew the cotton-growing business inside and out. One of them was Alf Stone. The other two were Stone's old friend, LeRoy Percy, and Oscar Johnston, an attorney from Clarksdale.[6]

Between them, the four men agreed to form a growers' cooperative to purchase and market their crops. The name of the organization they formed was the Staple Cotton Cooperative Association, usually referred to by its abbreviated name, Staplcotn. Through the spring and summer of 1920, they explained the business model to other growers and attracted partners. A market plan for the cooperative was in place by April, and soon ten planters from the Delta had pledged a thousand dollars apiece to get the organization going. By May 1921, 1,800 cotton growers from ten counties in the Delta had signed contracts with Staplcotn, pledging to sell their cotton to the cooperative for a price to be negotiated on their behalf. Staplcotn was incorporated in Memphis, Tennessee, on May 23, 1921. Fifteen members of the Staplcotn board of directors attended their first meeting two days later at the Gayoso Hotel. Stone was there, as well as at their second meeting on June 14, when they elected him vice president with a monthly salary of two hundred dollars. Stone was the first chief administrative officer for Staplcotn.[7]

6. Noel Workman, "Staplcotn: The First 75 Years," www.staplcotn.com/History/index.htm. Details in this paragraph come from chapter 2, "The Train Ride That Changed the Cotton Business." Stone would have been an obvious person for Bledsoe to approach because of Stone's well-known advocacy of cooperative ventures to advance economic development in the Delta ("Alfred H. Stone on Co-Operation," *Greenville Daily Democrat*, December 18, 1912). Stone continued his interest in agricultural cooperatives on a national level after Staplcotn's success ("2,000,000 Farmers to Unite in a Cooperative Move on a Nation-Wide Scale," *New York Times*, July 30, 1929).

7. Workman, "Staplcotn: The First 75 Years," chap. 2, "The Train Ride that Changed the Cotton Business," and chap. 3, "A New Way to Do Business." Staplcotn's service area included Coahoma, Quitman, Bolivar, Sunflower, Leflore, Washington, Issaquena, Sharkey, Humphreys, and Holmes Counties.

Staplcotn was not an instantaneous success. It had to contend with suspicion of its new way of marketing cotton and aggressive efforts by people who were being bypassed to sabotage the project by spreading false rumors. One thing the cooperative had going for it was the decision not to deal in all types of cotton but to focus instead on the long staple variety alone. This variety was more desirable in the mills because of its inherent advantage in the spinning process. Staplcotn capitalized on that decision by developing a classification system that became standard in the industry for the long staple product.

The boll weevil made growing the long staple variety of cotton more difficult because of the long time it takes to mature. Nevertheless, Staplcotn survived the lean years and began to prosper. Stone was an important part of the effort to make it work, and on February 1, 1923, he assumed duties as editor of the *Staple Cotton Review.* Although Stone eventually got out of the cotton-growing business, he would continue as editor of Staplcotn's newsletter for thirty-two years. He also served on the board of directors until he was named emeritus director for life in 1954, the year before he died.[8]

Although the Delta produced a large cotton crop for the 1925–26 season, the grade was low, and members of Staplcotn were still struggling to make ends meet when the biggest threat to their livelihood yet came rushing over their fields in 1927.[9] The levee broke on Thursday morning, April 21, at Mound Landing near Scott, Mississippi, the small town about thirteen miles north of Greenville and about three miles from where Captain Walter Wilson Stone had stepped off the ferry on his first visit to the Delta.[10] Before it was over, the overflow would become known as the greatest flood of the Mississippi River in recorded history.[11] More than 5.2 million acres of delta

8. Ibid., chap. 3 "A New Way to Do Business," and chap. 4, "Surviving Tough Times." Stone's appointment to emeritus status comes from "The Staplcotn Story," www.staplcotn.com/History/timeline.pdf.

9. A[lfred] H[olt] Stone, "As to 1925 Low Grades," *Staple Cotton Review* 5 (March 1927): 3.

10. Stone's father, Walter Wilson Stone, was still alive and living in Greenwood. Captain Stone died on June 4, 1930, in Greenville, Mississippi, where he had gone to attend a high school graduation ("Captain Stone Death Victim," *Jackson Daily Clarion-Ledger,* June 5, 1930). Captain Stone is buried in Greenwood Cemetery in Jackson, Mississippi (sec. 6, lot no. 92, [Grave Registration Form, MDAH]). His brother, Owen, and Owen's wife, Mary (Captain Stone's wife's sister), are buried in the same plot.

11. "Total of Homeless in Mississippi Valley Floods Shatters All Previous Disaster records," *Greenville Daily Democrat-Times,* May 17, 1927.

land flooded, and 246 people were drowned along with an estimated 165,000 head of livestock that could not escape the rapidly rising water. Over 333,000 people were made homeless, and water stood fifteen feet deep in places.[12]

The topography of the Delta had not changed since the flood of 1912, which meant that Dunleith was again in the mouth of a drainage basin formed by the convergence of Deer Creek and Bogue Phalia. Water could be seen collecting in low places at Dunleith on Friday afternoon, and the fields were covered by sunrise the next day.[13] The water continued to rise until April 27, when it reached an average depth of almost seven feet.[14] Refugees on the high ground between Leland and Arcola, the neck of the funnel, were cut off from the outside world without food or drinking water.[15] "As I look out across this island sea," Stone was quoted as saying, "I marvel at the foolishness which kept me here, in response to the same subtle dream of wilderness conquest, which brought my forefather three centuries ago."[16]

Growing cotton in the Mississippi Delta was a risky business, and it was hard for anyone, white or black, to get ahead. But not all was lost. Staplcotn had flood insurance, and the insurance adjuster inspected the 33,000 bales Staplcotn had stored in warehouses that flooded. Eventually the insurance company paid Staplcotn more than $2 million for its losses, which proved to be salvation for the members.[17] Money inspired hope, and Stone was determined to stick it out. "We shall weather the storm," he wrote to L. O. Crosby, a timber man who had been put in charge of the state's response to the flood. "We shall stay here and see it through."[18]

The flood of 1927 made it clear that the individual states bordering the Mississippi River were not capable by themselves of protecting their citizens

12. Workman, "Staplcotn: The First 75 Years," chap. 6, "The River's Last Great Flood."

13. [Alfred Holt Stone], "The Aftermath," *Staple Cotton Review* 5 (August 1927): 4–5.

14. Stone, "The Overflow," 2. The maximum depth of the water at Dunleith during the 1912 flood had reached five feet, four inches.

15. "Rescue Work Continues as More Land Submerges: Leland and Arcola Cut Off 'from the outside world without food or drinking water,'" *Greenville Daily Democrat-Times*, April 24, 1927. Although water began to recede in late April, it was July 10 before the road from Dunleith to Greenville was passable and another three days before the rest of the water was gone ([Stone], "The Aftermath," 5).

16. Workman, "Staplcotn: The First 75 Years," chap. 6, "The River's Last Great Flood." Stone was probably referring to the wilderness in Virginia where his forebearer William Stone settled after immigrating from England in 1628.

17. Ibid.; Stone, "The Overflow," 3–4.

18. Stone to Crosby, September 1, 1927, Red Cross Papers, NA, quoted in Barry, *Rising Tide*, 363.

from periodic overflows. Consequently, many of Mississippi's leaders went to Washington to lobby for passage of a federal flood-control bill. Stone was among this group. A member of the Flood Control Committee appointed by the United States Chamber of Commerce, Stone worked along with other influential advocates, including LeRoy Percy, who were successful in getting a federal flood-control measure passed in March 1928.[19] The new legislation marked the first time the federal government recognized that flooding along the Mississippi River was a national problem that had to be dealt with centrally rather than being left to the individual states.[20] The legislation authorized $300 million to strengthen the levees along the river. In addition, the federal act empowered the Levee Board in New Orleans, which had been largely ineffectual to that point, to help guard the Crescent City from overflows.[21]

The flood of 1927 was bad enough for the Delta, but the disaster created by the collapse of the stock market two years later inundated the entire state. By the time Governor Mike Conner was inaugurated in January 1932, Mississippi was in dire straits. There was only $1,326 in the state treasury to cover outstanding debts that totaled almost $6 million. Teachers in the public schools had not been paid in full for almost a year, and one of the first acts the state legislature passed when it convened that year was a concurrent resolution authorizing the state tax commissioner to borrow $750 so he could buy stamps to mail income tax notices to the taxpayers.[22]

Conner was convinced that a new source of revenue had to be found. The state's economic condition was such that increasing taxes on income and property would create more problems than it would solve. Conner favored a sales tax, but revenue measures had to originate in the legislature, and the idea of imposing new taxes, particularly a sales tax, met with stiff opposition in both houses. Merchants were against it, and there was little sympathy for

19. "Hoover's Opinions on Flood Problem Sought by Senate," *New York Times*, January 27, 1928; Barry, *Rising Tide*, 406–6. Stone was active in the national chamber with agricultural programs in general ("Farm Plans Carry in Business Vote," *New York Times*, October 18, 1928).

20. A[lfred] H[olt] Stone, "The Mississippi River National Problem: Memorandum Submitted to the Flood Control Committee of the House," *Staple Cotton Review* 5 (December 1927): 2–12.

21. "Katrina Coverage," katrinacoverage.com/2005/12/09/comment-with-historical-background.html.

22. William Winter, "Governor Mike Conner and the Sales Tax, 1932," *Journal of Mississippi History* 41 (August 1979): 213–15.

a sales tax among the public at large.[23] But the economic crisis did not abate, banks continued to close, and something had to be done.[24]

Conner knew that his work was cut out for him. Revenue measures had to pass with three-fifths majorities (60 percent), and opposition to the tax bill was well organized. Debate began in early March. Walter Sillers served as the floor leader for the tax bill in the House, and the issue came to a final vote on April 19. The measure needed eighty-two votes to pass, and when the roll was called, only eighty-one "ayes" were counted. But just before the clerk announced the results, Oscar Wolfe, the representative from Bolivar County in the Delta, rose to change his vote. Wolfe later said that he had decided to vote "aye" when he realized that the legislature was deadlocked and failure to pass the tax measure would be a disaster for the state.[25]

A new 2 percent sales tax went into effect May 1, 1932, and Governor Conner appointed Alf Stone Chairman of the Tax Commission to implement the measure.[26] Conner knew that Stone was the right man for the job. Stone understood the importance of making the new tax work, and he proved to be skillful in eliciting cooperation from the merchants who collected the sales tax and the public who had to pay it. "As I see it the situation which confronts the state is as critical as any in its history," Stone said when he was interviewed about his new appointment. "I am going into office with no pre-conceived ideas as to any aspects of its operations," he added. "I have made no obligations other than those which I acknowledged to Governor Conner and the legislature for affording me an opportunity of public service."[27]

23. For examples of the newspaper ads opposing the sales tax and marches organized to protest its passage, see V. B. Wheeless, "The Sales & Use Tax: Its Origin and Background in Mississippi through 1965" (typescript, 1966, MDAH), 8–16. The pages containing photocopies of newspaper clippings are not numbered.

24. "Sales Tax Fight Will Finally End with House Vote," *Jackson Daily Clarion Ledger,* April 19, 1932; Winter, "Governor Mike Conner," 214–19.

25. "Solons Speed Business after Voting Sales Tax; Early Adjournment Seen," *Jackson Daily Clarion-Ledger,* April 20, 1932; Wheeless, "Sales & Use Tax," 16; Winter, "Governor Mike Conner," 226–28.

26. Conner had wanted a 3 percent sales tax but failed to get it (Wheeless, "Sales & Use Tax," 19). In addition to appointing Stone tax commissioner, Conner reorganized the Tax Commission. Wheeless does not provide an explanation as to why Conner picked Stone for the job.

27. "Sales Tax Strikes Mississippians at Midnight's Gong," *Jackson Daily Clarion-Ledger,* May 1, 1932.

Stone did such a good job as state tax commissioner that he was appointed to four consecutive terms and served under six governors subsequent to Mike Conner.[28] There were several reasons for his longevity in office. For one, he was fair. For another, he was honest. And in addition to those important characteristics, Stone was an excellent administrator.[29]

The first crisis Alf Stone faced as the new tax commissioner was a dispute over the valuation of property used in the tax assessments. The assessed value of equivalent parcels land varied widely from county to county because the state did not have a uniform and impartial procedure for affixing values. Personal and political influence had corrupted the system. Stone tackled the problem head on, using his interpersonal skills to nudge assessors into compliance and his power as the tax commissioner to demand cooperation when his charm failed to have the desired effect. "I was neither a lawyer, an accountant nor an engineer," Stone recalled, "but I had both common honesty and common sense. With the full cooperation of my associates, although with inadequate funds, we determined to develop an efficient and a dependable valuation service." It took him two years to reform the system, but Stone prevailed, and the taxpayers of the state benefited from an equalization of the tax burden.[30]

With the controversial issue of tax valuation resolved, Stone turned his attention to the sales tax. He discovered that while collection of the new tax was proceeding without a serious hitch in most of Mississippi, it had encountered difficulties in counties that bordered neighboring states. Part of the problem was enforcing the collection of the 2 percent sales tax on purchases of less than fifty cents because coins small enough to pay the tax were not available. If a person bought something for twenty-five cents, the cashier would collect a penny for the sales tax, which was 4 percent of the purchase. No one liked paying a higher tax rate than allowed by law, particularly people in bordering states who received no benefits from the tax.

28. The six governors succeeding Mike Conner (1932–36) were Hugh Lawson White (1936–40); Paul Burney Johnson Sr. (1940–43); Dennis Murphree (1943–44); Thomas Lowry Bailey (1944–46); Fielding L. Wright (1946–52); and Hugh Lawson White (1952–56). Stone was reappointed tax commissioner in 1938, 1944, and 1950 ("Alf Stone Named Commission Head another 6 Years," *Jackson Daily News*, April 3, 1950; "Alf H. Stone Confirmed for Fourth Consecutive Term," *Jackson Clarion-Ledger*, April 6, 1950).

29. Wheeless, "Sales & Use Tax," 18. Wheeless observed that Stone, as the tax commissioner, "was to prove to be absolutely honest, absolutely fair, and as effective as humanly possible."

30. A[lfred] H[olt] Stone, *The Assessing of Public Utilities by the State Tax Commission* (Jackson: Mississippi State Tax Commission, 1935), 6–10.

To solve the problem, Stone imported an idea that had been tried in Oklahoma but modified it to make it more effective. The idea was to issue tokens as fractional currency in two denominations: one-mill or five-mill, which were equivalent to 1/10 and 1/2 of a cent, respectively. The one-mill tokens were made of aluminum; the five-mills token of brass. Stone made sure that each token bore the following inscription, "to make change for correct sales tax payment." The state tax commission continued to issue tax tokens until 1952, although it had to switch to fiber tokens during the Second World War and plastic tokens later. Tens of millions were bought and put into circulation. It was estimated that 53 million of the aluminum and brass tokens were in the system in 1940. Regardless of the final number, Alf Stone was in charge when each and every one was issued.[31]

Working closely with the banks, Stone set up a system for the distribution and redemption of the tokens, which were treated as if they were money. The availability of a fractional currency coupled with an extensive effort to educate retailers and customers increased the efficiency of tax collection in Mississippi. In addition, Stone oversaw the deployment of auditors to make sure that everyone was playing by the rules. What Stone liked about his plan was that "these tokens made possible a more equitable collection of the tax."[32] Although many states eventually used tokens to increase the efficiency of sales tax collection, Stone was one of the first tax administrators to adopt fractional currency as means of improving the collection of a sales tax.[33]

Despite the issuance of tax tokens and consistent enforcement of the sales tax, there was one loophole that Stone had trouble closing: bootleg beer. Technically, Mississippi was a dry state. Although the Twenty-first Amendment, which repealed the Eighteenth (Prohibition) Amendment, had become effective in December 1933, the Mississippi legislature had not voted on the measure, nor had it passed legislation authorizing the sale of liquor.

31 The figures come from columnist Jack Sunn's two columns about Mississippi's tax tokens in the *Jackson Daily News*, "Token of a Day Gone & Forgotten" and "Tax Token History," March 24 and April 4, 1980, respectively. More detail about the manufacture and procurement of the Mississippi tax tokens than most readers would care to know can be found in a typescript by V. B. Wheeless entitled "The Sales Tax Token in Mississippi" (typescript, 1974, MDAH). Photocopies of Sunn's columns and Wheeless's typescript can be found in the "Tax Token" subject file, MDAH.

32. A[lfred] H[olt] Stone, "Tokens Add Effectiveness to Mississippi Sales Tax," *National Association of Tax Administrators Report* (December 1936): 18.

33. The first tax tokens were issued in 1935. ("Sales Tax Tokens," users.pullman.com/fjstevens/tokens/taxinfo.html). Twelve states, including Mississippi, eventually issued fractional currency to increase the efficiency of sales tax collection.

In fact, it would not be until 1966 that Mississippi ratified the Twenty-first Amendment, the last state to do so, but the door to alcohol had been opened in 1934 by a law that allowed Mississippians to buy beer containing less than 4 percent alcohol by volume. Still in the death grip of the Great Depression, the legislators who allowed Mississippians to drink beer acted quickly to tax their brew. The revenue was to come from a sales tax on the beer itself and a privilege tax on the wholesalers, distributors, and retailers who sold it. Collection of this revenue fell to the state tax commission.[34]

Regulating wholesalers, distributors, and retailers was not a problem, but stopping untaxed beer from entering the state was. There was no easy way to determine whether a particular bottle or can served over the counter had come from a taxpaying distributor. Stone's solution was to force beer manufacturers to purchase bottle caps indicating that the sales tax had been paid at the source. In short, he set up a procedure for collecting the sales tax at the beginning, not the end, of the barley pop's hop from brew kettle to belly. "Manufacturers will have to buy caps especially designed to show that tax has been paid—there will be no more revenue stamps," he explained. "Both bottles and cans of beer will be required to use the 'revenue cap' and not only will it eliminate bootlegging, but it will bring more money to the state treasury and much more quickly, because the manufacturer has to buy the caps from the Tax Commission, paying cash!"[35]

Stone's innovations and excellent management skills brought him to the attention of tax administrators across the country. Just as he had been active in several professional organizations during his career as a racial theorist, Stone joined the ranks of national organizations concerned with the development and implementation of tax policies. In 1937, Stone was elected president of the National Association of Tax Administrators (NATA), which he helped organize,[36] and he was elevated to the same position in the National Tax Association (NTA) a year later.[37] It was on his way home from the NTA meeting in New York that Stone typed his recollections of life in the Delta. In

34. Tom Spight Hines, "Mississippi and the Prohibition Controversy" (master's thesis, University of Mississippi, 1960), 98, 104–7.

35. Untitled newspaper clipping in Stone's subject file, MDAH, from the *Webster [County] Progress*, December 15, 1938.

36. Stone was recognized for a lifetime of service to the NATA at their annual meeting in Yellowstone National Park in 1953 ("Tax Administrators Vote Alf Stone Due Recognition," *Jackson Daily News*, July 19, 1953). Stone was not able to attend due to a hernia operation ("3 State Officials Reported Better after Illnesses," *Jackson Daily News*, June 2, 1953).

37. Stone's presidency of the NTA comes from his obituary in the *Jackson Clarion-Ledger*, May 12, 1955. The NTA was founded in 1907 as "a nonpartisan, nonpolitical educational asso-

1943, he was named chairman of the board of the Federation of Tax Administrators (FTA).[38] In 1946, fourteen years after Stone had assumed the post of tax commissioner in a state that had only $1,326 in the treasury, it was estimated that Stone had overseen the collection of $237 million.[39]

Alf Stone's duties as the state tax commissioner did not prevent him from pursuing his passion for southern history. He had expressed an interest in history since his days as a student at the University of Mississippi when he joined the Mississippi Historical Society. Unfortunately, that iteration of the society had failed within four years, largely because its meetings were closed to the public and because its limited membership was too penurious for the organization to survive. But that state of affairs had changed in 1897, when Franklin L. Riley revived the society. Stone took an active role in the society's revival and remained active during the ten years during which the former Confederate general and state icon Stephen D. Lee served as the society's president (1898–1908).[40] When Lee died in 1908, another Confederate veteran, Richard Watson Jones, served as president until he died. Stone was handed the gavel as the society's third president in 1911.[41]

Alf Stone had climaxed his tenure as president of the Mississippi Historical Society with an address to the members on January 9, 1914. His address was entitled "Fact and Tradition in Southern History." In it, Stone mourned the passing of a truly southern identity. The society could be instrumental in reversing the trend, he argued, by becoming "a forum for the discussion of every fact and phase of southern history. And the only standard which should obtain in such discussions, and in the investigations preparatory thereto," he added, "is the single standard of truth."[42]

ciation that fosters study and discussion of complex and controversial issues in tax theory, practice and policy, and other aspects of public finance" ("About NTA," www.ntanet.org).

38. "Alfred Stone, Veteran Tax Official, Dies," *Jackson State Times*, May 12, 1955. The FTA was organized in 1937 by representatives from three independent organizations for tax administrators, one of which was the NATA. The purpose of the federation was "to improve the quality of state tax administration by providing services to state tax authorities and administrators" ("Facts about the Federation," www.taxadmin.org/FTA/ftafact1.html).

39. "Taxes, Greetings Fill His Coffers," *Memphis Commercial Appeal*, May 3, 1946.

40. Riley, "Work of the Mississippi Historical Society," 35–38.

41. Jones's Confederate service comes from a footnote in R[ichard] W[atson] Jones, "Confederate Cemeteries and Monuments in Mississippi," *PMHS* 8 (1904): 87.

42. Stone's address was reprinted in the *Memphis Commercial Appeal* on January 18, 1914, and forty-one years later, shortly after his death, in the *Journal of Mississippi History* (17 [July 1955]: 1–23).

Stone's work with the Mississippi Historical Society meant working closely with a good friend, Dunbar Rowland, who was the director of the Mississippi Department of History and Archives. In September 1933, a slot on the department's board of trustees became available when Judge Stone Deavours died. Stone's name quickly surfaced as a replacement. "He is an historian and economist, and has always taken a decided interest in the history of Mississippi and in the work of the Department," Rowland wrote in his endorsement of Stone to the other directors. Rowland also noted that Stone had been an active member of the Mississippi Historical Society for thirty years as well as serving as its president.[43]

The sitting directors voted unanimously to welcome Stone as a new member. "I do not know of anyone whom I would favor more highly as the successor to Judge Stone Deavours than Honorable Alfred Holt Stone," John F. Frierson wrote in his letter endorsing Stone's appointment. "No better nomination could have been made," Bishop Theodore Du Bose Bratton stated, and the compliments went on. "So there is your laurel wreath, my dear friend," Rowland wrote in his letter informing Stone of his election. "We congratulate ourselves that we are to have the benefit of your wise counsel during the coming years. You know what I have been trying to do all these years [with the Department of Archives and History], hence, it is hardly necessary for me to tell you how delighted I am personally and officially to have you associated with us."[44]

A year or so after Alf Stone joined the board of the MDAH, he deposited his collection of material entitled "The Negro and Cognate Subjects" at the MDAH. It is not clear when the collection arrived, but what we know for sure is that it was there in 1937.[45] It is possible that Stone made the decision to store the large collection at the MDAH when he let Dunleith go back to the bank in 1932[46] or when Mary Stone moved to Jackson to join her husband in 1934.[47] Whatever the case, the collection was initially on loan

43. Rowland to Judge R. H. Thompson, September 26, 1933, Dunbar Rowland Letter Books and Correspondence (box 4743), MDAH.

44. Rowland to Stone, October 10, 1933, Dunbar Rowland Letter Books and Correspondence (box 4743), MDAH.

45. Handwritten notation dated September 2, 1937, in the margin of a typed bibliography of hardbound books entitled "The Stone List" in Stone's subject file, MDAH.

46. Brieger, "Hometown Mississippi," 516; "Dunleith, Washington County," typescript in the Dunleith subject file, MDAH. Stone had renewed a deed of trust for Dunleith in June 1932 in the amount of $39,311.85, quite a sum to borrow in the depths of the Depression (Book 245, pages 71–72, WCCH).

47. Personal communication (e-mail) from Princella W. Nowell to Anne Lipscomb Web-

but eventually became a permanent part of the department's holdings. As a member of the board, Stone would have access to his collection whenever he needed it.

In the spring of 1939, Eri Douglass, director of the Federal Writers' Project for Mississippi, presented the MDAH with a typewritten volume entitled "Index of Alfred H. Stone Collection on the Negro and Cognate Subjects Loaned to the Department of Archives and History."[48] The compiler of the index was Norma Wilkins, and the bibliography that accompanied the index provided a complete list of articles, pamphlets, and monographs contained in the collection's 113 volumes. Unfortunately, the index itself was cursory and, in many cases, misleading. The biggest problem with the index was the choice of entries. One entry, for example, was "speeches," with no subdivision to identify specific topics. Another curious example of the indexing system Wilkins used was the inclusion of "Negro" as an entry in the index for a collection in which every item was purportedly about that subject. Nevertheless, the bibliography that Wilkins compiled yielded a definitive accounting of items in the Stone Collection and allowed the researcher to browse a list of over three thousand titles, if he or she had the time to leaf through 417 pages of single-spaced text in the two large, three-ring notebooks that held the document.[49]

Although Stone was not writing much anymore by the late 1940s, he was still sought after as a speaker. His favorite topic was the Constitution of 1890,

---

ster, July 12, 2005, in the author's possession. Entries for Stone in various editions of biographical directories (e.g., *Who's Who*) indicate that Alf and Mary lived at 306 North State Street until they moved to 817 Arlington Street in the Belhaven area sometime in the late 1940s or early 1950s.

48. During the Depression, the Federal Writers' Project "supported over six thousand journalists, freelance writers, novelists, poets, Ph.D.s, and other jobless persons experienced in putting words on paper" (Dixon Wecter, *The Age of the Great Depression* [New York: Macmillan, 1948], 260). The Writers' Project was controversial in several parts of the country, such as New York, because of the socialist views of some of the participants. However, the project in Mississippi appears to have operated successfully without controversy (Eri Douglass, "The Federal Writers' Project in Mississippi," *Journal of Mississippi History* 1 [April 1939]: 71–76). The primary accomplishment of the Mississippi Writers' Project during its existence from 1935 to the Second World War was the publication of two volumes in the American Guide Series: *Mississippi: A Guide to the Magnolia State* (New York: Viking, 1938; reprint, New York: Hastings House, 1949), and *Mississippi Gulf Coast: Yesterday and Today, 1699–1939* (Gulfport, [MS]: Gulfport Printing, 1939). The definitive history of the Federal Writers' Project is Jerre Mangione, *The Dream and the Deal: The Federal Writers' Project, 1935–1943* (Boston: Little, Brown, 1972).

49. "Historical News and Notices," *Journal of Mississippi History* 1 (July 1939): 200–201.

especially its sections that related to suffrage.[50] In these talks, he was fond of pointing out that the constitutional convention that drafted the document "did not legislate against the negro as a race." Rather, it "legislated against his racial characteristics." Stone would cite the poll tax as an example, explaining that it was not the two dollars that had an effect but, rather, the fact that the voter had to pay it two years in advance kept large numbers of African Americans from voting. But Stone did not attempt to disguise the true purpose of the constitution's "understanding clause." He admitted freely to his white audiences that the chancery clerks used that clause as a means of denying registration to African Americans regardless of their educational level. As far as the illiterate white man was concerned, Stone quoted a chancery clerk he knew when he asked about how the clerk handled that situation. "Easy enough," the clerk said, "if he's a white man, I just register him."[51]

As far as race relations were concerned, Stone's thinking had not changed much since he first articulated his belief that slavery was good, that reconstruction was a terrible mistake, and that you could not possibly understand the problems associated with two races living in close proximity unless you lived in the Deep South.[52] Nevertheless, the Depression had created a new set of challenges, and Stone did not like Franklin Delano Roosevelt's solutions. Stone recognized that the New Deal upset the balance of power between the federal government and the rights of individual states.[53] However, what motivated Stone to start writing again was President Harry Truman's Committee on Civil Rights, which criticized the treatment of African Americans in the southern states.[54] Stone defended the South by strengthen-

50. Various newspaper clippings from the *Jackson Daily News* (April 6 and 16, 1947; May 1, October 10, 1949) and the *Memphis Commercial Appeal* (June 17, 1948) in Stone's subject file, MDAH.

51. Stone, "The Constitution of 1890," typescript in Stone's "Articles and Speeches" subject file, MDAH.

52. For example, Stone's article entitled "The South in National Politics: The Political Effects of the War," which initially had been published in 1909 as a chapter in *The South in the Building of the Nation*, was reprinted in 1941 without revision as "Post Bellum Reconstruction, an American Experience" (*Journal of Mississippi History* 3 [July 1941]: 227–46).

53. Alfred Holt Stone, "Four Terms and Freedom—Why Not?" *Staple Cotton Review* 23 (January 1945): 1–7.

54. Truman's initiatives in regard to civil rights infuriated many southern politicians and led to the formation of a third party, the Dixiecrats, during the 1948 presidential race. Governor Fielding Wright, who later reappointed Stone to his fourth term as tax commissioner, led the Mississippi delegation's protest of Truman's policies at the Democratic convention and became the Dixiecrats' candidate for vice president (Jere Nash and Andy Taggert, *Mississippi Politics: The Struggle for Power, 1976–2006* [Jackson: University Press of Mississippi, 2006], 12–13).

ing his old arguments with new facts and figures. His work in this regard was an early salvo in the southern response to the modern civil rights movement. Although he would die within a year of the Supreme Court's ruling in *Brown v. Board of Education,* Stone's reaction to the federal government's intrusion into the states' regulation of race relations would serve as a rallying cry for people in Mississippi who favored massive resistance. His article "A Mississippian's View of Civil Rights, States Rights and the Reconstruction Background" was a prototype for that reaction. Initially published in 1948 in the *Staple Cotton Review,* it was reprinted six months later in the *Journal of Mississippi History* and reissued again privately in 1965, the year after Mississippi's "Freedom Summer."[55]

Alf Stone's longevity, both chronologically and administratively, was acknowledged on October 16, 1948, when employees of the State Tax Commission presented him with a cake bordered with seventy-eight dollar bills to commemorate his birthday. "'Uncle Alf' claims no credit for reaching such a ripe old age," read the caption under the picture of Stone with his birthday cake in the newspaper, "In fact, he says, anyone can get to be 78 years old 'if they just live long enough.'"[56] Within days, however, illness struck, and Stone was transported to the Touro Infirmary in New Orleans for an operation. His convalescence was lengthy, but he continued to work from his bed. In fact, the regional board of directors for the Farm Credit Administration traveled to New Orleans and held their meeting in his hospital room in order not to spoil Stone's record of perfect attendance that extended for almost thirty years.[57]

Alf Stone's next trip to the hospital occurred during the summer of 1953, when he underwent a hernia operation at Baptist Hospital in Jackson.[58] That operation was a success, but his health failed again in early 1955, and he was confined to his bed at home in the Belhaven area of Jackson. On May 9, he lapsed into a coma, only to rally briefly around two o'clock on Wednesday, May 11. He died about an hour later. Funeral services were scheduled in the

55. Alfred Holt Stone, "A Mississippian's View of Civil Rights, States Rights and the Reconstruction Background," *Journal of Mississippi History* 10 (October 1948): 181–239; reprint (Natchez, MS: Ketchings, 1965), [12] pages. The original title of the article in the *Staple Cotton Review* (26 [April 1948]: 1–32) was "The President's Program: Civil Rights, States' Rights, and the Reconstruction Background." A preface by W. M. Drake to the 1965 edition makes explicit the relationship between Stone's argument and resistance to integration.

56. "Stone Reaches 78th Birthday," *Jackson Daily News,* October 17, 1948.

57. "Alf H. Stone Is Practically Well," *Jackson Daily News,* December 6, 1948; "A. H. Stone Is Re-Appointed to Farm Credit Post," *Jackson Daily News,* December 30, 1952.

58. "3 State Officials Reported Better after Illness," *Jackson Daily News,* June 2, 1953.

First Presbyterian Church on State Street for Friday morning, and he was buried in the Ireys family plot in the Greenville Cemetery that afternoon.[59] His wife, Mary, joined him in 1969.[60]

Governor Hugh White said of Stone's passing, "No state has ever had a more faithful servant than Mississippi had in Alf Stone."[61] The board of trustees for the MDAH called a special meeting to honor their colleague. They remembered him as a "wise counselor, good friend, and great and noble Mississippian."[62]

The *Greenville Delta Democrat-Times* expressed the community's sadness over the news that one of their own had died. "Going into office with Governor Mike Connor [*sic*]," an editorial recalled, "Mr. Stone found the state's credit bad, teacher pay poor, and the whole economic structure floundering. As chairman of the newly-created State Tax Commission, he revised the tax structure and the sales tax was inaugurated. Significantly, he held this job through every administration since he first took over, although he took no part in politics except to vote."[63]

An editorial in the *Jackson Daily News* commemorated Stone's death as well. "To say that Alf Stone had within him the essentials of greatness is a poor and inadequate tribute to the man," it read. "Human greatness is something hard to define within a few phrases, but one had to be in the presence of Alf Stone only a few minutes to realize that he was a man set apart, a figure above the multitude. . . . It has been said that genius is merely capacity for hard work. If this be true Alf Stone was truly a genius, for he had a well-nigh limitless capacity for hard work."[64]

The editorial ended with poetry. "The pity of his passing is that we are not producing more like him in our state and nation, for he was a com-

59. "Alf Stone Dead; Native Son Was Tax Commissioner," *Greenville Delta Democrat-Times;* "Alfred Stone, Veteran Tax Official, Dies," *Jackson States Times;* and "Alf Stone Dies at Residence: Had Long Been in Ill Health," *Jackson Clarion-Ledger,* all May 12, 1955. Stone had been active in the First Presbyterian Church since moving to Jackson and was a key player in the church's fund-raising for a new sanctuary on the corner of North State and Belhaven streets in 1949 ("Presbyterian Men to Meet Tuesday to Plan Campaign," *Jackson Daily News,* October 31, 1949).

60. Mary Stone's grave marker, Greenville Cemetery.

61. "Alf Stone Dies, Rites Set Friday for Tax Official," *Jackson Daily News,* May 12, 1955.

62. "Minutes of the Meeting of the Board of Trustees of the Department of Archives and History," June 1, 1955, MDAH.

63. "Alf H. Stone," *Greenville Delta Democrat-Times,* May 13, 1955. Stone maintained his voter registration in Washington County, traveling there to vote on election days.

64. "Alfred Holt Stone," *Jackson Daily News,* May 12, 1955.

plete answer to that stirring demand in [Josiah Gilbert] Holland's beautiful poem—"[65]

> God give us men. A time like this demands
> Strong minds, great hearts, true faith, and ready hands;
> Men whom the lust of office does not kill;
> Men whom the spoils of office cannot buy;
> Men who possess opinions and a will;
> Men who have honor—men who will not lie;
> Men who can stand before a demagogue,
> And damn his treacherous flatteries without winking.
> Tall men, sun-crowned, who live above the fog
> In public duty and in private thinking.

65. Holland lived from 1819 to 1881 ("Josiah Gilbert Holland," www.2020site.org/poetry/jgh.html]). This poem sometimes goes by the titles "Wanted" or "God, Give Us Men!"

14 AND IN PRIVATE THINKING

As a public servant, Alf Stone was an innovator. As a racial theorist, he was a traditionalist. Stone stuck to ideas about the race problem that he had formed as a boy listening to his father and his father's friends when they gathered under a giant sycamore in front of the family house on Deer Creek.[1] Although Stone claimed repeatedly that he was receptive to other points of view, he did not let alternative explanations of the racial phenomena he presumed to understand affect his opinions, and despite his paeans to openness, Stone's mind remained closed when it came to issues involving the race problem.[2]

Stone's intransigence was not unusual given his time and place. Most people that Stone knew thought the same way he did and held to their convictions with a determination that matched his own. For that reason, Stone always found a sympathetic audience for what he said and became a spokesperson for people who shared his pessimistic assessment of the ability of African Americans to prosper and advance.

What made Stone different from other white southerners who subscribed to traditional views of the race problem was his prominence on a national level. Stone's articles and papers were widely read, and his writings influenced the thinking of some of the country's most prominent men and

1. A. H. Stone, "Some Recollections," chap. 1, p. 11. Stone's words were, "They would sit in its shade, on its gnarled and exposed roots, for what seemed to me hours at a time, and discuss crops, war, negro voters and negro labor, politics and reconstruction. It was in this rustic forum that I first learned of the problems of our people."

2. For a thoughtful analysis of Stone's receptivity to different points of view, see John David Smith, "Alfred Holt Stone and Conservative Racial Thought in the New South," *The Human Tradition in the New South*, ed. James C. Klotter (Lanham, MD: Rowman and Littlefield, 2005), 47–65.

women. This accomplishment was remarkable, given that Stone was an amateur competing with professionals, and although he was not professionally trained as were most of the social scientists with whom he associated, he was a member of their societies, a presenter at their meetings, and a confidant among their leaders.

In many ways, Stone's emergence as an authority on the race problem was reminiscent of another amateur, James Ford Rhodes, who gained widespread respect as a historian in the 1890s with the publication of his multivolume *History of the United States.*[3] Independently wealthy, Rhodes had pursued his passion for history without the accoutrement of an academic degree and, like Stone, gained distinction the old-fashioned way through "hard work and determination."[4] But there was more to Stone than hard work and determination. Stone was able to garner respect among his peers because he embraced the new wave of scientific thinking.

By the turn of the century, the scientific method had gained ascendancy among social theorists because it was considered to involve the objective assessment of large amounts of data in order to discover the "truth" of the phenomenon being studied. Dunbar Rowland, the first director of the Mississippi Department of Archives and History and one of Stone's close friends, embraced the scientific method and described it in a language that Stone understood and appreciated.[5] "History, in its modern acceptation, has been called the science of the progressive evolution or development of human society," Rowland observed in 1905. "The progress of historic thought and method has undergone a wonderful change since Leopold Von Ranke, the great German historian and the founder of the objective school, developed history as a systematic science. He taught that the aim of history was to narrate the truth and the truth only."[6]

Truth was the ultimate goal of the new approach, and Stone echoed

3. For Rhodes's racial views, see Cruden, *James Ford Rhodes,* 82, 166–67, and Smith, *Old Creed for the New South,* 116.

4. John David Smith, "James Ford Rhodes, Woodrow Wilson, and the Passing of the Amateur Historian of Slavery," *Slavery, Race, and American History,* by Smith, 16–17. According to Smith, Ford's *History of the United States from the Compromise of 1850,* published in 1892, was "one of the earliest analyses of slavery as an institution in the Old South." See also Smith, *Old Creed for the New South,* 115.

5. Rowland wrote, "We have been close friends for many years, in fact, from our student days at the University," in a letter to Judge R. H. Thompson dated October 2, 1933, in regard to Stone's nomination to the Mississippi Department of Archives and History's board of trustees (Dunbar Rowland Letter Books and Correspondence [box 4743], MDAH).

6. Dunbar Rowland, *The Mississippi Plan for the Preservation of State Archives: An Address De-*

Rowland's words when he wrote the preface to *Studies in the American Race Problem*. "It is a difficult thing to do—this telling of the truth, the narrating of facts, without at once creating the impression of a partisan temper or a biased mind," Stone wrote, adding, "Yet so far as this writer [Stone] knows[,] he has neither."[7]

Evidence that Stone adhered to the precepts of the scientific method can be found in a paper he read before the Washington County [Mississippi] Historical Association in 1910.[8] Stone began his talk by comparing the traditional approach, which he termed the "attractive" method, with the scientific method and its emphasis on the accretion of facts. He called the traditional approach "attractive" because it was exemplified by the literary brilliance of writers such as Thomas Carlyle and Thomas Macaulay.[9] Their works, which were catalogued as literature as well as history, stood in stark contrast with "a severely plain presentation of barren facts" by adherents of the scientific approach. Stone referred to them as the "'dry-as-dust' school of historians."

Despite an interest in literature that can be dated to his days as an undergraduate at the University of Mississippi, and despite his reference to scientific history as "dry-as-dust," Stone urged his audience to embrace the scientific view. "The place of the local historical society is to secure in the general scheme of developing and permanently recording the *facts* of American history," he told them.[10] To emphasize the point, Stone invoked the name of Herbert Baxter Adams, the father of scientific history whose program of professional training at Johns Hopkins had greatly influenced the contemporary generation of historians. "One great secret of Professor Adams's success," Stone observed, "was his ability to arouse interest and enthusiasm in local

livered by Invitation, before the General Assembly of Tennessee, and the Tennessee Historical Society, January 10, 1905 (Nashville, TN: Brandon Printing, [1905?]), 5.

7. Stone, *Studies in the American Race Problem*, ix.

8. Alfred Holt Stone, "The Functions of a Local Historical Society," typescript in Stone's "Articles and Speeches" subject file, MDAH. The quotations come from pages [1], [2], [4], & [5].

9. Thomas Carlyle was a British historian and essayist who gained prominence during the Victorian period. Carlyle "opposed analytic reasoning and quasi-scientific treatment of social questions," preferring instead "the more emotional and intuitive approach" ("Thomas Carlyle [1795–1881]," www.kirjasto.sci.fi/carlyle.htm). Thomas Macaulay was a British historian, poet, and essayist who is best remembered for his four-volume *History of England*. Macaulay did not hesitate to express opinions in his writings or to promote a political (Whig) agenda ("Thomas Babington Macaulay [1800–1859]," www.age-of-the-sage.org/history/historian/Thomas_Macaulay.html).

10. Emphasis added.

subjects." In that regard, local historians in the southern states would do well to learn from the example of historical societies in the Northwest and Midwest. "The libraries of the Eastern, New England and older Western states are filled with the chronicles of their counties and towns," Stone told his audience. "We have made a lot of history in the South, but we have been criminally negligent in preserving an authentic record of it." Southerners could only blame themselves, he added. "The chief, in fact, the one legitimate function of such an organization as ours, is the collection and preservation of the source material of which history may be written."

Stone's talk before the Washington County Historical Association may explain his persistence in collecting articles, pamphlets, and monographs for a history of black people. It also provides insight concerning the thoroughness with which he prepared the syllabus for the book he was planning to write for the Carnegie Institution.[11] The fact that Stone was not able to finish either book does not detract from his effort to do what scientific historians thought essential, "the most unwearied search after every scrap of proof that could throw light upon the topic."[12] Although Stone may have been surprised to know that he would be remembered best for the work he did in preparing for a book he did not write, his collection of material on the race problem that is currently held by the Mississippi Department of Archives and History is the most important product of his scholarship to survive.

Another characteristic of the scientific method that Stone tried to emulate was objectivity. Albert Bushnell Hart called this state of mind "dispassionate"; Stone referred to it as the absence of "a partisan temper." Stone's desire to be objective was evident. In 1907, for example, Stone ended his response to the discussants' comments regarding his "Race Friction" paper with: "Something has been suggested as to the rights and wrongs involved in the situation, and about what might be if only men would be just and honest, and so forth. My only reply is that I did not come here to discuss ethical questions. I am just now concerned only with the hard, stern, inexorable facts in the case."[13]

Stone thought that by adhering dispassionately to "facts" he could avoid the pitfalls inherent to a debate over human rights, fairness, and other slippery philosophical concepts. In this regard, Stone's assumption was similar to that held by Walter F. Willcox, who believed that his reliance on census

11. Stone expanded the syllabus and had it published as the *Material Wanted* pamphlet.

12. [Meriwether], "Scientific History," 301.

13. "Response by Mr. Stone," in "Discussion of the Paper by Alfred Holt Stone," 840.

data somehow protected him from the allegation of bias and allowed him to rise above the controversy generated by partisan arguments. "Believing himself and his data objective," Mark Aldrich has written, "Willcox imagined that he could maintain a middle position between white racists and such black leaders as the eminent sociologist W. E. B. Du Bois. His statistical truths, he believed, would eventually provide the common ground on which both sides could agree."[14]

Stone also attempted to chart a course on the common ground, midway between white racists, such as James K. Vardaman, and black troublemakers, such as T. Thomas Fortune, editor of the *New York Age*, who, Stone believed, stirred up black workers at Dunleith by providing them with unrealistic expectations.[15] Stone was convinced that although the two men, Vardaman and Fortune, were at opposite ends of the ideological continuum, they were responsible for a common outcome, racial unrest. As a cotton grower, Stone abhorred the prospect of racial unrest because anything that disrupted the status quo threatened the productivity of his work force. As the historian John David Smith has observed, "Stone's concern for African Americans, . . . always stemmed from his ingrained fear of social and economic change, not from altruism toward blacks. Sober, law-abiding industrious black workers should be encouraged, Stone maintained, because white southerners lived surrounded by blacks and depended upon their livelihoods."[16] The degree to which this assumption affected Stone's thinking can be seen clearly in a letter Stone wrote to his friend LeRoy Percy on May 21, 1907.[17]

To appreciate the insight this letter offers in regard to Stone's thinking requires that it be placed in a context that, as was usually the case, involved politics. Senator Hernando DeSoto Money from Mississippi had decided not to seek reelection in 1908, and several prominent politicians entered the race for Money's seat. Although the state legislature elected senators at the time, the Democratic nominee, who was sure to win the election in this one-party state, was selected in a statewide primary. The primary election to fill Money's seat was scheduled for early August 1907.[18]

14. Aldrich, "Progressive Economists and Scientific Racism," 3.

15. For criticism of Fortune's optimism in regard to black workers, see Stone, "Economic Future of the Negro," 203–4. For an example of Fortune's militancy, see T. Thomas Fortune, "The Afro-American," *Arena* (December 1890): 115–18. There is a copy of Fortune's article in TSC.

16. Smith, "Alfred Holt Stone and Conservative Racial Thought," 58.

17. Stone to Percy, May 21, 1907, PFP (box 2, folder 6).

18. Holmes, *The White Chief*, 177–95.

The most prominent politician to enter the race was James K. Vardaman, the popular governor in the third year of his four-year term. Vardaman's opponent was John Sharp Williams, an eight-term congressman from Mississippi and the Democratic minority leader in the U.S. House of Representatives. The two men were contrasts in style and demeanor. Vardaman was a newspaperman from a middle-class family who resorted to race-baiting to stir up his uneducated power base. Williams was a wealthy planter whose education and refinement appealed to the traditional kingmakers in Mississippi.[19] Not surprisingly, LeRoy Percy backed John Sharp Williams.[20]

Percy was concerned that Vardaman's race-baiting would create unrest among African Americans in the Delta. More specifically, he was concerned that black workers would flee if Vardaman's poor-white constituency acted on the threats implied by the governor's diatribes.[21] Although Percy thought that immigrants from Europe would eventually replace black workers in the Delta, he also knew that the transition would take time.[22] In the meantime, Percy believed that African Americans should be afforded the opportunity to receive an education in the public schools. Unlike Vardaman, who was convinced that educating black people made them inefficient laborers, Percy thought that withholding education from the black race acted like an anchor to drag down all southerners. Furthermore, Percy believed that education would allow black workers to find jobs in other parts of the country as European immigrants took their place.[23]

Percy found an opportunity to express his views on black education and, not incidentally, to support Williams's campaign, at the annual meeting of

19. Cresswell, *Rednecks, Redeemers, and Race,* 200.

20. Baker, *The Percys of Mississippi,* 31–32. For evidence that Percy was active as a consultant in Williams's campaign for the Senate, see his correspondence with John Sharp Williams in PFP, esp. Percy to Williams May 8, 1906, November 2, 1906, and June 22, 1907 (box 1, folders 9 and 13, box 2, folder 7, respectively).

21. The extent to which Percy was willing to go to prevent black workers from leaving the Delta was illustrated during the flood of 1927, when he blocked his son's attempt to evacuate African Americans from their refuge on top of the levees (Percy, *Lanterns on the Levee,* 249–69; Bertram Wyatt-Brown, *The House of Percy: Honor, Melancholy, and Imagination in a Southern Family* [New York: Oxford University Press, 1994], 241–42).

22. In 1907, the experiment with Italian workers at Sunnyside was still a going concern. Nevertheless, by 1910 the number of foreign-born people in Mississippi was smaller than it had been in 1870 (McMillen, *Dark Journey,* 258).

23. "Mississippi State Bar Association Banquet at Carroll Hotel Last Night," *Vicksburg Herald,* May 9, 1907. For Vardaman's antieducation stance, see Holmes, *The White Chief,* 121–23. Percy's faith in the experiment with Italian labor at Sunnyside was still intact during the spring of 1907.

the Mississippi Bar Association in Vicksburg on May 8, 1907. He was the last speaker on the program, and he attacked Vardaman's antieducation rhetoric head on. "The statement is daily heard that education ruins the negro as a laborer, and that we should not be burdened with educating him," Percy told the audience. "I want to enter my protest against both assertions. I deny that education impairs the usefulness of the negro; I deny the assertion that the South should not educate him."[24]

Stone read Percy's speech in the newspaper and offered support in a letter dated May 21. "I agree heartily with all you said," Stone wrote, "and am glad you said it," for Stone was as concerned about the disruptive effects of Vardaman's race-baiting as was Percy.[25] "It seems to me that the time has come when the sober and more thoughtful people of the South must take a definite stand against the tide of 'negrophobia' which has been sweeping over our section for the past few years," Stone wrote. "*I do not mean this in any theoretical or sentimental sense,—but as a cold-blooded, practical proposition.* Unrest, disquiet and fear among so large a section of our population as the negro now constitutes will inevitably react upon upon [*sic*] the white element in time. And fear and unrest must, and will, follow the everlasting stirring up of strife along the lines pursued by politicians seeking office on the strength of successful 'race-problem' agitation."[26]

Stone's letter to Percy speaks for itself and can be accepted as insight into Stone's assumptions concerning the race problem for several reasons. First, it captured a spontaneous exchange of views, as opposed to Stone's articles in popular magazines or professional journals, in which he had the opportunity to choose his words carefully. In fact, the letter reads like the continuation of a conversation that has been cut short by the press of another engagement. (A typescript of the letter can be found in appendix E.) Second, the letter was written to someone whom Stone knew well and who shared his point of view. The two men were close friends.[27] Third, every point Stone makes in the letter is consistent with what we already know. There are no surprises.

24. Leroy Percy, "A Southern View of Negro Education," *Outlook*, August 3, 1907, 731. This article is a transcript of the speech Percy gave at the Mississippi Bar Association on May 8, 1907. Williams defeated Vardaman in the primary by 648 votes out of the 118,344 cast (Holmes, *The White Chief*, 190).

25. Stone also agreed with Percy in regard to the practical necessity of providing an elementary education for black pupils (McMillen, *Dark Journey*, 77).

26. Stone to Percy, May 21, 1907 (box 2, folder 6) PFP; emphasis added. Note that Stone did not capitalize the initial letter in the word "negro."

27. When Percy was under assault in 1923 for his opposition to the Ku Klux Klan, Stone wrote a lengthy testimonial in regard to Percy's good character and paid to have it printed (Alfred H. Stone, *As to Senator Percy: A Personal Word* [Memphis, TN: n.p. (1923?)], 29 pages).

What makes the letter interesting is Stone's unguarded admission of the assumptions behind his racial theories. As such, Stone's letter to Percy offers a glimpse into Stone's mind that confirms much of what many of us may have suspected all along.

Stone's determination to maintain a "cold-blooded" approach to his data eventually backfired on him, for it imposed an intellectual rigidity that handicapped Stone when it came to considering opposing points of view. Had he been more flexible in his thinking, more willing to accommodate arguments that revealed flaws in his theories, the importance of what Stone had to say might have survived, and we would remember him today for something other than the big collection of research material he assembled in order to write a book.

Colyer Meriwether, editor of *Publications of the Southern History Association*, observed in 1907 that there were three stages when it came to applying the scientific method to the social sciences, particularly history. Seeking records of the past was the first, followed by the precise and accurate acknowledgment of the sources being used so that other historians could locate that material and weigh its significance independently. Up to this point, Meriwether noted, the scientific historian "can claim to be as impartial as any man with balance or crucible glass. But when he essays the third stage, that of interpretation, then the difficulty arises of reducing himself to an automaton with [a] brain. He has sought to avoid this obstacle by either exhibiting the other sides of any disputed question, and they are myriad in history, or by fairly recognizing another standpoint."[28] It was in his execution of the third stage that Stone failed. Stone collected material that was relevant to his argument, and he cited his sources accurately, but he had difficulty recognizing, much less accepting, opposing points of view.

Stone was not alone in his failure to let the objective analysis of data he collected shape ideas that were already well formed in his mind. It happened to the best of them, especially those studying the race problem. "Despite their [practitioners of 'scientific' history] determined efforts at detachment and scholarly rigor," John David Smith has written, "the first generation of historians of slavery succumbed to the dictates of the old slavery debate. Ideology ultimately prevailed over historical method."[29] Although ideology may have triumphed over methodology in most cases, it did not always do

Percy opposed the Klan because he was convinced that it would create unrest in the black community and thereby destabilize his labor force (Wyatt-Brown, *House of Percy*, 227–30).

28. [Meriwether], "Scientific History," 301.

29. Smith, *Old Creed for the New South*, 9.

so. Some social scientists changed their minds as they weighed the evidence. One of Stone's friends and colleagues, Carl Kelsey, serves as an example of how the application of the new approach could change a person's point of view.

Kelsey was a graduate student in sociology at the University of Pennsylvania when he visited Stone at Dunleith in 1902 while he was collecting data for his dissertation on the "Negro farmer." Kelsey and Stone hit it off, and Stone reviewed the dissertation in the *Publications of the Southern History Association* when it was published as a monograph. "It would not be extravagant praise to pronounce it the best piece of work yet done in its field," Stone wrote.[30]

What Stone liked about the dissertation was Kelsey's insistence that African Americans had inherited characteristics that made them different from persons of European descent. "He [Kelsey] has frankly recognized that the negro is a negro, and not a *black-skinned white man;* that as such he possesses certain racial traits and tendencies as much deserving of study, if he is to be wisely helped, as are the characteristics of any other race that we might wish to know."[31] Stone also liked the way Kelsey exonerated slavery from causing the shortcomings observed among African Americans following emancipation. "Mr. Kelsey does another sensible thing," Stone wrote in his review; "he recognizes the fact that it is ridiculous and unscientific to shut our eyes to the negro in Africa, untouched by American slavery, and charge to that institution whatever is bad or unpromising in the negro of the South."[32]

Stone retained his high esteem for Kelsey's dissertation three years later when he wrote new material for *Studies in the American Race Problem,* calling it "the best study of the Negro farmer that I have seen."[33] But Kelsey's think-

30. Stone, Review of *The Negro Farmer,* 239. The year of Kelsey's visit to Dunleith comes from Stone to Willcox, November 6, 1902, WFWP. Both men were born in 1870.

31. Emphasis added. Stone's allusion to Frederick L. Hoffman's influential book on "racial traits and tendencies" was not a coincidence. The assumption that black people were innately inferior to white folk was almost universally accepted in white America at the time, and Hoffman's book was one of the pillars in that ideological framework.

32. Stone, review of *The Negro Farmer,* 239–40. Kelsey's words were, "The North is slowly learning that the Negro is not a dark-skinned Yankee, and that thousands of generations in Africa have produced a being very different from him whose ancestors lived an equal time in Europe" (Carl Kelsey, *The Negro Farmer: A Thesis Submitted to the University of Pennsylvania in Partial Fulfillment of the Requirements for the Degree of Ph. D.* [Chicago: Jennings and Pye, 1903], 67). There is a copy of Kelsey's dissertation with portions underlined by Stone, including this quotation, in TSC.

33. Stone, "Foundations of Our Differences," 47.

ing was changing, even though Stone may not have been aware of it at the time. Although Kelsey praised Stone as "a frank and honest student" of the race problem in his review of *Studies in the American Race Problem* after it was published, he hedged his endorsement. Although "No one need accept Mr. Stone's conclusions[,]" Kelsey wrote, "no honest man can fail to appreciate the importance of his arguments."[34]

Kelsey's qualified praise for Stone's book may have reflected the shift his thinking had taken in terms of the importance of heredity in determining black traits and tendencies. "The time has come for all students to cease quibbling about the relative importance of heredity and environment—both absolute essentials to the life of every human being, and capable therefore of no such comparisons," Kelsey had remarked as a discussant for a paper on social Darwinism at the first annual meeting of the American Sociological Society in 1906. By then, Kelsey was convinced that the effect of heredity was limited to "those physical characteristics which come to us through the germ cells of the parents." He was convinced that other characteristics, such as industry, intelligence, and integrity were shaped by the environment. "Another result of our studies," Kelsey noted, "is to weaken the belief in superior and inferior races. It now seems very probable that there is an approximate equality of mental ability among the various races, and that race differences are the result of different environments."[35]

By 1909, Kelsey had developed a theory of social causation, which he described in a paper entitled "Influence of Heredity and Environment upon Race Improvement." "In a word, we live and think too much in vicious circles," he wrote.

> Men and women live and work under bad conditions. The children are poorly nourished and sadly neglected. Low ideals are inculcated—result, inefficiency, poverty, vice, crime. In another group opposite conditions prevail, opposite results follow. Popular opinion of the successful group say heredity—blood tells; that of the other says environment, exploitation, lack of opportunity. I know of no better way of contrasting the philosophy of the so-called upper and lower worlds.

34. Kelsey, review of *Studies in the American Race Problem*, 217–18.

35. Carl Kelsey, "Comments as a Discussant Regarding D. Collin Wells, *Social Darwinism*," *American Journal of Sociology* 12 (March 1907): 711. Kelsey was influenced by the work of Franz Boas, an anthropologist from Europe who argued that race did not account for differences observed in the mental capacity and social behavior of groups of people (Degler, *In Search of Human Nature*, 61–62, 85–86). See also John Cartwright, "A Review of *Evolution and Human Behavior*," home.comcast.net/~neoeugenics/cartw.htm.

> To such loose thinking an increasing protest is arising. Unconscious, perhaps, of its full significance, many of those now grappling with social problems are condensing their statement of causes into the one word, "maladjustment." In a word, we create the evil as well as the good.[36]

Kelsey and Stone had moved far apart theoretically by 1909, for Stone's position concerning the innate characteristics of African Americans had not changed since Kelsey had visited him at Dunleith in 1902. This is not to say that Stone's racial theories were the same in 1909 as they had been in 1902. They were not; they were more sophisticated. But Stone had wrapped layers of sophistication around a core of assumptions that had not changed, despite Stone's exposure to a wide range of thoughts, ideas, and evidence. For example, the economic situation in the Delta had changed with the advent of the boll weevil, and Stone took that unwelcome circumstance into account. But Stone's assessment of the ability of black people in the Delta to meet the new challenge was as pessimistic as it had been when he predicted their expulsion as a labor force when Italian immigrants had been brought in to raise cotton at Sunnyside. "Under the new order, the white man has in his greater intelligence and keener appreciation of the necessity of mending his ways, an advantage over the negro," Stone wrote in an article for the *Annals of the American Academy of Political and Social Science* in 1909. "The masses of the latter are handicapped by a concentration of thought upon immediate desires, with a contemptuous disregard of either future good or future evil." If anything, the economic situation created by the boll weevil exacerbated the inherited shortcomings of the black laborer. "In the present condition of the plantation negroes of the cotton states, we have a situation which is in part the inevitable result of a thoroughly vicious economic system, and in part attributable to what seem to be the more or less fixed characteristics of this class of labor."[37]

Not only did Stone continue to hold to a theory of racial differences that was heavily slanted toward the heritability end of the continuum, he continued to describe slavery as a liberating experience for African Americans.

36. Carl Kelsey, "Influence of Heredity and Environment upon Race Improvement: An Introductory Paper upon the Significance of the Problem," *Annals of the American Academy of Political and Social Science* 34 (July 1909): 7–8. By 1908, Kelsey was active in the child welfare movement at the University of Pennsylvania and played a role in establishing what was to become one of the most prestigious social work programs in the nation ("The Penn Approach: The Significance of Penn in Social Work History," www.sp2.upenn.edu/about/approach/). Kelsey died in 1953. The years of his life, 1870–1953, overlap Stone's almost exactly.

37. Stone, "Negro Labor and the Boll-Weevil," 169–70, 172.

"Slavery transformed the savage negro into a civilized man; it taught him to work, and showed him what could be accomplished by the labor of his hands; and then it left him as a free man with almost a monopoly of the field in which he had been employed as a slave," Stone wrote in a chapter he authored in 1909 for *The South in the Building of the Nation.* If you were to accept Stone's logic, slavery was actually a generous investment program for black people brought to America as slaves. "In 1865," he observed, "no other body of negroes in the world occupied as advantageous a position economically as those in the Southern States."[38] Their failure to capitalize on the advantages that accrued with emancipation was due to the innate shortcomings African Americans had inherited as people of African descent. The idea that slavery may have played a role in shaping a generation of people who were ill-prepared to cope with the responsibilities of freedom found no expression in Stone's point of view.

Kelsey and Stone had started on common ground; they both assumed that inherited traits and tendencies accounted for the superiority of one race, the white one, and the inferiority of the other, the black one. But Kelsey had expanded his theories to encompass social factors as he studied the race problem. Stone stuck closely to theories that depended almost entirely on the dominance of inherited characteristics.

It is tempting to invoke education and geography as explanations of why Kelsey changed and Stone did not. However, pointing out that Kelsey was professionally trained and from the North and that Stone was an amateur who hailed from the South does not tell us anything that we do not already know. Moreover, the viability of this explanation breaks down quickly when it is observed that Walter F. Willcox was from Massachusetts and had excellent professional credentials, yet he held tenaciously to a theory of racial differences that was similar to Stone's. Furthermore, the Reverend Quincy Ewing was a southern man and amateur sociologist, yet he was able to escape from the racial straightjacket that sought to restrain him.

The key to understanding Stone's intellectual rigidity was a persistence in avoiding information that contradicted his beliefs, a shortcoming that was obscured by Stone's insistence that he had an open mind. Throughout his writings, Stone assured the reader that he was eager to consider alternative points of view. For example, he used a quotation from Marcus Aurelius as the epigragh for *Studies in the American Race Problem:* "If any one is able to convince me and show that I do not think or act right, I will gladly change, for

38. Stone, "American Negro Origins," in *South in the Building of the Nation,* 10: 177–78.

I seek the truth, by which no man was ever injured. But he is injured who abides in his error and ignorance." In the preface to the book, Stone stated, "My only hope is that my influence, however small, may be cast upon the side of better understanding, which shall make for mutual respect and tolerance for our several points of view."[39] Despite his bold front in this regard, Stone repeatedly ignored or actively avoided evidence that was contrary to what he already believed.

In some instances, Stone's distancing himself from his critics was understandable, even if by today's standards doing so appears to have been overly defensive. His exchanges with W. E. B. Du Bois are an example. Du Bois offered frank and discriminating criticism of Stone's theories, and Stone ignored it. But, from Stone's perspective, Du Bois was not in a position to criticize him. Although Du Bois had received his undergraduate degree from Fisk University in Nashville, Tennessee, and although he had conducted most of his research on African Americans in southern states, Du Bois was still a Yankee, and a mulatto at that. Furthermore, Du Bois was an academic, an intellectual, and a visionary. Stone was convinced that Du Bois did not understand the race problem as it existed in the Deep South among the unwashed and uneducated masses of black people who worked on plantations like Dunleith.

Dealing with feedback from Booker T. Washington was a different story. Washington was from the South, born a slave. Although Washington obviously had white blood in his veins, he knew what he was talking about when it came to black agricultural workers.[40] Furthermore, Washington endorsed black subservience to the white race, the very thing that Stone insisted was essential to peaceful race relations in the South. In fact, Stone had praised Washington in an editorial for the *Greenville Times* in 1900. "During his whole career he [Washington] has preached to his people the abandonment of politics, the dignity of labor and the supreme importance of cultivating, by proper conduct, the friendship and respect of the dominant race," Stone had observed. For that reason, "he [Washington] deserves the good will and the thanks of the white people of the South."[41] Yet, Stone ignored evidence that Washington tried to show him because it contradicted Stone's point of view. Just how and why can be seen in an exchange of letters in 1905.

Stone initiated this particular round of correspondence with Washington

39. *SARP*, vii (epigraph), xi (preface).

40. For evidence that Stone lumped Washington and Du Bois into the same racial category, see Stone, "The Mulatto Factor," in *SARP*, 430.

41. "Booker T. Washington," *Greenville Times*, September 22, 1900.

on March 29, when he wrote to request copies of the Tuskegee Institute's annual report for the past four years, which Washington provided by return mail.[42] Stone thanked Washington for the reports and asked for one of his pamphlets, *The Successful Training of the Negro*.[43] Washington responded to Stone's letter by sending a copy of the pamphlet and took the opportunity to confront Stone about a statement Stone had made that bothered him. Washington complained of a paper Stone had authored in which Stone had accused Washington of glossing over black shortcomings.[44] Washington disagreed. He may not have dwelled on the shortcomings of black people when he addressed a white audience because he wanted to focus on that race's responsibilities in dealing with the race problem. "In my addresses to my own race," Washington assured Stone, "I never fail to dwell upon the very points that are brought out in your paper."[45]

So far so good, but it was what Washington wrote next that relates to the question of Stone's openness to data that contradicted his beliefs. "I wish very much that you might see an experiment that is being conducted within four or five miles of this institution," Washington wrote to counter Stone's emphasis on black shortcomings.

> I was talking with the man who has charge of it this morning, and he tells me that he has this year 65 or 67 families on the plantation. I asked him how many moved away this year, and he said one, and I asked him how many last year and he said two at his request, and the year before only one left. This man has the greatest faith in the capacity and the character of the colored people upon his plantation, and because the people know that he has faith in them and trusts them, they stand by him.

Washington ended the letter with, "I am very sorry that we have never been able to meet each other, and hope at some time that we can."[46]

42. Stone to Washington, March 29, 1905, and Washington to Stone, March 31, 1905, both in BTWM (reel 261).

43. Booker T. Washington, *The Successful Training of the Negro* (New York: Doubleday, Page, 1903), [25] pages. Stone's copy of this pamphlet is in TSC. His letter to Washington was dated April 5, 1905 (BTWM [reel 261]).

44. The paper to which Washington referred was Stone's "A Plantation Experiment," which was published in the *Quarterly Journal of Economics* in February 1905. This article was reprinted in 1908 as a chapter in *Studies in the American Race Problem*. At the very end of the article (146–47), Stone quoted a statement Washington had released to the press in November 1904.

45. Washington to Stone, April 6, 1905, BTWP (8: 248–49). There is also a legible copy of the letter in BTWM (reel 261).

46. Ibid.

Stone waited for six weeks until he had time to respond at length. He began by stating that he was "speaking with utmost candor" and defended his assessment of the African Americans' shortcomings as a labor force. "My personal experience strongly inclines me to the conclusion," he wrote, "that the great mass of our colored population lacks the element of stability that is essential to the highest degree of economic efficiency." Nevertheless, Stone wanted to know more about the operation Washington had referred to in his letter. "I am deeply interested in all plantation experiments, and do not harbor the delusion that I 'know it all,' and I should be glad to have the name and address of the party to whom you refer. I would be delighted to correspond with him, and have no doubt that I might profit by some suggestion from him."[47]

The party to whom Washington referred was William V. Chambliss, a Tuskegee graduate who managed a plantation near the Tuskegee Institute for the Southern Improvement Company. The plantation was a working cotton farm of about 4,500 acres in Macon County, Alabama. The property had been purchased by a group of northern investors who were interested in supporting Washington's effort to demonstrate that African Americans could succeed as independent farmers if provided with decent land and affordable financing. To that end, the investors divided the plantation into small tracts averaging about forty to eighty acres apiece and sold them to reliable black farmers on easy terms. The lease-purchase arrangement was for seven years, and the tenants paid rent yearly, either as a share of the crop or cash. Just as was the case at Dunleith, the Southern Improvement Company contract stipulated that the black lessees would submit to strict supervision in the cultivation and harvesting of their cotton crop.[48]

There were several similarities between the Southern Improvement Company experiment in Macon County, Alabama, and Stone's experiment in the Mississippi Delta. Both were cotton-growing concerns, although the operation in Alabama had added a sawmill, a cotton gin, a gristmill, and a brick-making facility by 1903. Both were located in regions known for

47. Stone to Washington, May 19, 1905, BTWM (reel 261).

48. Descriptions of the Southern Improvement Company's operation comes from "Home Buying" and "Hampton's Outgrowth's," both in *Southern Workman* 32 (February 1903): 124–25 and 33 (January 1904): 7–8, respectively. See also Anderson, "The Southern Improvement Company," 113–18. The identification of the manager comes from Robert E. Zabawa and Sarah T. Warren, "From Company to Community: Agricultural Community Development in Macon County, Alabama, 1881 to the New Deal," *Agricultural History* 72 (Spring 1998): 466.

their dark, rich soil.[49] At around 3,500 acres, Dunleith was comparable in size to the Southern Improvement Company plantation.[50] During the years of Stone's experiment at Dunleith, 1899–1903, there were from fifty-eight to seventy-five families working for him. During the first four years of the Southern Improvement Company experiment, 1900–1904, there were sixty-five to sixty-seven families. Of these, Stone lost an average of 20 percent of his work force each year. The average rate of attrition at the Southern Improvement Company plantation was less than 2 percent.[51]

Stone abandoned his experiment in 1904 because of the turnover of his labor force and went exclusively to a sharecropping arrangement without the favorable terms he had offered previously for his black workers. The Southern Improvement Company experiment continued to do well until the boll weevil infested the fields and made the business less profitable. The northern investors sold out, having made a good return on their money in the intervening years and a large profit on the property, which had been improved substantially by the tenants. Unfortunately, the tenants did not fare as well. The thin edge of profitability for the small farmer kept the tenants on the Southern Improvement Company property from getting clear of debt and paying off their lease. Consequently, most of them did not realize the expectation of owning the land they worked before the boll weevil made their financial situation too precarious to continue. However, not one farmer had defaulted on his lease agreement with Southern Improvement Company over the first nine years of its existence.[52]

Washington was proud of the Southern Improvement Company operation in Macon County and more than willing to compare the results of his experiment with those of Stone's. To that end, Washington included a balance sheet for the Southern Improvement Company in his reply to Stone on July 10, 1905. The company had been in existence for four years, Washington

49. For the type of soil found in Macon County, see "Alabama's Black Belt," irhr.ua.edu/blackbelt/intro.html.

50. The estimate of Dunleith's acreage comes from Mrs. T. P. [Betty Paschal] O'Connor, *My Beloved South* (New York: Putnam's, 1914), 302. A deed of trust executed on June 6, 1932, puts Dunleith's acreage at 3,900 (Book 245, pages 71–72, WCCH). It is likely that Stone acquired additional acreage during the intervening years. O'Connor's book also contains a colorful, if sensational, description of life on Dunleith around 1910.

51. Zabawa and Warren, "From Company to Community," 465–66. The data for Dunleith come from a table at the end of "A Plantation Experiment," in *Studies in the American Race Problem*, 148.

52. Anderson, "Southern Improvement Company," 118–24.

noted, and had returned a 7 percent dividend to the investors in 1904. "I had a talk with the manager [Chambliss] only yesterday," Washington remarked, "and he tells me that he has more persons applying to him for opportunities to work and buy farms than he can accommodate." Washington's claims were not mere braggadocio, for he was sharing proprietary information with Stone. "I shall have to ask you to be very sure to keep this information private as I have no right to give out these figures," Washington wrote at the end of his letter. "The manager of the company has let me have them for your eye only:—Please return this to me when you are through with it."[53]

Stone's response was short and to the point. "It is impossible for me intelligently to criticize the showing, but I presume I misunderstood the nature of the experiment to which you referred in a former letter," he wrote after thanking Washington for the balance sheet, which he returned. "I received the impression that it was purely an agricultural enterprise, in which I see I was in error." But Stone did not explain why a cotton plantation in the Black Belt of Alabama that diversified its operation was not comparable to a cotton plantation in the Mississippi Delta, given that both "experiments" had been run concurrently and thus had been subject to the same fluctuations of labor, climate, and market forces. Nevertheless, Stone thanked Washington for sharing the information with him. "I am always glad to know of the success of any such undertaking, whether agricultural or industrial, particularly if it be something in the nature of a co-operative enterprise," Stone wrote. "I shall be glad to correspond with the manager of The Southern Improvement Co., at a later date, and know that I shall be interested in whatever observations he may make."[54]

As far as we know, Stone never did write Chambliss, nor did he visit Tuskegee. "I want to make an effort to visit the Institute during this session," Stone wrote to Washington a year later in a request for the official report of the commemoration of Tuskegee's twenty-fifth anniversary. "Will you kindly suggest the best time for such a visit, if I find it possible to make it at all."[55] Washington was eager to have Stone visit, and someone, possibly Washington, penciled the notation "Stone—get here!" in the margin of Stone's letter. "I write to say that it will give us very great pleasure if you can come to Tuskegee, say from May 14 to 18," Washington wrote in his reply. "I shall be at the school at the time and it will afford me very great pleasure to

53. Washington to Stone, July 10, 1905, BTWM (reel 261).
54. Stone to Washington, July 14, 1905, BTWM (reel 261).
55. Ibid., April 13, 1906, BTWM (reel 273).

offer you the opportunity to look through the work we are doing here. I very much hope that it will be possible for you to come at that time."[56]

May 14 to 18 came and went without a word from Stone. Finally, on May 19, Stone replied. "I regret that I find it impossible to run down at or about the time suggested by you. I am hoping to be able to arrange for a trip this winter, or next spring, which shall take in all the more important colored industrial schools in the South."[57] He never made it. The winter and spring of 1907 were a busy time for Stone while he was assembling a research team for his work with the Carnegie Institution and working on *Studies in the American Race Problem*. It was a shame, because Tuskegee was on the main rail line between Atlanta, Georgia, and Jackson, Mississippi, if Stone had traveled by way of Montgomery, Alabama.[58] He may not have been able to make it on those dates, but Stone was traveling back and forth between Mississippi and Washington two and three times a year during this period, and it seems as if he could have gotten off the train at Tuskegee for a visit, had he wanted.

Booker T. Washington certainly thought so. "He [Stone] and I have had considerable correspondence on the subject of the progress of the Negro and I know him pretty well," Washington informed Colonel Wright at the Carnegie Institution in April 1907, when he wrote to complain of Stone's appointment as a division head. "I have tried over and over again to get him to come to Tuskegee or to go to Hampton or some other institution and see something of the effects of rational education on the life of the Negro, but I have never succeeded in getting him to go where he could see for himself what is being accomplished."[59]

Stone's failure to visit Tuskegee does not reflect favorably on his oft-stated desire to seek the truth. His complaint that the Southern Improvement Company operation was not "purely agricultural" appears to have been an excuse to avoid dealing with Washington's success as compared to Stone's failure. Certainly, a businessman with Stone's experience could have appreciated the advantage of taking a closer look at a project that was generating a 7 percent return per annum for its investors. It was true that the Southern Improvement Company operation had diversified by adding a sawmill, a cotton gin,

56. Washington to Stone, April 19, 1906, BTWM (reel 273).

57. Stone to Washington, May 19, 1906, BTWM (reel 273).

58. Rail lines for the period can be determined from maps available online at the Library of Congress's "Railroad Maps, 1828–1900" Web site, rs6.loc.gov/ammem/gmdhtml/rrhtml/rrhome.html.

59. Washington to Wright, March 26, 1907, BTWP (9: 233).

a gristmill, and a brick-making facility, but these investments improved the operation's self-sufficiency and were factors in its success. As far as a school building and a teacher for the tenants' children were concerned, Stone may have decided that adding a school at Dunleith would have been a wise investment too.

Despite the numerous similarities between the two experiments, there was one fundamental difference: workers for the Southern Improvement Company operation had signed a lease, whether it was to be paid with cash or a share of the crop, that would allow them to purchase the land they worked and the homes they lived in over a period of seven years. This powerful incentive did not exist at Dunleith, and black workers who labored there did so with the expectation that they would be working as tenants indefinitely, unless they could save some money and move on to establish their independence elsewhere.

The promise of self-sufficiency through land ownership is a powerful motivator, as had been illustrated when the attempt to import Italian workers at Sunnyside failed, in part, because that option was not open to them.[60] At the Southern Improvement Company plantation in Alabama, ownership was a crucial factor in its success, as one of the workers explained to a correspondent who attended the dedication in 1906 of a portrait of Alexander Purves, one of the Southern Improvement Company investors. "In talking with one of these men, whose home I visited," the correspondent reported, "I learned that for twenty-six years he had been paying out money annually for rent and had nothing to show for it, but that with the chance which Mr. Purves had given him he now owns forty acres of land, upon which is a comfortable, three-room house with a kitchen; he has a garden and a good stable for his stock. There is a fence around his house and garden. He has also sufficient food for his wife and children. In about two years, with no unforeseen misfortunes, he will have finished paying for his property and will have a home which he can call his own. His children attend school daily and his wife attends the mothers' meetings which are held in the school building weekly by one of the school-teachers. This man's life has been made happy by Mr. Purves's having come in touch with it, as is true of the others for whom he made it possible to secure homes."[61]

The contractual arrangements at the Southern Improvement Company

60. Alberto Pecorini, "The Italian as an Agricultural Laborer," *Annals of the American Academy of Political and Social Science* 33 (March 1909): 159, 163; Wyatt-Brown, "Leroy Percy and Sunnyside," 93.

61. J. H. Palmer, "The Unveiling of a Picture," *Southern Workman* 35 (March 1906): 188.

plantation and Dunleith reflect different ways of obtaining a return on an investment. As such, it might be said that the difference was a matter of preference. Stone preferred to protect his investment by retaining title to his property, while the Southern Improvement Company investors preferred to cash out at the end of seven years with a profit. The preference for wealth invested in land as opposed to wealth held as cash is a decision based on personal and financial factors. Yet, there was more to Stone's decision to keep his workers in a perennial state of tenancy. It was a reflection of his desire to maintain the traditional relationship between cotton planter and field hand, the model that planters had relied on in antebellum times. The Southern Improvement Company investors operated under a different investment model, a model for agricultural development that emerged in the South after the demise of slavery.

Which model fit best depended on the assumptions the landowner made. In Stone's case, it is easy to understand why he would have discounted the desire for self-sufficiency as a way to motivate his black workers. What could he expect from a race of people whom he knew to be listless, lazy, and improvident? As far as Stone was concerned, the only reason black families left his plantation was to evade his supervision. From that perspective, their desire for independence was a character flaw. For the Southern Improvement Company investors, the desire for independence was a potent force to be harnessed.[62]

Stone was stuck in the past.[63] He could not let go of the notion that using unskilled workers under close supervision was the only way to grow cotton. Although mechanization, which would eventually displace the antebellum tradition, was still four decades in the future,[64] there were many options available to Stone that fell between the two extremes. Nevertheless, he persisted in looking at the world through the eyes of his father's generation.[65]

In 1909, John Dewey, the eminent professor of philosophy at Columbia University, offered his views on the process of social and intellectual change

62. Cresswell has noted that black mobility was the one weapon at the disposal of African Americans that "gave them leverage with employers and landlords" (*Rednecks, Redeemers, and Race*, 49).

63. Wyatt-Brown has described Stone's friend LeRoy Percy, who shared Stone's point of view, as "too arrogant to bend" ("Leroy Percy and Sunnyside," 93).

64. Holley, *The Second Great Emancipation*, passim.

65. For an extended discussion of variations in the transition from the old plantation system to modern methods of agriculture, see Charles S. Aiken, *The Cotton Plantation South since the Civil War* (Baltimore: Johns Hopkins University Press, 1998), especially chap. 3, "The Demise of the Plantation" (63–96).

in a lecture on the influence of Charles Darwin. "Old ideas give way slowly; for they are more than abstract logical forms and categories," Dewey stated in reference to the fundamental shift in thinking that had occurred as a result of Darwin's theory of evolution.

> They [old ideas] are habits, predispositions, deeply ingrained attitudes of aversion and prejudice. Moreover, the conviction persists—though history shows it to be a hallucination—that all the questions that the human mind has asked are questions that can be answered in terms of the alternatives [philosophy or science] that the questions themselves present. But in fact intellectual progress usually occurs through sheer abandonment of questions together with both of the alternatives they assume—an abandonment that results from the decreasing vitality and a change of urgent interest. *We do not solve them: we get over them. Old questions are solved by disappearing, evaporating, while new questions corresponding to the changed attitude of endeavor and preference take their place.*[66]

Stone believed, as did Willcox and many other intelligent and well-intentioned thinkers who delved into the complexities of the race problem at the turn of the century, that the application of the new methods of science to questions concerning the proper relationship of the black and white races in America would provide definitive answers that would chart the course of race relations for the future. But the questions Stone and the other scientific racists asked failed to take into account that the world had fundamentally changed, and that their asking the wrong questions over and over would never give them the right answers.

The Reverend Ewing recognized this pervasive rejection of postemancipation reality when he asserted that the heart of the race problem could be summarized in one basic question, "How to maintain the social, industrial, and civic inferiority of the descendants of chattel slaves."[67] To ask the same question in economic terms that Stone would have appreciated, How could planters in the South perpetuate the existence of a labor force that was obedient, productive, and willing to work for low wages?[68]

66. John Dewey, "The Influence of Darwinism on Philosophy," in *The Influence of Darwin on Philosophy and Other Essays*, by Dewey (Amherst, NY: Prometheus Books, 1997), 19; emphasis added.

67. For a similar assessment, see Phillips, "The Central Theme of Southern History," 30–43.

68. Stone said as much in his lecture at the University of Virginia in 1914 ("Responsibility of the Southern White Man to the Negro," 11, 16–18). The fact that he expressed these views

Alfred Holt Stone asked this question repeatedly, even after the failure of the labor experiment at Sunnyside made it clear that European immigrants were not going to replace African Americans. "As there is no likelihood of an influx of immigration to the South," Stone observed in 1910, "the question remains one of speculating upon the length of time the plantation system, based upon Negro labor, can be maintained. Under this system the Negro's part in southern agricultural development will remain merely the part played by any labor, working under supervision in any line of industry. Where another furnishes the capital, the brains, and the direction, the part played by labor is not more than that of a tool or a machine."[69]

The new order of things that emerged as the country entered the twentieth century cried out for answers, but not for answers to the old questions that had dominated southern discourse in the century before. Rather, it demanded that new questions be asked, questions that reconceptualized race relations in the United States in response to the basic shift that had occurred in the nation's political, economic, and social reality. Alfred Holt Stone's efforts to shed light on the "American race problem" did not acknowledge this fundamental shift. He kept trying to force a round peg into a square hole, but instead of reshaping the hole when the peg did not fit, he blamed the peg.

---

five years after he had returned full-time to the business of growing cotton is further evidence that his views in regard to the race problem had not changed. See also Smith, "High Authority of Failed Prophet?" 209–12.

69. Stone, "Negro in Agricultural Development," 14. Stone's reverence for slavery and the plantation system that employed it and resistance to the economic and social changes that had occurred as a result of emancipation can be read in his unpublished memoirs at MDAH ("Some Recollections," chap. 1, sec. 2 [pp. 10–19]).

Appendix A

VERIFICATION OF STONE'S AUTHORSHIP OF EDITORIALS IN THE *GREENVILLE TIMES*

Editorials in the *Greenville Times* at the turn of the twentieth century were not signed, and Stone's name does not appear on the masthead during the period of his editorship. Nevertheless, there is compelling evidence to suggest that he wrote most of the editorials that appeared in the *Greenville Times* during the latter part of 1900 and first part of 1901.

To start, Stone claimed on several occasions that he served as editor of the *Greenville Times*. For example, the biographical sketch that he wrote for himself in *The Official and Statistical Register of the State of Mississippi, Centennial Edition, 1917* states that "from June, 1900, to June, 1901, he [Stone] edited the Greenville Times.[1] On January 10, 1902, Stone told Willcox that "For a year I was editor of the Greenville Times,—giving up the place a few months since, in order to be able to devote all my leisure time to the work above referred to."[2]

The editor of the *Greenville Times* in 1900 was H. T. Crosby. The masthead read "H. T. Crosby, Editor and M'gr." through the August 25 issue of that year. However, the masthead was altered in the next issue (September 2) to read "H. T. Crosby . . . Business Manager." Apparently, Stone assumed his editorial duties around this time. Regardless of the actual date, the first of nine editorials concerning the race problem that can be attributed to Stone appeared on September 22.[3]

1. Dunbar Rowland, *The Official and Statistical Register of the State of Mississippi, Centennial Edition, 1917* (Madison, WI: Democrat Publishing, 1917), 942–43. The typescript for this sketch with Stone's proof corrections can be found in his subject file at MDAH.

2. Stone to Willcox, January 10, 1902, WFWP.

3. "Booker T. Washington," *Greenville Times*, September 22, 1900. The last of these editorials appeared on May 11, 1901.

The evidence linking the authorship of these editorials to Stone is both circumstantial and substantive. Many of the circumstantial connections are noted in the text. However, one of the most convincing is the reference to Walter F. Willcox in an editorial that appeared on February 9, 1901. "One of the most careful students of negro criminality is Dr. W. F. Willcox, of Cornell University," it reads. "He cannot be accused of prejudice, for, as he says of himself, he was born and reared 'within the shadow of Bunker Hill Monument.'"[4] The quotation in the editorial is from a footnote to an article entitled "Negro Criminality" from the *American Journal of Social Sciences.* Seven years later Stone reprinted the article, with the footnote, as a chapter in *Studies in the American Race Problem.*[5] Willcox may have been well connected in academic circles, but it doubtful that his work was well known in the Mississippi Delta prior to his association with Stone.

There is also substantive evidence to verify Stone's authorship of these editorials. For example, on November 3, 1900, an editorial was published in the *Greenville Times* that read as follows: "A combination of Spanish and Indian produces the *mestizo,* who, while combining some of the mental and physical characteristics of both parents, belongs, ethnologically speaking, properly to the race of neither."[6] Three years later, Stone wrote an article for the *Atlantic Monthly* that contained this language: "Just as the crossing of the Spaniard upon the Indian has given us the *mestizo* of Central American and Mexico, so the blending of white and Negro blood has given us a type which combines some of the racial characteristics—good and bad—of both its progenitors."[7]

Another example of parallel composition can be found in an editorial published on December 1, 1900:

> Taking the New York Age, edited by T. Thomas Fortune, and the Colored American, of Washington, edited by Edward E. Cooper, as types of the highest development of the negro paper, we fail to see how the most enthusiastic believer in the intellectual ability of the negro race is to wring much comfort out of a review of this field of its mental endeavor. . . . We find in them such articles as Fortune's attack on Charles Dudley Warner,

4. "Education as a Solution," *Greenville Times,* February 9, 1901.

5. Walter F. Willcox, "Negro Criminality," in *SARP* 443–75. A version of the quotation "within sight of Bunker Hill" appears on page 443.

6. "A Lame Conclusion," *Greenville Times,* November 3, 1900; emphasis added.

7. "The Mulatto Factor in the Race Problem," *Atlantic Monthly* 91 (May 1903): 658–62; reprinted in *SARP,* 425–39. The quotation is from page 431 in the book (emphasis added).

> echoed in Cooper's "the Afro-American's condition would be greatly improved if more men of the caliber of Charles Dudly [*sic*] Warner were transported to the 'Great Beyond.'"[8]

Stone's article in the *Atlantic Monthly* contained this language:

> One of the leading race papers in the country, published at the nation's capital, in enumerating certain things which it would like to see occur, as being beneficial to the Negro, included "the death of a few more men like *Charles Dudley Warner,*" and this merely because that good man and true friend of the Negro had, shortly before his death, reached and expressed conclusions concerning Negro higher education at variance with opinions he had formerly entertained.[9]

Charles Dudley Warner was past president of the American Social Science Association and a respected editor with proslavery sentiments.[10] Stone was familiar with Warner's work and referred to him in his first letter to Willcox. "Did you notice Chas. Dudley Warner's address [talk], at the [American Social Science] Association's meeting in Washington?"[11]

It would be understandable if the reader questions whether the authorship of nine brief editorials warrants this much attention. In response, it can be noted that the importance of these editorials is not their content—for Stone restated all of the points they contained at greater length and with more detail in later writings—but their timing. The fact that Alfred Holt Stone was writing about these issues with conviction and knowledge as early as the fall of 1900 gives credence to his claim that he had been a serious student of the race problem for many years before he began publishing articles on the topic. In that regard, it is safe to say that Stone's interest in racial theories was neither superficial or transient, and this evidence of his dogged determination helps us appreciate the persistence with which Stone assembled his collection of material on "the Negro and cognate subjects."

8. "Negro Journalism," *Greenville Times,* November 3, 1900.
9. "The Mulatto Factor," 434.
10. Smith, *Old Creed for the New South,* 91.
11. Stone to Willcox, May 18, 1900, WFWP.

Appendix B

VERIFICATION OF STONE'S ANNOTATIONS

More than 110 articles and pamphlets in the Stone Collection at the Mississippi Department of Archives and History (MDAH) have been marked by someone wanting to identify specific passages for future use. Occasionally, these notations are accompanied by a brief comment, but usually they stand alone. The person making the notations applied them consistently. For example, all of the notations were made using a soft pencil, never an ink pen.

The most common notation was a single vertical line in the margin to indicate an important passage. Sometimes, a sentence or part of a sentence in the passage thus noted was underlined. Occasionally, the person making the notations would use double or even triple vertical lines for added emphasis. The only other marginal notation was the pound sign (#), which apparently indicated a particularly important passage.

These notations occurred in various combinations. One vertical line seemed to communicate something akin to "Look at this passage again." Double vertical lines with underlining in the text identified a passage as a candidate for becoming a quotation. A combination of two or three vertical lines with a pound sign seemed to cry out, "Don't forget this passage when you are writing that article!" Given these guidelines, the reader can determine what seemed to interest the person making these notations. For example, he or she seemed particularly interested in miscegenation and consistently noted passages dealing with this topic, whether they favored or criticized mixed-race unions.

There are several lines of evidence to suggest that Alfred Holt Stone was the person who marked these articles and pamphlets. Accessibility is the obvious place to start. After all, the collection at the MDAH was Stone's collection, and he was probably the only person who had access to it while it was being assembled. The fact that these notations occur as early as volume 3 and as late as volume 111, the second to last volume, suggests that whoever made them did so over time rather than after the fact.

A more direct line of evidence is the fact that Stone's signature appears, in pencil, on the front page of many pamphlets. In addition, comments in the margins were invariably in his hand, which can be verified by comparing the handwriting of these notes with that found in his letters to Walter F. Willcox.

More direct evidence can be found when Stone refers directly in other writings to an article in the collection that has been marked. "Are you familiar with two papers by J. Stahl Patterson on 'Increase and Movement of the Colored Population'?" Stone asked Willcox in a letter dated April 24, 1906. "I did not know them when you wrote me in regard to the paper on which you were engaged on the probable negro increase. In fact [I] have just come across them. They are in the Popular Science Monthly for September and October, 1881. He [Patterson] makes several allusions to negro gains through intermarriage between negro men and white women. He does not seem to advance any definite opinion as to the future."[1] Patterson's article, in two parts, is in the Stone Collection at the MDAH, and several passages, especially those related to miscegenation, have been marked.[2]

There are also several examples of a direct link between material noted or underlined in the collection and quotations Stone used in *Studies in the American Race Problem.* One example occurs on page 71, where Stone states: "There is a lot of common sense in Mr. Kelsey's suggestion that one trouble with too many Northern men who cannot understand Southern conditions is that they insist on regarding the Negro as a 'dark skinned white man.'" A published copy of Kelsey's dissertation is in the Stone Collection at MDAH, and many passages have been marked, specifically a sentence in the second to last chapter entitled "The Outlook." "The trouble is that we at the North are unable to disabuse ourselves of the idea that the Negro is a dark skinned Yankee and we think, therefore, that if all is not as it should be that something is wrong, that somebody or some social condition is holding him back."[3]

Another example occurs on page 211 in *Studies in the American Race Problem,* where Stone opened his essay "Race Friction" with a description of

1. Stone to Willcox, April 24, 1906, WFWP.

2. J. Stahl Patterson, "Increase and Movement of the Colored Population," *Popular Science Monthly,* September 1881, 665–75; and October 1881, 784–90. The article is number 4 in volume 80, TSC.

3. Kelsey, *The Negro Farmer;* emphasis in original.

a ceremony celebrating the integration of public schools in Boston, an event that would not be widely known in the Deep South more than fifty years later. As far as can be determined, Stone's description of the ceremony comes from a twenty-four-page pamphlet published in 1856 to commemorate the occasion, which is in the collection and is marked.[4]

A third example comes from pages 359–60 of *Studies in the American Race Problem*, where Stone quotes T. Thomas Fortune at length in regard to the political naiveté of African Americans. Stone acknowledges his source—an article by Fortune—in a footnote, and the article cited is in the Stone Collection at the MDAH with the relevant quotation marked.[5] There can be little doubt that Stone used items in his collection for his research and marked them accordingly.

4. *Triumph of Equal School Rights in Boston: Proceedings of the Presentation Meeting Held in Boston, Dec. 17, 1855; Including Addresses by John T. Hilton, Wm. C. Nell, Charles W. Slack, Wendell Phillips, Wm. Lloyd Garrison, Charles Lenox Remond* (Boston: R. F. Wallcut, 1856), 24 pages. The article is number 14 in volume 111, TSC.

5. T. Thomas Fortune, *The Negro in Politics* (New York: Ogilvie and Rowntree, 1886), 61 pages. The article is number 15 in volume 111, TSC.

Appendix C

RECONCILING STONE'S BOOKS WITH HIS COLLECTIONS

Linking Stone's collections to Stone's books can be confusing because he attempted to assemble two collections and started three book projects, of which only one collection survives today and only one book was published. The collection that survives was compiled for one of the books he failed to write. The collection that did not survive was for another book that Stone failed to complete. The book that he did publish, *Studies in the American Race Problem,* was an anthology of his work covering a period of seven years. Consequently, some of the chapters in this work predated his collection efforts, while others were written at a time when Stone could use material in the collection that has survived. Confusing? The picture can be reconciled by approaching Stone's collection efforts and book projects chronologically.

Stone's first book project was a "history of the Negro race," a project he referred to as "my life work" early in his career as a racial theorist. The Stone Collection of articles, pamphlets, and monographs that is housed today at the Mississippi Department of History and Archives (MDAH) was assembled for this book, and that collection reflects the scope of his ambitious project. For example, a "history of the Negro race" would encompass the black experience in Africa, and that may be one reason why there are almost 160 titles in the collection dealing with that continent. A "history of the Negro race" would also have to deal with their deportation as slaves to the Caribbean, which helps explain why there are over 90 titles in the collection touching on that geographic region.

The international flavor to the Stone Collection at the MDAH sets it apart from the collection associated with the book he was obligated to write for the Carnegie Institution, which focused exclusively on the role of African Americans in the economic development of the United States. Another difference between the two collections is that the Stone Collection at the MDAH consists entirely of printed material. By comparison, Stone wanted manuscripts for the Carnegie project. One other factor that separates the MDAH collection from Stone's work for the Carnegie Institution is the publication dates of the material in the MDAH collection. That collection effort appears to have been initiated as early as 1905, more than a year before Stone became associated with the Carnegie Institution.

These factors reinforce the contention that the Stone Collection at the MDAH was the result of his plan to write a "history of the Negro race," even though he was not able to complete that project. Although Stone scaled down the scope of the work as he went, recognizing his naiveté in selecting so broad a topic, it was the sheer magnitude of the project that served as a rationale for the huge collection of material he assembled.

Thus, there were two attempts to collect material and two book projects. One of the collections exists, one does not, and neither book was published. Where does the third book, *Studies in the American Race Problem*, fit in? Of the nine chapters Stone contributed to that work, five were written prior to 1905 and four in 1907. Footnotes in the four chapters written after 1905 clearly indicate that Stone used material from the collection that is now housed at the MDAH. The earlier chapters do not contain references to material in that collection.[1] Furthermore, there is no indication in any of the chapters that Stone had access to the manuscript material he was trying to collect for the Carnegie Institution. Consequently, it appears that although he failed to complete his magnum opus (a "history of the Negro race"), he used items he was collecting for that work in articles and addresses as they became available.

1. For example, the footnotes in "Race Problems Contrasts and Parallels," which was adapted from addresses made in 1905, refer to standard sources that would have been readily available at the time, such as Booker T. Washington's *Future of the American Negro* and W. E. B. Du Bois' *Souls of Black Folk*. By comparison, "The Foundations of Our Differences," an extension of "Race Problems Contrasts and Parallels," was written in 1907 expressly for *Studies in the American Race Problem*. Footnotes in that chapter cite several titles in the MDAH collection. One example appears in a footnote on page 58, which refers to a report from a committee in the Massachusetts legislature (Theodore Lyman, *Free Negroes and Mulattoes* [Boston: True and Green, 1822], 16 pages). Stone's reference to this obscure pamphlet suggests that he used the MDAH collection in writing this chapter.

Appendix D

SHOULD THE *N* IN *NEGRO* BE CAPITALIZED?

There was no consensus in the United States at the turn of the twentieth century in regard to the question of whether the initial letter in the word *Negro* should be capitalized.[1] The first edition of the *Chicago Manual of Style,* published in 1906, offered forty-nine rules for capitalizing words but said nothing about capitalizing the word *Negro,* or the name of any other race, for that matter.[2] Nevertheless, this stylistic issue was not a trivial matter that could be easily ignored because deciding whether to capitalize *Negro* communicated a particular point of view regarding persons of African descent.

Although both Walter F. Willcox and Alfred Holt Stone had consistently spelled *negro* with a lowercase *n* in their writings prior to March 1908, they decided to capitalize *Negro* in their forthcoming book, *Studies in the American Race Problem.* Why did they change their minds? An exchange of correspondence in 1907 between Willcox and H. B. Frissell, principal of the Hampton Normal and Agricultural Institute, offers a glimpse of Willcox's, and possibly Stone's, thinking in regard to the issue and may provide insight as to the rationale behind their decision to depart from their previous practice in regard to the word *Negro.*

Willcox wrote H. B. Frissell on April 11 and raised the question of whether the first letter in *Negro* should be capitalized.

1. See footnote 65 in chapter 11 to place this appendix in a context.

2. *Manual of Style: Being a Compilation of the Typographical Rules in Force at the University of Chicago Press* (Chicago: University of Chicago Press, 1906), 1–20. The title was officially changed to *The Chicago Manual of Style* with the thirteenth edition in 1982. A facsimile of the first edition can be read online at "The Chicago Manual of Style Online," www.chicagomanualofstyle.org/about15_facsimile.html.

> The question has recently been raised whether it is better to print the word "negro" with a capital "N." So far as I have observed the usage of good writers in the matter, I find that the majority of writers belonging to that race, but not all of them, use the capital, and the majority of writers belonging to the white race, but not all of them, use the small letter. The usage in England, so far as I have observed[,] is uniformly in favor of the small letter, and the assumption from the derivation and analogy seems to point to that conclusion. But I understand that a strong preference is felt by many leading colored men for the capital[,] and I have no serious objection to using it if that be the case. Will you be kind enough to tell me what your judgment is, and so far as you can, the reasons for it?[3]

Frissell responded to Willcox's letter on April 16, 1907:

> I referred the matter of the capitalization of the word "Negro" to my Commandant of Cadets, Major R. R. Moton. This is what he says: "Negro should be written with a capital letter. It is the designation of a race—the only race not spelled with a capital letter. The reasoning employed by the people who do not use the capital is, that it is a descriptive adjective and other races take their name from the countries in which they live. But so is 'Missouri' a descriptive adjective, meaning muddy. The name 'Smith' first meant the trade of the man to whom it belonged. Numerous other cases might be mentioned."
>
> I am also enclosing for [*sic*] Major Moton another paragraph on the subject.

The attachment consisted of a single sheet without a title or attribution, although it apparently had been written by Major Moton.[4]

> Since the objection to the writing of the word Negro seems to come mainly from the fact that the noun was derived from an adjective, it may be worth while to point out that a number of similar instances of proper nouns derived from adjectives are in common use. For instance—Alba Longa (means simply "long white") is a pretty good case, yet this is always

3. All three letters can be found in the "F" folder, box 15, WFWP. Although salutations and signature blocks have been omitted, the typescripts of the three letters with one attachment have not been edited or abridged.

4. Robert Russa Moton graduated from the Hampton Institute in 1890 and served as "Commandant in charge of military discipline" for twenty-five years. In 1915, Moton left Hampton to assume the presidency of the Tuskegee Institute following the death of Booker T. Washington. He served as president of Tuskegee until 1935, when he retired because of ill health ("Dr. Robert Russa Moton," www.tuskegee.edu/Global/story.asp?S=1079853).

> capitalized. The same is true of Albion, the ancient name of Britain. And certainly Nigritos [*sic*], the name of one of the Philippine tribes, ought to satisfy the most critical. Again the original of the name Ethiopian means simply "burnt faces," a fairly good point, yet this word is never written without the initial capital letter.

Willcox responded to Frissell in a letter dated April 19, 1907.

> Thank you for your notes of April 16th about the capitalization of "Negro." That my position hitherto on the subject may be a little clearer let me say one word of explanation.
>
> There are two main methods of classifying the races of mankind, one by the skin color, and the other by habitat or geographical location. The former, and much the oldest, introduced by Blumenbach, divides the race of mankind into white, the yellow, the black or negro, the red and the brown. The geographical classification divides mankind into the Caucasian, Indo European or Indo German, and Mongolian, and African, the American or Indian, and the Malay. I think the best usage calls for capitals in all terms of the second group and for small letters in all terms of the first group, and the reason for the distinction is obvious. The origin and ~~general~~ original significance of geographic terms like "Missouri" or "Alba Longa" seems to me unimportant.
>
> Turning from reasoning to authority, which in a matter of this sort must be decisive, I find the great weight of authority, including so far as I have observed all English writers, spell "negro" as they do "white" as a racial term with a small letter. The weight of argument and of authority therefore seems to me clearly on the side of the small letter, but if this usage is distasteful to a large body of my fellow countrymen, even though their reasoning may seem to me unsound[,] I am ready to change.

Not everyone was as willing as Willcox to change in regard to the issue of capitalizing *Negro*. The dean of American historians, J. Franklin Jameson, wrote a blistering response to W. E. B. Du Bois on June 22, 1910, because the black scholar had complained to Jameson about his refusal to capitalize *Negro* in the influential *American Historical Review*, which Jameson edited.[5]

5. Jameson to Du Bois, June 22, 1910, JFJP. The origin of Du Bois' complaint was probably the paper he authored in the July 1910 issue of the journal (W. E. Burghardt Du Bois, "Reconstruction and Its Benefits," *American Historical Review* 15 [July 1910]: 781–99), throughout which the first letter of *Negro* is lowercased. Apparently, Du Bois' letter to James on June 13 was prompted by his review of page proofs for the article.

> On my return from Maine and Indiana I find here your letter of June 13. It astonishes and afflicts me very much, that I, grandson of an old Abolitionist [*sic*], brought up to know no differences between black and white, and still—so far as I can see—without a trace of that race prejudice which I see exist but cannot comprehend, should be thought to inflict upon a professional colleague, contributing to this journal, anything which he can regard as a personal insult. The word "negro" is not a proper noun, not an indication of nationality but of physical traits. We have always held in this journal that it should no more be capitalized than white man, brown man, or red man, or than the Spanish *negro*, black. So far as I have been able to look into the matter without going to the Library of Congress (which is impossible because I leave town tonight), the authority of dictionaries runs in the same way, and apparently their opinion is based on the same argument. It would be making no more of the white man to capitalize that phrase; it is making no less of the negro to leave the word not capitalized. The question is simply one of typography. I cannot admit that the maintenance of uniform practice with regard to such matters is a small thing. To print "the church" in all our articles for fifteen years and then suddenly and in one article alone capitalize "the Church" would be nothing short of ridiculous and would justly expose us to the charge of having no typographical standards. The typography of this journal stands, I believe rather high in the scale, and we value our reputation for consistency as well as for accuracy. We should never be above making improvements, but we have never agreed to alter our standards for one article alone, at the request of any contributor. I shall hope that it will ultimately appear to you that we have not treated you in any other manner than that in which we have treated every other contributor from the beginning of the journal.[6]

Today, the *Chicago Manual of Style* treats the word *Negro* as the name of a specific racial group, which requires that the first letter be capitalized.[7] For terms like *black* or *white* that are "based loosely on color," Chicago recommends lowercase but notes that an author's preference should be followed.

6. Notice that Jameson capitalizes *Abolitionist*, treating it as a proper noun, although it could be argued that *Abolitionist* is descriptive of personal beliefs just as *Negro* describes physical characteristics.

7. *The Chicago Manual of Style*, 15th ed. (Chicago: University of Chicago Press, 2003), 325–26.

Appendix E

STONE'S LETTER TO LEROY PERCY

Personal[1]
Frankfort, Ky., May 21/07
Dear Mr. LeRoy:

Yours of 15th to hand. The [Vicksburg Daily] Herald may not be forwarded from Washington, but I have seen a copy containing your talk before I left.[2] In fact, I had fully intended to write you in regard to it, but I had to leave by a fixed date, to meet Mary in Louisville, and I was so crowded leaving the last few days that I postponed the matter to a more convenient season.[3]

I agree heartily with all you said, and am glad you said it. The force and effect of such utterances depend on the personality of the man behind them,—and I would rather have them come from you than from any other man in the State. It seems to me that the time has come when the sober and more thoughtful people of the South must take a definite stand against the tide of "negrophobia" which has been sweeping over our section for the past few years. I do not mean this in any theoretical or sentimental sense,—but as a cold-blooded, practical proposition. Unrest, disquiet and fear among so large a section of our population as the negro now constitutes will inevitably react upon upon [*sic*] the white element in time. And fear and unrest must, and will, follow the everlasting stirring up of strife along the lines pursued

1. Box 2, folder 6, PFP, MDAH.

2. The *Vicksburg Daily Herald* printed excerpts from Percy's address to the Mississippi State Bar Association in its May 9, 1907, issue.

3. Stone had left Greenville for Washington, D.C., on March 18 before proceeding to Madison, Wisconsin, for the summer (Stone to Willcox from Dunleith, March 12, 1907, WFWP). Stone was on his way to Madison when he wrote this letter.

by politicians seeking office on the strength of successful "race-problem" agitation.

But I did not mean to write you a letter,—though I wish that I could discuss the whole question with you in person. I only meant to write a hurried line of appreciation of the position which you have ~~under~~taken. I am glad to know that your remarks were so well received. Just between us, I wish that Capt. "Mac" had given them a stronger & more emphatic editorial endorsement.[4] Some time I want to discuss with you the matter of broaching this aspect of the subject with him.

I leave in a few days for Madison Wisc., where I shall work this summer. With love to wife Camille, & best regards for yourself,

Sincerely your friend, Alfred H. Stone

4. Captain J. S. McNeily was the editor of the *Vicksburg Daily Herald* (LeRoy Percy to "Dear Capt. Mac:," March 9, 1906, box 1, folder 8, PFP, MDAH). See also Wyatt-Brown, *House of Percy,* 173.

Bibliography

MANUSCRIPTS

Adams, Herbert B., Papers. Johns Hopkins University, Baltimore, MD.
Carnegie Institution archives, Washington, DC.
Dodd, William E., Papers. Library of Congress, Washington, DC.
Du Bois, W. E. B., Papers (microfilm). University of Massachusetts, Amherst.
Jameson, J. Franklin. Papers. Library of Congress, Washington, DC.
Percy Family Papers. Mississippi Department of Archives and History, Jackson.
Stone Collection, Alfred H. Mississippi Department of Archives and History, Jackson.
———. University of Mississippi, Oxford.
———. Literary papers. Mississippi Department of Archives and History, Jackson.
———. Miscellaneous typescripts and newspaper clippings. Mississippi Department of Archives and History, Jackson.
Stone, Walter Wilson. Miscellaneous papers. Mississippi Department of Archives and History, Jackson.
Washington, Booker T., Papers (microfilm). Library of Congress, Washington, DC.
Willcox, Walter F., Papers, Library of Congress, Washington, DC.

GOVERNMENT DOCUMENTS

Congressional Globe, 33rd Cong., 1st sess.
Deed books, Washington County Court House, Greenville, MS.
Mississippi Constitutional Convention. *Journal of the Proceedings of the Constitutional Convention, of the State of Mississippi. Begun at the City of Jackson on August 12, 1890, and Concluded November 1, 1890.* Jackson: E. L. Martin, 1890.
Mississippi Legislature, 1916–18. *Joint Report of the Senate and House Committees Appointed at the Session on 1916 to Consider the State's Revenue System and Fiscal Affairs.* Jackson: Tucker Printing, [1918].

Record of Wills. Washington County Court House, Greenville, MS.

U.S. Congress, Select Committee. *Mississippi in 1875. Report of the Select Committee to Inquire into the Mississippi Election of 1875, with the Testimony and Documentary Evidence.* Washington, DC: Government Printing Office, 1876.

U.S. Department of Agriculture, Bureau of Agricultural Economics, Office of Farm Management, Reports, Speeches, and Articles Relating to Farm Management, RG 83, National Archives, College Park, MD.

U.S. Department of Justice, RG 60, National Archives, College Park, MD.

War of the Rebellion: A Compilation of the Official Records of the Union and Confederate Armies. 128 vols. Washington, DC: Government Printing Office, 1880–91.

DISSERTATIONS

Hines, Tom Spight. "Mississippi and the Prohibition Controversy." Master's thesis, University of Mississippi, 1960.

Merideth, Mary Louise. "The Mississippi Woman's Rights Movement, 1889–1923: The Leadership Role of Nellie Nugent Somerville and Greenville in Suffrage Reform." Master's thesis, Delta State University, 1974.

Simpson, Robert Reynolds. "The Origin of State Departments of Archives and History in the South." Ph.D. diss., University of Mississippi, 1971.

NEWSPAPERS AND SERIALS

Chicago Tribune, 1902
Cincinnati Enquirer, 1904
Corinth (Mississippi) Herald, 1902
Greenville Daily Democrat, 1912–13
Greenville Daily Democrat-Times, 1917–24
Greenville Delta Democrat-Times, 1955–62
Greenville Democrat, 1894–96
Greenville Times, 1875–1907
Greenwood Commonwealth, 1902–4
Hinds County (Mississippi) Gazette, 1875
Jackson Clarion, 1865–87
Jackson Clarion-Ledger, 1889–93, 1949–81
Jackson Daily Clarion-Ledger, 1930–32
Jackson Daily News, 1947–80
Jackson Daily Times, 1876
Jackson State Ledger, 1891
Manufacturer and Builder, 1893
Memphis Commercial Appeal, 1914–48
Natchez Daily Democrat, 1893

New Orleans Daily Picayune, 1881–99
New Orleans Times-Democrat, 1899
New York Times, 1907–28
New York Times Book Review, 1909
Port Gibson (Mississippi) Reveille, 1890
Southern Workman, 1903–6
Southwestern Presbyterian, 1900
Staple Cotton Review, 1927–55
Vicksburg Daily Herald, 1907
Washington Post, 1903
Webster [County] (Mississippi) Progress, 1938
Yazoo City (Mississippi) Herald, 1903

ARTICLES, CHAPTERS, PAMPHLETS, AND TYPESCRIPTS

Adams, Charles Francis. "Reflex Light from Africa." *Century Magazine*, May 1906, 101–11.

———. *"The Solid South" and the Afro-American Race Problem. Speech of Charles Francis Adams at Academy of Music, Richmond, Va., Oct. 24, 1908*. Boston: n.p., 1908.

Adams, Herbert Baxter. "The Study and Teaching of History." *Publications of the Mississippi Historical Society* 1 (1898): 73–84.

Aikman, William. *The Future of the Colored Race in America: Being an Article in the Presbyterian Quarterly Review, of July, 1862*. New York: Anson D. F. Randolph, 1862. 35 pages.

Aldrich, Mark. "Progressive Economists and Scientific Racism: Walter Willcox and Black Americans, 1895–1910." *Phylon* 40 (1st quarter 1979): 1–14.

"Alfred H. Stone Recalls Railroad Pioneers." *Illinois Central Magazine* 27 (September 1938): 3–5.

Alston, Lee J., and Robert Higgs. "Contractual Mix in Southern Agriculture since the Civil War: Facts, Hypotheses, and Test." *Journal of Economic History* 42 (June 1982): 327–53.

American Economic Association. "The Fourteenth Annual Meeting." *Publications of the American Economic Association*, 3rd ser, 3 (February 1902): 42–58.

———. "Geographical Distribution of Members and Subscribers." *Publication of the American Economic Association*, 3rd ser., 3 (February 1902): 40–41.

———. "Papers and Proceedings of the Twelfth Annual Meeting, Ithaca, N.Y., December 27–29, 1899." *Publications of the American Economic Association*, 3rd ser., 1 (February 1900): 7–42.

———. "Supplement, Hand Book of the AEA, 1906." *Publications of the American Economic Association*, 3rd ser., 7 (August 1906): 1–48.

———. "The Twenty-fourth Annual Meeting." *American Economic Review* 2 (March 1912): 1–4.

Anderson, James D. "The Southern Improvement Company: Northern Reformers' Investment in Negro Cotton Tenancy, 1900–1920." *Agricultural History* 52 (January 1978): 111–31.

Baker, Thomas Nelson. "An Explanation." *Outlook*, June 27, 1903, 528–29.

Barringer, P[aul] B[randon]. *The American Negro: His Past and Future.* 3rd ed. Raleigh, NC: Edwards and Broughton, 1900.

Bassett, John Spencer. "Herbert Baxter Adams." In *Dictionary of American Biography*, 1: 69–71. New York: Scribner's, 1928.

———. Review of *Studies in the American Race Problem*, by Alfred Holt Stone. *American Historical Review* 14 (July 1909): 837–38.

Berthoff, Rowland T. "Southern Attitudes toward Immigration, 1865–1914." *Journal of Southern History* 17 (August 1951): 328–60.

Bettersworth, John K. "Mississippi Historiography: Research Materials and Researchers." *Mississippi Quarterly* 10 (Summer 1957): 138–45.

Bingham, Robert. *An Ex-Slaveholder's View of the Negro Question in the South.* N.p., 1900. Reprint from the European edition of *Harper's Monthly Magazine*. 16 pages.

Blackmar, F. W. Review of *Studies in the American Race Problem*, by Alfred Holt Stone. *American Historical Review* 14 (May 1909): 837–39.

Blaine, James G., L.Q.C. Lamar, James A. Garfield, Alexander H. Stephens, Wendell Phillips, Montgomery Blair, and Thomas A. Hendricks. "Ought the Negro to Be Disenfranchised? Ought He Have to Have Been Enfranchised?" *North American Review* 268 (March 1879): 225–83.

Brieger, James F. "Hometown Mississippi." 2nd ed. Photocopy of a typescript, 1980. Mississippi Department of Archives and History. 557 pages.

Brooks, Walter H. "The Evolution of the Negro Baptist Church, 1851." *Journal of Negro History* 7 (January 1922): 11–22.

Brough, Charles Hillman. "Work of the Commission of Southern Universities on the Race Question." *Annals of the American Academy of Political and Social Science* 49 (September 1913): 47–57.

Calhoon, S. S. "The Causes and Events that Led to the Calling of the Constitutional Convention of 1890." *Publications of the Mississippi Historical Society* 6 (1902): 105–10.

Campbell-Kelly, Martin. "Punched-Card Machinery." In *Computing before Computers*, edited by William Aspray, 122–55. Ames: Iowa State University Press, 1990.

Carnegie, Andrew. *The Negro in America: An Address Delivered before the Philosophical Institution of Edinburgh, 16th October 1907.* Philadelphia: E. A. Wright Bank Note Co. for the Committee of Twelve for the Advancement of the Interests of the Negro Race, [1907?]. 32 pages.

Chamberlain, D[aniel] H[enry]. *Ex-Governor Chamberlain on the Negro Problem.* Charleston, SC: News and Courier, 1904. 32 pages.

Coody, A. S. "Repair of and Changes in the Old Capitol." *Journal of Mississippi History* 11 (April 1949): 89–96.

Coulter, E. Merton. "What the South Has Done about Its History." *Journal of Southern History* 2 (February 1936): 3–28.

[Croly, David Goodman]. *Miscegenation: The Theory of the Blending of the Races, Applied to the American White Man and Negro.* New York: H. Dexter, Hamilton, and Co., 1864. 72 pages.

"Development of Historical Work in Mississippi." *Publications of the Southern History Association* 6 (July 1902): 335–40.

Dewey, John. "The Influence of Darwinism on Philosophy." In *The Influence of Darwin on Philosophy and Other Essays,* by Dewey, 1–19. Amherst, NY: Prometheus Books, 1997.

"Discussion of the Paper by Alfred Holt Stone, 'Is Race Friction between Blacks and Whites in the United States Growing and Inevitable?'" *American Journal of Sociology* 13 (May 1908): 820–40.

Douglass, Eri. "The Federal Writers' Project in Mississippi." *Journal of Mississippi History* 1 (April 1939): 71–76.

Douglass, Frederick. *Address by Hon. Frederick Douglass, Delivered in the Metropolitan A.M.E. Church, Washington, D.C. Tuesday, January 9th, 1894, on the Lessons of the Hour, in which He Discusses the Various Aspects of the So-Called, But Mis-Called, Negro Problem.* Baltimore: Thomas and Evans, 1894. 36 pages.

———. *The Race Problem. Great Speech of Frederick Douglass, Delivered before the Bethel Library and Historical Association, in the Metropolitan A.M.E. Church, Washington, D.C., October 21, 1890.* N.p., [1890?]. 16 pages.

Du Bois, W. E. Burghardt. *The Atlanta Conferences.* Atlanta University Leaflet, no. 16. Atlanta: n.p., 1902. 14 pages.

———. *The Conservation of Races,* Occasional Papers, no. 2. Washington, DC: American Negro Academy, 1897. 16 pages.

———. "The Economic Future of the Negro." *Publications of the American Economic Association,* 3rd ser., 7 (February 1906): 219–42.

———. "The Future of the Negro Race in America." *East and the West* (January 1904): 6–7.

———. "The Negro and Crime." *Independent* 51 (May 18, 1899): 1355–57.

———. *The Negro Artisan: Report of a Social Study Made under the Direction of Atlanta University; Together with the Proceedings of the Seventh Conference for the Study of Negro Problems, Held at Atlanta University, on May 27, 1902.* Atlanta University Publications, no. 17. Atlanta: Atlanta University Press, 1912.

———. "The Negro as He Really Is." *World's Work,* June 1901, 848–66.

———. "The Negro in the Black Belt: Some Social Sketches." *Bulletin of the Department of Labor* ([May 1899]): 401–18.

———. "The Negro Landholder of Georgia." Special issue, *Bulletin of the Department of Labor* 35 (July 1901): 647–777.

———. "The Negroes of Farmville, Virginia: A Social Study." *Bulletin of the Department of Labor* 14 (January 1898): 1–38.

———. "Reconstruction and Its Benefits." *American Historical Review* 15 [July 1910]: 781–99.

———. "The Training of Negroes for Social Power." *Outlook*, October 17, 1903, 409–14.

"The Economic Future of the Negro: Discussion." *Publications of the American Economic Association*, 3rd ser., 7 (February 1906): 295–334.

Ewing, Quincy. "The Heart of the Race Problem." *Atlantic Monthly* 103 (March 1909): 389–97.

———. "How Can Lynching Be Checked in the South?" *Outlook*, October 12, 1901, 359–61.

Fetter, Frank A. "Social Progress and Race Degeneration." *Forum* (October 1899): 228–38.

Fleming, William H. *Slavery and the Race Problem in the South, with Special Reference to the State of Georgia: Address of Hon. Wm. H. Fleming, before the Alumni Society of the State University, Athens, June 19, 1906*. Boston: Estes [1906?]. 66 pages.

Fortune, T. Thomas. "The Afro-American." *Arena*, December 1890, 115–18.

Fry, Henry Peck. *The Voice of the Third Generation: A Discussion of the Race Question for the Benefit of Those Who Believe that the United States is a White Man's Country and Should Be Governed by White Men*. Chattanooga, TN: self-published, 1906. 32 pages.

Gatewood, Willard B. "Sunnyside: The Evolution of an Arkansas Plantation, 1840–1945." In *Shadows over Sunnyside: An Arkansas Plantation in Transition, 1830–1945*, edited by Jeannie M. Whayne, 3–24. Fayetteville: University of Arkansas Press, 1993.

Genovese, Eugene D. "The Negro Laborer in African and the Slave South." *Phylon* 21 (4th quarter 1960): 343–50.

George, J[ames] Z. *The Federal Election Bill: Speech of Hon. J. Z. George, of Mississippi, in the Senate of the United States, December 10, 1890*. Washington, DC: n.p., 1890. 50 pages.

Goodloe, H. H. "The Negro Problem." *Southern Magazine* 7 (April 1874): 373–76.

Hampton, Wade. "The Race Problem." *Arena*, July 1890, 132–38.

Harris, Mrs. L. H. "A Southern Woman's View." *Independent* 51 (May 18, 1899): 1354–55.

Harris, Wiley P. "Kansas and Nebraska: Speech of Hon. W. P. Harris, of Mississippi, in the House of Representatives, April 24 and 26, 1854." *Congressional Globe*, 33rd Cong., 1st sess., appendix, 547–50.

Hart, Albert Bushnell. "Imagination in History." *American Historical Review* 15 (January 1910): 227–51.

Henry, George. *An Address to the Hon. General Assembly of the State of Rhode Island, of 1880, and to the Editor of the Providence Journal, On the Repeal of the Law Prohibiting the Intermarriage of Whites and Blacks*. Providence: J. A. and R. A. Reid, 1881. 12 pages.

Hoffman, Frederick L. "The Race Traits and Tendencies of the American Negro." Special issue, *Publications of the American Economic Association* 11 (January–March–May 1896): 1–329.

Hopkins, John H. *Bible View of Slavery*. N.p., [1861?]. 16 pages.

Humphreys, Benjamin G. *In Defense of the State of Mississippi against the Charge of Peonage: Speech of Hon. Benjamin G. Humphreys on Mississippi in the House of Representatives, Monday, March 2, 1908*. Washington, DC: n.p., 1908. 15 pages.

Ireys, Henry T. "Autobiography." In *Memoirs of Henry Tillinghast Ireys: Papers of the Washington County Historical Society, 1910–1915*, edited by William D. McCain and Charlotte Capers, 6–36. Jackson: Mississippi Department of Archives and History, 1954.

———. "County Seats and Early Railroads of Washington County. *Publications of the Mississippi Historical Society* 14 (1914): 267–300.

Jefferson, Mark. Review of *Studies in the American Race Problem*, by Alfred Holt Stone. *Bulletin of the American Geographical Society* 43, no. 4 (1911): 298–99.

Jervey, Theo[dore] D. Review of *American Negro Slavery: A Survey of the Supply, Employment, and Control of Negro Labor as Determined by the Plantation Regime*, by Ulrich Bonnell Phillips. *American Historical Review* 25 (October 1919): 117–18.

Johnson, Donald. "W. E. B. Du Bois, Thomas Jesse Jones and the Struggle for Social Education." *Journal of Negro History* (Summer 2000): 71–95.

Jones, R[ichard] W[atson]. "Confederate Cemeteries and Monuments in Mississippi." *Publications of the Mississippi Historical Society* 8 (1904): 87–119.

Kaplan, Sidney. "The Miscegenation Issue in the Election of 1864." *Journal of Negro History* 34 (July 1949): 274–343.

Kelsey, Carl. "Comments as a Discussant Regarding D. Collin Wells, *Social Darwinism*." *American Journal of Sociology* 12 (March 1907): 711.

———. "Influence of Heredity and Environment upon Race Improvement: An Introductory Paper upon the Significance of the Problem." *Annals of the American Academy of Political and Social Science* 34 (July 1909): 3–8.

———. Review of *Studies in the American Race Problem*, by Alfred Holt Stone. *Annals of the American Academy of Political and Social Science* 33 (March 1909): 217–18.

Kilpatrick, William H[eard]. *Preserving Southern History Material: An Address before the Southern Club of Columbia University, July 31, 1923*. [New York]: Columbia University Press, 1923. 7 pages.

Layton, Blake. *The Boll Weevil in Mississippi: Gone, But Not Forgotten* [Starkville?]: Mississippi State University Extension Service, n.d. 13 pages.

Lucas, Albert. "American Jewish Relief in the World War." *Annals of the American Academy of Political and Social Science* 79 (September 1918): 221–28.

Lyman, Theodore. *Free Negroes and Mulattoes*. [Boston: True and Green, 1822]. 16 pages.

[Lyon, James A.]. "Slavery, and the Duties Growing Out of the Relation." *Southern Presbyterian Review* 16 (July 1863): 1–54.

Mabry, William Alexander. "Disfranchisement of the Negro in Mississippi." *Journal of Southern History* 4 (August 1938): 318–33.

McAfee, Cleland Boyd. Review of *Studies in the American Race Problem*, by Alfred Holt Stone. *Journal of the Royal African Society* 8 (January 1909): 145–53.

McLean, William C. "Remarks." In *Proceedings of a Reunion of the Survivors of the Constitutional Convention of 1890 on the Thirty-Seventh Anniversary of the Adoption of the Constitution Held in the Senate Chamber of the New Capitol at Jackson, Mississippi, November 1, 1927*, 27–30. Jackson: Mississippi Department of Archives and History, 1928.

Memorials of the Life and Character of Wiley P. Harris, of Mississippi. Jackson, MS: Clarion Printing, 1892. 40 pages.

[Meriwether, Colyer]. "Scientific History." *Publications of the Southern History Association* 11 (September–November 1907): 297–311.

Milani, Ernesto R. "Peonage at Sunnyside and the Reaction of the Italian Government." In *Shadows over Sunnyside: An Arkansas Plantation in Transition, 1830–1945*, edited by Jeannie M. Whayne, 39–48. Fayetteville: University of Arkansas Press, 1993.

"A Model Italian Colony in Arkansas." *American Monthly Review of Reviews* (September 1906): 361–62.

Morse, W. Eugene. "'Judge' William Hemingway, 1869–1937." *Journal of Mississippi History* 36 (November 1974): 338–51.

Murphy, Edgar Gardner. "An Address at Tuskegee." N.p., [1900?]. 7 pages.

———. "Backward or Forward?" *South Atlantic Quarterly* 8 (January 1909): 19–38.

———. *The White Man and the Negro at the South: An Address Delivered under the Invitation of the American Academy of Political and Social Science, the American Society for the Extension of University Teaching, and the Civic Club of Philadelphia, in the Church of the Holy Trinity, Philadelphia, on the Evening of March 8th, A.D. 1900*. N.p., [1900?]. 55 pages.

"The Negro Problem." *National Criterion* (September 1902): 11–13.

"A Negro's View of the Race Problem." *Outlook*, June 13, 1903, 392.

"The Nicaragua Canal Convention at New Orleans." *Manufacturer and Builder* (April 1893): 78–79.

"O. B. Crittenden & Company." In *The 39th Anniversary Edition: Greenville Times*. Greenville, MS: Times Print and Publishing, 1910. Reprint, Greenville, MS: Office Supply Co., 1969. 1 page.

Palmer, J. H. "The Unveiling of a Picture." *Southern Workman* 35 (March 1906): 187–88.

Patterson, J. Stahl. "Increase and Movement of the Colored Population." *Popular Science Monthly*, September 1881, 665–75; and October 1881, 784–90.

Payne, Caledonia Jackson, comp. "Old Greenville Cemetery, Greenville, Mississippi, 1880–1982." Photocopy of a typescript, 1983. Mississippi Department of Archives and History. 114 pages.

Pecorini, Alberto. "The Italian as an Agricultural Laborer." *Annals of the American Academy of Political and Social Science* 33 (March 1909): 155–66.

Percy, Leroy. "A Southern View of Negro Education." *Outlook*, August 3, 1907, 730–32.

Phillips, Ulrich B. "The Central Theme of Southern History." *American Historical Review* 34 (October 1928): 30–43.

Powers, [James] LeGrand. "The Negro in the Yazoo-Mississippi Delta: Discussion." *Publications of the American Economic Association*, 3rd ser., 3 (February 1902): 273–78.

"Question of Race Diversely Viewed." *New York Times Book Review*, January 9, 1909. 1 page.

Ralls, J. R. *The Negro Problem. An Essay on the Industrial, Political and Moral Aspects of the Negro Race in the Southern States. As Presented under the Late Amendments to the Federal Constitution.* Atlanta: Jas. P. Harrison and Co., 1877. 116 pages.

"Report of the Proceedings of the Fourth Annual Meeting, April 18 and 19, 1901." *Publications of the Mississippi Historical Society* 4 (1901): 2–14.

Rice, Stuart A. "Walter Francis Willcox." 18 *American Statistician* (December 1964): 26.

Riley, Franklin L. "The Work of the Mississippi Historical Society." *Publications of the Mississippi Historical Society* 10 (1909): 35–45.

Ripley, William Z. "Races in the United States." *Atlantic Monthly* 102 (December 1908): 745–59.

Roudebush, George S. *An Address Delivered on the One Hundredth Anniversary of the Organization of the Presbytery of Mississippi Held in Pine Ridge Church, October 11, 1916.* Jackson, MS: Tucker Printing, 1916. 12 pages.

Rowland, Dunbar. *The Mississippi Plan for the Preservation of State Archives: An Address Delivered by Invitation, before the General Assembly of Tennessee, and the Tennessee Historical Society, January 10, 1905.* Nashville, TN: Brandon Printing, [1905?]. 14 pages.

———. "Wiley P. Harris." In *Courts, Judges, and Lawyers of Mississippi, 1798–1935*, 267–70. Jackson: Press of Hederman Bros. for State Department of Archives and History and the Mississippi Historical Society, 1935.

Royce, Josiah. "Race Questions and Prejudices." *International Journal of Ethics* (April 1906): 265–88.

Scarborough, W. S. "The Race Problem." *Arena*, October 1890, 560–67.

Schieffelin, William Jay. Review of *Studies in the American Race Problem*, by Alfred Holt Stone. *Political Science Quarterly* 24 (December 1909): 701–2.

Scomp, Henry A. "Can the Race Problem Be Solved?" *Forum* (December 1889): 365–76.

Seaman, L. *What Miscegenation Is! And What We Are to Expect Now that Mr. Lincoln is Re-Elected.* New York: Waller and Willetts, [1865?]. 8 pages.

Sharfstein, Daniel J. "The Secret History of Race in the United States." *Yale Law Journal* 112 (March 2003): 1473–510.

Shofner, Jerrell H. "Mary Grace Quackenbos: A Visitor Florida Did Not Want." *Florida Historical Quarterly* 58 (January 1980): 273–90.

Smith, John David. "Alfred Holt Stone and Conservative Racial Thought in the New South." In *The Human Tradition in the New South*, edited by James C. Klotter, 47–65. Lanham, MD: Rowman and Littlefield, 2005.

———. "Alfred Holt Stone: Mississippi Planter and Archivist/Historian of Slavery." *Journal of Mississippi History* 45 (November 1983): 262–70. Reprinted in *Slavery, Race, and American History: Historical Conflict, Trends, and Method, 1866–1953*, by Smith, 55–60. Armonk, NY: M. E. Sharpe, 1999.

———. "High Authority or Failed Prophet? Alfred Holt Stone and Racial Thought in Jim Crow America." *Journal of Mississippi History* 68 (Fall 2006): 195–211.

———. "The Historiographic Rise, Fall, and Resurrection of Ulrich Bonnell Phillips." *Georgia Historical Quarterly* 65 (Summer 1981): 138–53. Reprinted in *Slavery, Race, and American History: Historical Conflict, Trends, and Method, 1866–1953*, by Smith, 44–52. Armonk, NY: M. E. Sharpe, 1999.

———. "James Ford Rhodes, Woodrow Wilson, and the Passing of the Amateur Historian of Slavery." *Mid-America* 64 (October 1982): 17–24. Reprinted in *Slavery, Race, and American History: Historical Conflict, Trends, and Method, 1866–1953*, by Smith, 16–22. Armonk, NY: M. E. Sharpe, 1999.

———. "'Keep 'Em in a Fire-Proof Vault': Pioneer Southern Historians Discover Plantation Records." *South Atlantic Quarterly* 78 (Summer 1979): 376–91. Reprinted in *Slavery, Race, and American History: Historical Conflict, Trends, and Method, 1866–1953*, by Smith, 133–43. Armonk, NY: M. E. Sharpe, 1999.

———. "'No negro is upon the program': Blacks and the Montgomery Race Conference on 1900." In *A Mythic Land Apart: Reassessing Southerners and Their History*, edited by Smith and Thomas H. Appleton, 125–50. Westport, CT: Greenwood Press, 1997.

———. "Ulrich Bonnell Phillips's *Plantation and Frontier Documents: 1649–1863:* The Historian as Documentary Editor." *Georgia Historical Quarterly* 77 (Spring 1993): 123–43. Reprinted in *Slavery, Race, and American History: Historical Conflict, Trends, and Method, 1866–1953*, by Smith, 154–64. Armonk, NY: M. E. Sharpe, 1999.

———. "W. E. B. Du Bois and Ulrich Bonnell Phillips: Symbolic Antagonists of the Progressive Era." *Centennial Review* 24 (Winter 1980): 88–102. Reprinted in *Slavery, Race, and American History: Historical Conflict, Trends, and Method, 1866–1953*, by Smith, 23–32. Armonk, NY: M. E. Sharpe, 1999.

"Special Libraries Association: Fortieth Anniversary, 1909–1949." *Special Libraries* 40 (April 1949): 126.

Stephenson, M[atthew] F. *Negro Equality: Or the Effects of the Fifteenth Amendment to the Constitution of the United States, on the Morals of the People. Proofs of the Falsity of the Doctrine Are from the Bible, from Physiology, Anatomy, Chemistry, and Their Cognate Sciences*. Augusta, GA: F. M. Singer and Co., 1870, 40 pages.

[Stone, Alfred Holt]. "The Aftermath." *Staple Cotton Review* 5 (August 1927): 4–5.

———. "'And the Waters Prevailed.'" *Staple Cotton Review* 5 (July 1927): 3.

———. "'As a Man Thinketh in His Heart, So Is He': Address to the Graduating Class of 1937 of Millsaps College, June 8, 1937." Typescript, 1937. Mississippi Department of Archives and History. 20 pages.

———. *The Assessing of Public Utilities by the State Tax Commission.* Jackson: Mississippi State Tax Commission, 1935. 26 pages.

———. "As to 1925 Low Grades." *Staple Cotton Review* 5 (March 1927): 2–3.

———. *As to Senator Percy: A Personal Word.* Memphis, TN: n.p., [1923?]. 29 pages.

———. "The Basis of White Political Control in Mississippi." *Journal of Mississippi History* 6 (October 1944): 225–36.

———. "The Constitution of 1890: A Speech by the Hon. Alfred H. Stone before the Jackson Lions Club, May 30, 1947." Typescript, 1947. Mississippi Department of Archives and History. 14 pages.

———. "The Cotton Factorage System of the Southern States." *American Historical Review* 20 (April 1915): 557–65.

———. "A Disclaimer." *Nation,* May 14, 1903, 393.

———. "The Early Slave Laws of Mississippi." *Publications of the Mississippi Historical Society* 2 (1900): 133–45.

———. "The Economic Future of the Negro: The Factor of White Competition." *Publications of the American Economic Association,* 3rd ser., 7 (February 1906): 243–94. Reprinted in *Studies in the American Race Problem,* by Stone, 149–208. New York: Doubleday, Page, 1908.

———. "Fact and Tradition in Southern History." *Journal of Mississippi History* 17 (July 1955): 1–23.

———. "Foundations of Our Differences." In *Studies in the American Race Problem,* by Stone, 40–78. New York: Doubleday, Page, 1908.

———. "Four Terms and Freedom—Why Not?" *Staple Cotton Review* 23 (January 1945): 1–7.

———. "The Functions of a Local Historical Society." Typescript, 1910. Mississippi Department of Archives and History. [13] pages.

———. "The Future of the Race Problem in America." *International* 3 (October 1908): 147–56.

———. "Is Race Friction between Blacks and Whites in the United States Growing and Inevitable?" *American Journal of Sociology* 13 (March 1908): 676–97. Reprinted under the title "Race Friction" in *Studies in the American Race Problem,* by Stone, 211–41. New York: Doubleday, Page, 1908.

———. "The Italian Cotton Grower: The Negro's Problem." *South Atlantic Quarterly* 4 (January 1905): 42–47.

———. "Italian Cotton-Growers in Arkansas." *American Monthly Review of Reviews* 35 (February 1907): 209–13.

———. *Material Wanted for an Economic History of the Negro.* N.p., n.d. 16 pages.

———"A Mississippian's View of Civil Rights, States Rights and the Reconstruction Background." *Journal of Mississippi History* 10 (October 1948): 181–239. First published in *Staple Cotton Review* 26 (April 1948).

———. "The Mississippi River National Problem: Memorandum Submitted to the Flood Control Committee of the House." *Staple Cotton Review* 5 (December 1927): 2–12.

———. "Mississippi's Constitution and Statutes in Reference to Freedmen, and Their Alleged Relation to the Reconstruction Acts and War Amendments." *Publications of the Mississippi Historical Society* 4 (1901): 143–226.

———. "More Race Problem Discussion." *Publications of the Southern History Association* 11 (January 1907): 20–37.

———. "More Race Problem Literature." *Publications of the Southern History Association* 10 (July 1906): 218–27.

———. "Mr. Roosevelt, the South, and the Negro." In *Studies in the American Race Problem,* by Stone, 242–350. New York: Doubleday, Page, 1908.

———. "The Mulatto Factor in the Race Problem." *Atlantic Monthly* 91 (May 1903): 658–62. Reprinted in *Studies in the American Race Problem,* by Stone, 425–39. New York: Doubleday, Page, 1908.

———. "The Negro and Agricultural Development." *Annals of the American Academy of Political and Social Sciences* 35 (January 1910): 8–15.

———. "The Negro Farmer in the Mississippi Delta." *Southern Workman* 32 (October 1903): 457–60.

———. "The Negro in Politics." In *Studies in the American Race Problem,* by Stone, 351–421. New York: Doubleday, Page, 1908. Portions reprinted as "The Negro Problem in the United States: The Psychological Influence of Reconstruction," in *Readings in Social Problems,* edited by A. B. Wolfe, 665–76. Boston: Ginn, 1916.

———. "The Negro in the South: American Negro Origins." In *The South in the Building of the Nation,* 10: 166–83. Richmond, VA: Southern Historical Publication Society, 1909.

———. "The Negro in the Yazoo-Mississippi Delta." *Publications of the American Economic Association,* 3rd ser., 3 (February 1902): 233–72. Reprinted in *Studies in the American Race Problem,* by Stone, 81–124. New York: Doubleday, Page, 1908.

———. "Negro Labor and the Boll-Weevil." *Annals of the American Academy of Political and Social Sciences* 33 (March 1909): 167–74.

[———]. "The Overflow." *Staple Cotton Review* 5 (May 1927): 2–4.

———. *A Personal Word to Our Taxpaying Friends with a Christmas Letter to Our Staff.* Jackson: [Mississippi State Tax Commission], 1941, 2 pages.

———. "A Plantation Experiment." *Quarterly Journal of Economics* 19 (February 1905): 270–87. Reprinted in *Studies in the American Race Problem,* by Stone, 125–48. New York: Doubleday, Page, 1908.

———. "Race Problem Contrasts and Parallels." In *Studies in the American Race Problem*, by Stone, 3–39. New York: Doubleday, Page, 1908.

———. "The Responsibility of the Southern White Man to the Negro." In *Lectures and Addresses on the Negro in the South*, 5–18. Charlottesville, VA: Michie, [1915?].

———. Review of *A Documentary History of American Industrial Society*, edited by John R. Commons, Ulrich B. Phillips, Eugene A. Gilmore, Helen L. Sumner, and John B. Andrews. *American Historical Review* 16 (October 1910): 137–39.

———. Review of *The Interest of America in International Conditions*, by A. T. Mahan. *American Historical Review* 16 (April 1911): 652–54.

———. Review of *The Negro Farmer*, by Carl Kelsey. *Publications of the Southern History Association* 8 (May 1904): 239–40.

———. Review of *The Negro in the New World*, by Harry H. Johnston. *American Historical Review* 16 (April 1910): 650–52.

———. Review of *The Souls of Black Folk*, by W. E. B. Du Bois. *Publications of the Southern History Association* 7 (September 1903): 395–96.

———. Review of *The Southern South*, by Albert Bushnell Hart. *American Political Science Review* 4 (November 1910): 609–14.

———. Review of *The United States in Our Own Time: A History from Reconstruction to Expansion, Being an Extension of "The History of the Last Quarter,"* by Benjamin Andrews. *American Historical Review* 9 (July 1904): 843–44.

———. "Some Recent Race Problem Literature." *Publications of the Southern History Association* 8 (November 1904): 451–61.

———. "Some Recollections of a Southern Planter." Typescript, 2 chaps., [1938]. Mississippi Department of Archives and History. The chapters are 19 and 22 pages, respectively, and are paginated separately.

———. "The South in National Politics: The Political Effects of the War." In *The South in the Building of the Nation*, 4: 553–78. Richmond, VA: Southern Historical Publication Society, 1909. Reprinted as "Post Bellum Reconstruction, an American Experience." *Journal of Mississippi History* 3 (July 1941): 227–46.

———. "Tokens Add Effectiveness to Mississippi Sales Tax." *National Association of Tax Administrators Report* (December 1936): 18, 21.

Stone, Alfred Holt, and Julian H. Fort. *The Truth about the Boll Weevil: Being Some Observations on Cotton Growing under Boll Weevil Conditions in Certain Areas of Louisiana, Texas and Mississippi.* [Greenville, MS]: n.p. for the First National Bank, 1910. 34 pages.

Stone, W[alter] W[ilson]. "Greenville in 1865." In *Memoirs of Henry Tillinghast Ireys: Papers of the Washington County Historical Society, 1910–1915*, edited by William D. McCain and Charlotte Capers, 78–81. Jackson: Mississippi Department of Archives and History, 1954.

———. "Some Post-War Recollections." In *Memoirs of Henry Tillinghast Ireys: Papers of the Washington County Historical Society, 1910–1915*, edited by William D. McCain and Charlotte Capers, 231–89. Jackson: Mississippi Department of Archives and History, 1954.

Straton, John Roach. "Will Education Solve the Race Problem?" *North American Review* 170 (June 1900): 785–801.

Sutherland, H. L. "The Destiny of the American Negro." *Memphis Medical Monthly* (December 1905): 1–10.

Sydnor, Charles S. "Letters from Franklin L. Riley to Herbert B. Adams, 1894–1901." *Journal of Mississippi History* 2 (April 1940): 100–110.

Thornwell, J. H. *The Rights and the Duties of Masters. A Sermon Preached at the Dedication of a Church, Erected in Charleston, S.C., for the Benefit and Instruction of the Colored Population.* Charleston, SC: Steam-Power Press of Walker and James, 1850. 51 pages.

Tillinghast, Joseph Alexander. "The Negro in Africa and America." Special issue, *Publications of the American Economic Association*, 3rd ser., 3 (May 1902): 6–330.

Tillman, Benjamin R. *The Race Problem: Speech of Hon. Benjamin R. Tillman, of South Carolina, in the Senate of the United States, February 23–24, 1903.* Washington, DC: Allied Printing, 1903. 32 pages.

Tilton, Theodore. *The Negro: A Speech by Theodore Tilton, at Cooper Institute, New York, May 12, 1863, at the Anniversary of the American Anti-Slavery Society.* 2nd ed. New York: n.p. for the Anti-Slavery Society, 1863. 16 pages.

Triumph of Equal School Rights in Boston: Proceedings of the Presentation Meeting Held in Boston, Dec. 17, 1855; Including Addresses by John T. Hilton, Wm. C. Nell, Charles W. Slack, Wendell Phillips, Wm. Lloyd Garrison, Charles Lenox Remond. Boston: R. F. Wallcut, 1856. 24 pages.

Van Tassel, David D. "John Franklin Jameson." In *Keepers of the Past*, edited by Clifford L. Lord, 81–96. Chapel Hill: University of North Carolina Press, 1965.

Vardaman, James K. "A Governor Bitterly Opposes Negro Education." *Leslie's Weekly*, February 4, 1904, 104.

"A Vermonter." *Review of a "Letter from the Right Rev. John H. Hopkins, D.D.LL.D. Bishop of Vermont on the Bible View of Slavery."* Burlington, VT: Free Press, 1861. 28 pages.

Wade, Alice, and Katherine Branton. "Marriage Records from the Greenville 'Times.'" In Early Records of Mississippi: Issaquena County, Washington County," 3: 79–93. Typescript, 1983. Mississippi Department of Archives and History.

Walthall, E[dward] C. *The Federal Election Bill. Speech of Hon. E. C. Walthall, of Mississippi, in the Senate of the United States, Friday, December 12, 1890.* Washington, DC: n.p., 1890. 27 pages.

Washington, Booker T. "The Montgomery Race Conference." *Century Magazine*, August 1900, 630–32.

———. *The Successful Training of the Negro.* New York: Doubleday, Page, 1903. [25] pages.

———. "Will Education Solve the Race Problem? A Reply." *North American Review* 171 (August 1900): 221–32.

Wheeless, V. B. "The Sales & Use Tax: Its Origin and Background in Mississippi through 1965." Typescript, 1966. Mississippi Department of Archives and History.

———. "The Sales Tax Token in Mississippi." Typescript, 1974. Mississippi Department of Archives and History. 19 pages.

Willcox, Walter F. "Census Statistics of the Negro." *Yale Review* (November 1904): 274–86.

———. "Negro Criminality." In Alfred Holt Stone, *Studies in the American Race Problem*, 459–63.

———. "Negro Education Not a Source of Crime." *Leslie's Weekly*, March 17, 1904, 252.

———. "The Probable Increase of the Negro Race in the United States." *Quarterly Journal of Economics* 19 (August 1905): 545–72.

Willcox, Walter F., W. E. B. Du Bois, H. T. Newcomb, W. Z. Ripley, and A. H. Stone. "The Economic Position of the Negro." *Publications of the American Economic Association*, 3rd ser., 6 (February 1905): 216–21.

Williams, Frank B. "The Poll Tax as a Suffrage Requirement in the South, 1870–1901." *Journal of Southern History* 18 (November 1952): 469–96.

Williamson, Orvil A. "The North and the Negro Problem." *New York Times*, April 9, 1923.

Winter, William. "Governor Mike Conner and the Sales Tax, 1932." *Journal of Mississippi History* 41 (August 1979): 213–30.

Wright, Carroll D. "An Economic History of the United States." *Publications of the American Economic Association*, ser. 3, 6 (May 1905): 160–64.

Wyatt-Brown, Bertram. "From Piety to Fantasy: Proslavery's Troubled Evolution." In *Yankee Saints and Southern Sinners*. Baton Rouge: Louisiana State University Press, 1985, 155–82.

———. "Leroy Percy and Sunnyside: Planter Mentality and Italian Peonage in the Mississippi Delta." *Arkansas Historical Quarterly* (Spring 1991): 60–84. Reprinted in *Shadows over Sunnyside: An Arkansas Plantation in Transition, 1830–1945*, edited by Jeannie M. Whayne, 77–94. Fayetteville: University of Arkansas Press, 1993

Wynes, Charles E. "The Reverend Quincy Ewing: Southern Racial Heretic in the 'Cajun' Country." *Louisiana History* 7 (Summer 1966): 220–28.

Yeatman, James E. *Suggestions of a Plan of Organization for Freed Labor, and the Leasing of Plantations along the Mississippi River, under a Bureau or Commission to be Appointed by the Government*. St. Louis, MO: Western Sanitary Commission, 1864. 8 pages.

Zabawa, Robert E., and Sarah T. Warren. "From Company to Community: Agricultural Community Development in Macon County, Alabama, 1881 to the New Deal." *Agricultural History* 72 (Spring 1998): 459–86.

BOOKS

Aiken, Charles S. *The Cotton Plantation South since the Civil War.* Baltimore, MD: Johns Hopkins University Press, 1998.

Aptheker, Herbert, ed. *Selections, 1877–1934.* Vol. 1, *The Correspondence of W. E. B. Du Bois.* Amherst: University of Massachusetts Press, 1973.

Bailey, Hugh C. *Edgar Gardner Murphy: Gentle Progressive.* Coral Gables, FL: University of Miami Press, 1968.

Baker, Lewis. *The Percys of Mississippi: Politics and Literature in the New South.* Baton Rouge: Louisiana State University Press, 1983.

Baker, Ray Stannard. *Following the Color Line: An Account of Negro Citizenship in the American Democracy.* New York: Doubleday, Page, 1908. Reprint, Williamstown, MA: Corner House, 1973.

Barry, John M. *Rising Tide: The Great Mississippi Flood of 1927 and How It Changed America.* New York: Simon and Schuster, 1997.

Biographical and Historical Memoirs of Mississippi: Embracing an Authentic and Comprehensive Account of the Chief Events in the History of the State and a Record of the Lives of Many of the Most Worthy and Illustrious Families and Individuals. 2 vols. Chicago: Goodspeed Publishing Co., 1891.

Blum, Edward J. *Reforging the White Republic: Race, Religion, and American Nationalism, 1865–1898.* Baton Rouge: Louisiana State University Press, 2005.

Brandfon, Robert L. *Cotton Kingdom of the New South: A History of the Yazoo Mississippi Delta from Reconstruction to the Twentieth Century.* Cambridge: Harvard University Press, 1967.

Brands, H. W. *T. R.: The Last Romantic.* New York: Basic Books, 1997.

Carnegie Institution of Washington. *Year Book No. 5, 1906.* Washington, DC: Judd and Detweiler for the Institution, 1907.

———. *Year Book No. 6, 1907.* Washington, DC: Judd and Detweiler for the Institution, 1908.

———. *Year Book No. 7, 1908.* Washington, DC: Judd and Detweiler for the Institution, 1909.

———. *Year Book No. 8, 1909.* Washington, DC: Judd and Detweiler for the Institution, 1910.

———. *Year Book No. 11, 1912.* Washington, DC: Judd and Detweiler for the Institution, 1913.

———. *Year Book No. 12, 1913.* Washington, DC: Judd and Detweiler for the Institution, 1914.

———. *Year Book No. 13, 1914.* Washington, DC: Judd and Detweiler for the Institution, 1915.

Casson, Herbert N. *The History of the Telephone.* Chicago: McClurg, 1910.

Chafee, Zechariah. *Free Speech in the United States.* Cambridge: Harvard University Press, 1946.

Chandler, Walter, ed. *Journal and Speeches of Greene Callier Chandler.* N.p., 1953. 245 pages.

Cobb, James C. *The Most Southern Place on Earth: The Mississippi Delta and the Roots of Regional Identity.* New York: Oxford University Press, 1992.

Cresswell, Stephen. *Multiparty Politics in Mississippi, 1877–1902.* Jackson: University Press of Mississippi, 1995.

———. *Rednecks, Redeemers, and Race: Mississippi after Reconstruction, 1877–1917.* Jackson: University Press of Mississippi, 2006.

Cruden, Robert. *James Ford Rhodes: The Man, the Historian, and His Work.* Cleveland, OH: Press of Western Reserve University, 1961.

Cutler, James Elbert. *Lynch-Law: An Investigation into the History of Lynching in the United States.* New York: Longmans, Green, 1905. Reprint, New York: Negro Universities Press, 1969.

Degler, Carl N. *In Search of Human Nature: The Decline and Revival of Darwinism in American Social Thought.* New York: Oxford University Press, 1991.

Dillon, Merton L. *Ulrich Bonnell Phillips: Historian of the Old South.* Baton Rouge: Louisiana State University Press, 1985.

Du Bois, W. E. Burghardt. *Dusk of Dawn: An Essay toward an Autobiography of a Race Concept.* New York: Harcourt, Brace, 1940.

———, ed. *Economic Cooperation among Negro Americans. Report of a Social Study Made by Atlanta University, under the Patronage of the Carnegie Institution of Washington, D.C.; Together with the Proceedings of the Twelfth Conference on the Study of the Negro Problems, held at Atlanta University, on Tuesday, May the 28th, 1907.* Atlanta: [Atlanta University Press], 1907.

———, ed. *The Negro Church: Report of a Social Study Made under the Direction of Atlanta University; Together with the Proceedings of the Eighth Conference for the Study of the Negro Problems, Held at Atlanta University, May 26th, 1903.* Atlanta University Publications, no. 8. Atlanta: Atlanta University Press, 1903.

———, ed. *The Negro in Business: Report of a Social Study Made under the Direction of Atlanta University; Together with the Proceedings of the Fourth Conference for the Study of the Negro Problems, Held at Atlanta University, May 30–31, 1899.* Atlanta University Publications no. 4. Atlanta: [Atlanta University Press], 1899.

———. *The Philadelphia Negro: A Social Study.* New York: Lippincott, 1899.

———. *Some Efforts of American Negroes for Their Own Social Betterment.* Atlanta: Atlanta University Press, 1898.

Evans, Eli N. *Judah P. Benjamin: The Jewish Confederate.* New York: Free Press, 1988.

———. *The Provincials: A Personal History of Jews in the South.* New York: Atheneum, 1973.

Federal Writers' Project of the Works Progress Administration (Mississippi). *Mississippi: A Guide to the Magnolia State.* New York: Viking, 1938. Reprint, New York: Hastings House, 1949.

———. *Mississippi Gulf Coast: Yesterday and Today, 1699–1939*. Gulfport, [MS]: Gulfport Printing, 1939.

Fortune, T. Thomas. *The Negro in Politics*. New York: Ogilvie and Rowntree, 1886.

Fredrickson, George M. 1971. *The Black Image in the White Mind: The Debate on Afro-American Character and Destiny, 1817–1914*. Hanover, NH: Wesleyan University Press, 1987.

Fulkerson, H[orace] S[mith], *The Negro; As He Was; As He Is; As He Will Be*. Vicksburg, MS: Commercial Herald, 1887. 119 pages.

Gossett, Thomas F. *Race: The History of an Idea in America*. Rev. ed. New York: Oxford University Press, 1997.

Gould, Stephen Jay. *The Mismeasure of Man*. New York: Norton, 1981.

Hair, William Ivy. *Carnival of Fury: Robert Charles and the New Orleans Race Riot of 1900*. Baton Rouge: Louisiana State University Press, 1976.

Harlan, Louis R. *Booker T. Washington: The Wizard of Tuskegee, 1901–1915*. New York: Oxford University Press, 1983.

Harlan, Louis R., and John W. Blassingame. *The Booker T. Washington Papers*. 14 vols. Urbana: University of Illinois Press, 1972.

Harris, J. William. *Deep Souths: Delta, Piedmont, and Sea Island Society in the Age of Segregation*. Baltimore: Johns Hopkins University Press, 2001.

Harris, William C. *The Day of the Carpetbagger: Republican Reconstruction in Mississippi*. Baton Rouge: Louisiana State University Press, 1979.

Higham, John. *History: Professional Scholarship in America*. 1965. Baltimore: Johns Hopkins University Press, 1983.

Historical Catalogue of the University of Mississippi, 1849–1909. Nashville, TN: Marshall and Bruce, 1910.

History of Mississippi: The Heart of the South. 3 vols. Spartanburg, SC: Reprint Co., 1978.

Holley, Donald. *The Second Great Emancipation: The Mechanical Cotton Picker, Black Migration, and How They Shaped the Modern South*. Fayetteville: University of Arkansas Press, 2000.

Holmes, William F. *The White Chief: James Kimble Vardaman*. Baton Rouge: Louisiana State University Press, 1970.

Horsman, Reginald. *Josiah Nott of Mobile: Southerner, Physician, and Racial Theorist*. Baton Rouge: Louisiana State University Press, 1987.

Howell, H. Grady. *For Dixie Land I'll Take My Stand! A Muster Listing of All Known Mississippi Confederate Soldiers, Sailors and Marines*. 3 vols. [Madison, MS]: Chickasaw Bayou Press, 1998.

Hurst, Jack. *Nathan Bedford Forrest: A Biography*. New York: Knopf, 1993.

Johannsen, Robert W., ed. *Reconstruction, 1865–1877*. New York: Free Press, 1970.

Johnsen, Julia E[mily]. *Selected Articles on the Negro Problem*. New York: Wilson, 1921.

Kelsey, Carl. *The Negro Farmer: A Thesis Submitted to the University of Pennsylvania in*

Partial Fulfillment of the Requirements for the Degree of Ph. D. Chicago: Jennings and Pye, 1903.

Kirwan, Albert D. *Revolt of the Rednecks: Mississippi Politics, 1876–1925.* New York: Harper and Row, 1951.

Landon, Michael De L. *The University of Mississippi School of Law: A Sesquicentennial History.* Jackson: University Press of Mississippi for the University of Mississippi School of Law, 2006.

Lemann, Nicholas. *Redemption: The Last Battle of the Civil War.* New York: Farrar, Straus, and Giroux, 2006.

Lewis, David Levering. *W. E. B. Du Bois: Biography of a Race, 1868–1919.* New York: Holt, 1993.

Logan, Rayford W. *The Negro in American Life and Thought: The Nadir, 1877–1901.* New York: Dial Press, 1954.

Mangione, Jerre. *The Dream and the Deal: The Federal Writers' Project, 1935–1943.* Boston: Little, Brown, 1972.

Manual of Style: Being a Compilation of the Typographical Rules in Force at the University of Chicago Press. Chicago: University of Chicago Press, 1906.

McCain, William D., ed. *The Story of Jackson: A History of the Capital of Mississippi, 1821–1951.* 2 vols. Jackson: Hyer, 1953.

McMillen, Neil R. *Dark Journey: Black Mississippians in the Age of Jim Crow.* Urbana: University of Illinois Press, 1989.

Morison, Samuel Eliot, and Henry Steele Commager. *The Growth of the American Republic.* 2 vols. New York: Oxford University Press, 1962.

Morris, Edmund. *Theodore Rex.* New York: Random House, 2001.

Murphy, Edgar Gardner. *The Basis of Ascendancy: A Discussion of Certain Principles of Public Policy Involved in the Development of the Southern States.* New York: Longmans, Green, 1909.

Myrdal, Gunner, Richard Sterner, and Arnold Rose. *An American Dilemma.* New York: Harper, 1944.

Nash, Jere, and Andy Taggert. *Mississippi Politics: The Struggle for Power, 1976–2006.* Jackson: University Press of Mississippi, 2006.

Nowell, Princella W. *A Closer Look: A History and Guide to the Greenville Cemetery.* Jackson, MS: Hederman, 2003.

Percy, William Alexander. *Lanterns on the Levee: Recollections of a Planter's Son.* New York: Knopf, 1941.

Phillips, Ulrich Bonnell. *American Negro Slavery: A Survey of the Supply, Employment and Control of Negro Labor as Determined by the Plantation Regime.* New York: Appleton, 1918.

Ponder, Jerry. *The 9th Missouri Infantry Regiment, C.S.A., and the 12th Missouri Regiment, C.S.A.* Doniphan, MO: Ponder Books, 1996.

Race Problems of the South: Report of the Proceedings of the First Annual Conference Held under the Auspices of the Southern Society for the Promotion of the Study of Race Con-

ditions and Problems in the South at Montgomery, Alabama, May 8, 9, 10, A.D. 1900. Richmond, VA: Johnson, 1900.

Read, Hollis. *The Negro Problem Solved; or Africa as She Was, as She Is and as She Shall Be. Her Curse and Her Cure.* New York: A. A. Constantine, 1864.

Roberts, Gene, and Hank Klibanoff. *The Race Beat: The Press, the Civil Rights Struggle, and the Awakening of a Nation.* New York: Knopf, 2006.

Rowland, Dunbar. *The Official and Statistical Register of the State of Mississippi, Centennial Edition, 1917.* Madison, WI: Democrat Publishing, 1917.

Saikku, Mikko. *This Delta, This Land: An Environmental History of the Yazoo-Mississippi Floodplain.* Athens: University of Georgia Press, 2005.

Skates, John Ray. *Mississippi's Old Capitol: Biography of a Building.* Jackson: Mississippi Department of Archives and History, 1990.

Smith, John David. "Religion and "The Negro Problem." Pt. 2 of *The Biblical and "Scientific" Defense of Slavery,* vol. 6 of *Anti-Black Thought, 1863–1925.* New York: Garland, 1993.

———. *An Old Creed for the New South: Proslavery Ideology and Historiography, 1865–1918.* Westport, CT: Greenwood Press, 1985.

Stephens, Frank F. *A History of the University of Missouri.* Columbia: University of Missouri Press, 1962.

Stiles, T. J. *Jesse James: Last Rebel of the Civil War.* New York: Knopf, 2002.

Stone, Alfred Holt. *Studies in the American Race Problem.* New York: Doubleday, Page, 1908.

Sullivan, Mark. *Our Times: United States, 1900–1925.* Vol. 3, *Pre-War America.* New York: Scribner's, 1930.

Thompson, Holland. *The New South: A Chronicle of Social and Industrial Evolution.* New Haven, CT: Yale University Press, 1919.

Upchurch, Thomas Adams. *Legislating Racism: The Billion Dollar Congress and the Birth of Jim Crow.* Lexington: University of Kentucky Press, 2004.

Viles, Jonas. *The University of Missouri: A Centennial History.* Columbia: University of Missouri, 1939.

Waldrep, Christopher. *Vicksburg's Long Shadow: The Civil War Legacy of Race and Remembrance.* New York: Rowman and Littlefield, 2005.

Warner, Ezra J. *Generals in Blue: Lives of the Union Commanders.* Baton Rouge: Louisiana State University Press, 1964.

———. *Generals in Gray: Lives of the Confederate Commanders.* Baton Rouge: Louisiana State University Press, 1959.

Washburn, Wilcomb E. *The Cosmos Club of Washington: A Centennial History, 1878–1978.* Washington, DC: Cosmos Club, 1978.

Wecter, Dixon. *The Age of the Great Depression.* New York: Macmillan, 1948.

Wharton, Vernon Lane. *The Negro in Mississippi, 1865–1890.* New York: Harper and Row, 1965.

Williamson, Joel. *The Crucible of Race: Black-White Relations in the American South since Emancipation.* New York: Oxford University Press, 1984.

Willis, John C. *Forgotten Time: The Yazoo-Mississippi Delta after the Civil War.* Charlottesville: University Press of Virginia, 2000.

Winters, John D. *The Civil War in Louisiana.* Baton Rouge: Louisiana State University Press, 1963.

Wyatt-Brown, Bertram. *The House of Percy: Honor, Melancholy, and Imagination in a Southern Family.* New York: Oxford University Press, 1994.

Index